introduction to ENGINEERING

introduction to ENGINEERING

Roger Mayne
Department of Mechanical and Aerospace Engineering
State University of New York at Buffalo

Stephen Margolis
Department of Electrical and Computer Engineering
State University of New York at Buffalo

McGraw-Hill Book Company

New York   St. Louis   San Francisco   Auckland   Bogotá   Hamburg
London   Madrid   Mexico   Montreal   New Delhi
Panama   Paris   São Paulo   Singapore   Sydney   Tokyo   Toronto

introduction to
ENGINEERING

Copyright © 1982 by McGraw-Hill, Inc.
All rights reserved. Printed in the United States of America.
Except as permitted under the United States Copyright Act of 1976,
no part of this publication may be reproduced or distributed
in any form or by any means, or stored in a data base or retrieval system,
without the prior written permission of the publisher.

4567890 KPKP 8987

ISBN 0-07-041137-9

This book was set in Memphis Medium by Progressive Typographers.
The editors were Julienne V. Brown and J. W. Maisel;
the designer was Jo Jones;
the production supervisor was Leroy A. Young.
The drawings were done by Felix Cooper.
The cover photograph of *Pont du Gard* was taken by Jean Roubier;
the photograph of the Solar Reflector courtesy NASA.
Kingsport Press, Inc. was printer and binder.

Library of Congress Cataloging in Publication Data
Mayne, Roger, date
Introduction to engineering.

Includes bibliographical references and index.
1. Engineering. I. Margolis, Stephen.
II. Title.
TA157.M34   620   81-14323
ISBN 0-07-041137-9   AACR2

To our families . . .
        who fill our present and give us a past and a future
        transcending our own mortality

CONTENTS

Preface······xiii

PART ONE THE PROFESSION

CHAPTER 1 PATHS TO AN ENGINEERING CAREER······3

1.1 **What Is Engineering?**······3
Definitions • Engineering as a Profession •
Salary and Other Motivations

1.2 **The Engineering Disciplines**······13
Electrical Engineering • Mechanical Engineering •
Civil Engineering • Chemical Engineering •
Industrial Engineering • Other Fields of
Engineering

1.3 **Education and Employment**······24
Educational Opportunities • The Engineering
Functions • Other Opportunities

1.4 **Professional Development**······36
Historical Perspective • The Role of Professional
Engineering Societies • References

PART TWO CALCULATIONS AND COMPUTING

CHAPTER 2 INTRODUCTION TO ENGINEERING CALCULATIONS   **51**

  2.1 **Numbers and Arithmetic**   **51**
      Significant Figures • Scientific Notation • Expressing Results of Calculations • Problems

  2.2 **Dimensions and Units**   **60**
      Dimensions • Le Système International d'Unités (SI) • Customary American Units—The U.S. Customary System (USCS) • Converting Units • Dimensional Analysis • Problems

  2.3 **Graphs**   **83**
      Basic Rules in Plotting Graphs • Logarithmic Scales • Developing Equations from Graphs • Problems

  2.4 **Statistical Concepts**   **97**
      The Mean • Histograms • The Normal Distribution • The Least-Squares Fit • Problems

CHAPTER 3 COMPUTERS   **115**

  3.1 **Engineers and Computers**   **116**
      What Do Engineers Do with Computers? • Problems

  3.2 **Historical Perspective: From Evolution to Revolution**   **117**
      Problems

  3.3 **Computer Basics**   **139**
      Organization and Design • Number Systems • The Binary System • Problems

  3.4 **Computer Arithmetic: Addition and Subtraction**   **148**
      Addition • Subtraction • Complements in Decimal • Complements and Subtraction in Binary

  3.5 **Computer Arithmetic: Hardware Fundamentals**   **152**
      Physical Representation of Binary Digits • Logic and Logic Circuits • Problems

  3.6 **Using Digital Hardware to Perform Arithmetic Operations**   **157**
      Operations • The Half-Adder • The Full Adder • Using the Full Adder to Subtract • Levels of Integration • Problems

3.7 **Programming Computers**   162
    The Programming Process • From Problem Statement to Flowchart • Checking the Procedure: Sample Calculations • Fortran Program • Basic Program • Machine-Language Program • Problems • References • Suggested Readings

PART THREE   BASIC TOPICS IN ENGINEERING THEORY

CHAPTER 4   STATICS   173

    4.1 **Forces**   173
        Cables • Pinned Joints • Trusses • Problems

    4.2 **Moments**   189
        Problems

    4.3 **General Comments**   194
        Statically Indeterminate Problems • References

CHAPTER 5   ELECTRIC CIRCUITS   197

    5.1 **Current and Resistance**   198

    5.2 **Simple Circuits**   199
        Voltage and Current Laws • Series Resistors • Parallel Resistors • Problems

    5.3 **Applications of Variable Resistance**   212
        Problems

    5.4 **Electromechanical Devices**   215
        The Galvanometer • Ammeters • Voltmeters • Electric Motors • Problems

    5.5 **Remarks**   224
        References

CHAPTER 6   THE THERMAL-FLUID SCIENCES   227

    6.1 **Mechanics of Fluids**   228
        Pressure in Fluid Columns • Manometers • Liquid Flow in Pipes • Problems

    6.2 **Conservation of Matter**   237
        A Combustion Problem • Combustion in Practice • Problems

6.3 Conservation of Energy 246
An Equation for Conservation of Energy • Applying the Conservation-of-Energy Equation • Energy Conservation in Steady Flow • Problems

6.4 Heat Transfer 259
Conduction • Convection • Experimental Determination of h • Calculating Equilibrium Temperature • Problems

6.5 General Comments 271
References

CHAPTER 7 DYNAMIC SYSTEMS 273

7.1 Solving a Transient Response Problem 274
Describing the System Mathematically • A Simple Numerical Solution • Problems

7.2 First-Order Systems 282
Calculating Equilibrium Values and Rate of Change • Problems

7.3 Second-Order Systems 287
Transient Response of a Second-Order System • General Nature of Second-Order Behavior • Problems

7.4 Electric Systems 299
Capacitance • Inductance • Dynamics in Electric Circuits • Problems

7.5 Frequency Response 306
Background • Frequency Response of First-Order Systems • Frequency Response of Second-Order Systems • Problems

7.6 Summary 320
References

PART FOUR DESIGN AND APPLICATIONS

CHAPTER 8 ENGINEERING DESIGN: AN OVERVIEW 323

8.1 The Design Process 325
An Outline of the Process • Application of the Design Process • Problems

  8.2 **Evolution of an Engineering Design**   **344**
    A Development of Automatic Riveting Machines

  8.3 **Aluminum as an Electric Conductor: A Problem in Engineering Design**   **353**
    Background · Design Summary for Transmission Cable · Aluminum House Wiring

  8.4 **Safety as a Vital Part of Performance Evaluation**   **363**
    References

CHAPTER 9   MATERIALS, NUTS, AND BOLTS   367

  9.1 **Materials**   **367**
    Mechanical Properties · Problems

  9.2 **Operation of Bolts and Screws**   **376**
    Theory of Threaded Fasteners · A Threaded Connection · Problems · References

CHAPTER 10   ELECTRIC CONNECTIONS WITH WIRE-BINDING SCREWS   387

  10.1 **The Need for Reliability in Electric Connections**   **387**

  10.2 **Electric Contact between Metal Surfaces**   **388**
    Experimental Results · Aluminum and Its Oxide, Alumina · Stress Relaxation in Aluminum

  10.3 **An Actual Detailed Design**   **391**
    Problems · References

CHAPTER 11   STRUCTURES AND BRIDGES   403

  11.1 **Modes of Failure**   **403**
    Yielding in Tension and Compression · Buckling · Bending · Problems

  11.2 **Truss Structures**   **409**
    Design Calculations · Problems

  11.3 **Some Actual Bridges**   **415**
    Truss Bridges · Suspension Bridges · Arch Bridges · References

CHAPTER 12   ELECTRIC FILTERS   421

  12.1 **A Tuned Circuit**   **422**
    Principles of AM Radio · Design of a Tuning Circuit · Analyzing the Circuitry · Selecting L and C · Problems

12.2 **Removing a High-Frequency Signal**   **430**
    The Filter-Design Problem • Choosing R and C •
    Problems • References

CHAPTER 13   THE PRODUCTION OF OIL   **441**

13.1 **Background**   **442**
    How Long Will the Oil Last? • Problems

13.2 **Some Design Calculations**   **446**
    Analysis of Required Pressure • Mixtures of
    Oil, Gas, and Water • Problems

13.3 **Tertiary Recovery by Steamflooding**   **453**
    Some Results of the Computer Study •
    Problems • References

Appendix   **460**

Answers to Selected Problems   **464**

Index   **467**

PREFACE

A student beginning the study of engineering should be introduced to the profession, to engineering methods, and to engineering theory and its application. This book is an outgrowth of a course which we and our colleagues have taught for several years at the State University of New York at Buffalo with those goals in mind.

The material presented here begins with a description of engineering opportunities and education. This is followed by an introduction to engineering calculations and then by sections on digital computers—their history, operating principles, and applications in engineering. A broad range of basic engineering theory is then presented, ranging from statics to electric circuits to thermodynamics and system dynamics. The book concludes with several chapters on design and applications which illustrate the use of engineering theory. Throughout this book we have presented challenging topics. We have attempted to make them understandable to the typical entering student by providing detailed discussion and examples for the more difficult concepts. At a few places in the text, concurrent course work in calculus and computer programming has been assumed, but most of the material requires only high school trigonometry as a prerequisite.

In writing this book we have tried to include both the scientific and mathematical aspects of engineering as well as the societal and other less quantifiable aspects of the profession. We believe, however, that an introductory engineering course should present some of the details of engineering and should reflect the material taught in engineering curriculums. Examples of engineering theory should be included along with illustrations of how the theory can be applied to problems in the real world of engineering. In this way students can get an initial taste of engineering and learn to appreciate the complexity and ingenuity which are part of modern technology.

A wide range of topics is presented here to provide varied examples of engineering theory and applications. The topics have been selected because each is of quite general importance while together they represent the spectrum of engineering specialties. Naturally, it might not be feasible to cover all of the topics in a single course, so instructors can select two or three application areas for discussion during the course which are compatible with their own backgrounds. The course would then include topics in computation, computers, and basic theoretical material which lead up to a discussion of the selected applications. The use of a project to involve the student in actually applying the theory then provides a complete exposure to the engineering process.

At the State University of New York at Buffalo (SUNYAB), the introductory engineering course (initiated by W. N. Gill and continued under the sponsorship of Dean George C. Lee) carries two semester hours of credit. The following sequence of material from this text has often been taught in that course: an introduction to the profession (Chap. 1); dimensions, units, graphs, and statistical concepts (Chap. 2); the basics of statics (Sec. 4.1); material properties (Sec. 9.1); bridges and structures (Chap. 11). These sections are followed by selected topics in electric resistance circuits (Chap. 5); dynamic systems (Chap. 7); and electric filters (Chap. 12). The course concludes with a general discussion of thermal-fluid topics (Chap. 6).

A separate two-hour course in computer programming is also taken by entering engineering students as part of our freshman offerings at SUNYAB. In that course digital computers are discussed (Chap. 3) and FORTRAN programming is taught from a separate text.

ACKNOWLEDGMENTS

The authors acknowledge with thanks the assistance of the many people who have contributed to the development of this book. We especially thank Professors R. C. Michelini and R. Ghigliazza of the Istituto di Meccanica Applicata alle Macchine at the University of Genoa, Italy, who helped to provide the congenial environment where the book was begun and where several chapters were written. We extend thanks also to Mr. T. H. Speller, Sr., founder of the General Electro Mechanical Corporation, who furnished the material on Gemcor which appears in Sec. 8.2; and to Henry J. Nowak and George Horan of the Niagara Mohawk Power Corporation, who provided background information on electric utility connector practice for Sec. 8.3 and Chap. 10. In addition we would like to acknowledge the assistance of Dr. V. Venkatraman and the late Stephen Hurst, who initially developed many of our illustrations, the work of Hilary Shreter on the computer examples of Chap. 3, the comments of Professor H. Reismann who taught from an early draft of the text, and the many students whose suggestions and patience with various portions of the book have helped to improve it.

Finally, we would like to thank the editorial staff at McGraw-Hill for making the writing of this book a truly enjoyable experience.

Roger Mayne
Stephen Margolis

introduction to
ENGINEERING

PART ONE

THE PROFESSION

The purpose of this book is to present a broad view of the engineering profession and the work done by engineers. We have organized the book into four parts which very naturally reflect four different aspects of engineering. Parts II and III are intended to survey the special knowledge and skills developed by practicing engineers and Part IV illustrates the way engineering methods are applied to real problems.

Part I, which you are about to begin, describes the general nature of engineering and the roles that engineers play in modern industrial practice. In reading this material we hope that you will not only learn about engineering opportunities but will also establish a professional attitude as you begin to plan your career in engineering.

1 PATHS TO AN ENGINEERING CAREER

The study of engineering opens the door to a wide variety of opportunities in our modern technological society. Our purpose in this opening chapter is to give you an overview of the opportunities found in both industry and other branches of society. We also describe the various educational possibilities in engineering and technology so that you can understand the options that are available to you. Of course, there are definite relations between the educational background you establish and the role that you may ultimately play in the engineering and business community. We discuss these relationships quite thoroughly so that you may consider your educational plans carefully and be certain that they are compatible with your career objectives.

The chapter begins with a general discussion of engineering and the responsibilities of the engineer as a professional. A description of engineering opportunities and education follows. The chapter concludes with a discussion of the various engineering associations and societies that are playing an expanding role in today's technology. Learning more about the engineering association closest to your interests would be a first step in your development of a professional orientation.

1·1 WHAT IS ENGINEERING?

Since you are about to embark on a career in engineering, you surely would like to know just what engineering is. The field is so broad and the roles of engineers so varied that it is sometimes quite difficult to give a single simple answer.

DEFINITIONS

Two definitions of engineering often quoted are:

> "The art of directing the great sources of power in nature for the use and convenience of man."[1]

> "The art of the practical application of scientific and empirical knowledge to the design, production or accomplishment of various parts of construction projects, machines and material of use or value to man."[2]

The work of engineers today (Secs. 1.2 and 1.3) extends beyond power, construction projects, machines, and material. Nonetheless, so many of today's engineers *are* concerned with these topics and the topics are fundamental, so that they continue to be an important part of the basic courses in almost all engineering curriculums.

A broader and somewhat more recent definition of engineering states that

> Engineers are paid by society to work on systems dealing with problems whose solutions are of interest to that society. These systems seem to group conveniently into (*a*) systems for material handling, including transformation of and conservation of raw and processed materials; (*b*) systems for energy handling, including its transformation, transmission, and control; and (*c*) systems for data or information handling, involving its collection, transmission, and processing.[1]

These three definitions were written years ago but are still typical of the many available definitions of engineering and indicate the more significant characteristics of engineering and also of engineers. First, engineering is directed toward applications that benefit society. Second, engineering is based on scientific and mathematical principles. Third, engineering involves value and cost. Fourth, engineering is an art. Let's now look at each of these characteristic features and see how they apply today.

Engineering Is Directed toward Applications that Benefit Society

An engineer's work is directed toward an application, something that is useful, convenient, or valuable to society: this clearly remains true today. The bridges, roads, structures, and communication systems designed by engineers are obviously *intended* to be valuable to human beings—despite possibly harmful side effects, as when a road divides a community. But engineers do want to serve, not harm, humankind.

A concern for application differentiates the engineer somewhat

---

[1] From the charter of the British Institution of Civil Engineers, 1828.
[2] From the charter of the American Society of Civil Engineers, 1852.

Engineering is a career for the future. The remaining frontiers for humanity are now engineering challenges. At left is Dam-Atoll, a revolutionary turbine device designed to produce electric power from ocean waves. At right is the Teleoperator Retrieval System designed to operate with NASA's space shuttle for maneuvering and adjusting payloads in earth orbit. Both pictures are artist's conceptions. (Left, Lockheed-California Company; right, NASA.)

from the scientist, whose principal goal is knowledge, not application. There are certainly exceptions to this. Just as we sometimes find scientists working on applications, we can also find engineers working on science. In fact, the same person may occasionally play both roles. However, scientists customarily seek knowledge for its own sake and engineers apply scientific principles for the benefit of society. For example, medical science determined that heartbeats could be initiated by an applied electric signal. It was through electrical engineering that practical pacemakers were developed to help people with heart defects live longer, more useful, lives.

Engineering is Based on Scientific and Mathematical Principles

Engineering curriculums which lead to a Bachelor of Science degree in engineering typically require four courses in mathematics beginning with calculus, two courses in chemistry, and two courses in physics. In addition, another four or more courses in engineering science (solid mechanics, fluid mechanics, materials, etc.) are also required. Thus, students pursuing an engineering degree spend at least one-fourth of their time studying mathematics and science. In fact, most detailed engineering course work does not begin until the junior year, or the later part of the sophomore year, after students have developed a proper scientific and mathematical background.

Engineering clearly continues to be rooted in science and mathematics. However, this should not lead the reader to believe that everything in engineering has an answer based on rigorously derived mathematical formulas. Much of engineering relies on empirical information—information that may be gathered by experience or simply by experiment. Often it is not completely quantitative but must be used by engineers when basic scientific theory is either incomplete or insufficient for the task at hand. Engineering practice began over 2500 years ago, so the centuries have furnished a foundation of experience for many aspects of modern engineering. Science and mathematics have helped engineers understand and expand this empirical information and have provided them with tools for efficient and accurate engineering solutions, while eliminating the trial-and-error approach. At times, however, there is simply no substitute for experience.

The design of interior walls in housing construction is an example of the use of empirical information. Building codes normally specify interior walls using 2-inch by 4-inch joists on 16-inch centers. This design is not based on a detailed mathematical analysis by an engineer. The design simply works well. Anything much lighter seems to be flimsy, and anything with more weight and strength is not usually worth the cost.

More-complex engineering problems often require a combination of practical but empirical knowledge with well-understood scientific and mathematical principles. Such problems include

- Protection of metal structures from corrosion
- Lubrication of bearings
- Mass production of integrated electronic circuitry
- Analysis of effects of pollutants on the environment
- Study of fatigue failure of materials
- Efficient refining of crude oil

Most practicing engineers could easily add another half-dozen examples to this list from their own work. Engineers continually use their theoretical knowledge based on scientific and mathematical principles together with empirical knowledge based on experience.

Engineering Involves Value and Cost

Twenty centuries ago the Roman engineer Marcus Vitruvious Pollio wrote that engineers know how to exert "thrifty and wise control of expense in the works." In today's cost-conscious society, this has become increasingly true. The costs of labor, materials, and production for any product directly influence the customer who makes comparisons with competing products and who decides if the value of an item justifies its cost. Thus value and cost have now become a primary concern of the engineer. Nearly every engineer must be aware of costs and most engineering managers are continually searching for ways to cut them.

Courses in engineering economy and value analysis have now become commonplace in engineering programs. The modern equivalent of the statement by our Roman engineering friend is that "an engineer can do for one dollar what any fool can do for ten."

Engineering Is an Art

By dictionary definition an "art" is a skill or craft that involves study and learning, experience, and practice as well as ingenuity and imagination. This definition is nearly a description of the engineering profession, furnishing a list of the requirements for people who are dealing with the engineering problems of our society. As a typical example, engineers who are now designing earthquake-proof buildings must have a basic understanding of mechanics, vibrations, and materials learned in the study of engineering. They must have experience with up-to-date construction techniques and standard practice in building design obtained only on the job. Finally, they must be able to combine this background to create or synthesize new designs which can be built safely and economically through acceptable modifications of standard procedures. Because there is no predetermined sequence of steps that produces a correct design, the synthesis process forces engineers to use their ingenuity and imagination. It is the ability to develop inventive ideas and to convince people of their value that identifies an outstanding engineer.

Part IV of this book is devoted to design and applications in engineering, or, in other words, to the combination of theory and practice that represents the art of engineering today.

Although engineering classes at the freshman and sophomore levels are often in lecture format, upper level classes and laboratories allow more personal attention and provide hands-on experience. Students here are studying human performance on a treadmill. (*State University of New York at Buffalo.*)

ENGINEERING AS A PROFESSION

In choosing to study engineering, you have also chosen to enter a profession. Before selecting engineering, you may have considered other professional occupations such as law, medicine, accounting, dentistry, or nursing, or scientific fields such as chemistry, physics, or biology. In what follows, we will examine what engineering has in common with these professions and how it differs from them.

Many definitions of the word "profession" emphasize the *specialized knowledge* and the *intensive preparation* that a profession requires. Others—as in the phrase "a real professional"—refer to character traits in terms of competence and craftsmanship.

In *The Engineer and His Profession* [2], John D. Kemper suggests that the labor-management relations act of 1947 (called the Taft-Hartley Act) defines profession as it relates to the engineer:

> The term "professional employee" means any employee engaged in work (i) predominantly intellectual and varied in character as opposed to routine mental, manual, mechanical or physical work; (ii) involving the consistent exercise of discretion and judgment in its performance; (iii) of such a character that the output produced or the result accomplished cannot be standardized in relation to a given period of time; (iv) requiring a knowledge of an advanced type in a field of science or learning customarily acquired by a prolonged course of specialized intellectual instruction and study in an institution of higher learning or a hospital, as distinguished from a general academic education or from an apprenticeship.

Let us adopt this as a working definition of professional employee and see what the legal language implies about the day-to-day work and working conditions of the engineer. You will notice that much of what we write assumes that the engineer is an employee. This is in contrast to lawyers, physicians, dentists, and certified public accountants—who are frequently self-employed or are members of partnerships or group practices. The fact that many engineers are employees is one of the unique aspects of the profession, one that is sometimes a cause of tension between the demands of the profession and the demands of the employer.

Professional Conduct

Let's now take a closer look at the items in the Taft-Hartley definition above. In particular we will see how you can begin developing a professional attitude while you are still a student.

Engineering is "predominantly intellectual and varied" The word "predominantly" is the key here. Professional engineers should expect to spend most (but not all) of their time on work of an intellectual and varied character. Routine mental, manual, mechanical, or physical work may be part of the job, but engineers—as professionals—should attempt to make the intellectual and varied parts of their work predominate.

As a student, you may be required to do routine mental tasks, such as working out solutions to problems in mechanics or circuit theory. While you do this, your fellow students in law school and medical school are doing similar tasks: outlining legal precedents or doing routine blood analyses. When you become an established professional, you may expect to delegate the more routine mental work of this type to others or in some cases to a computer program. While still in school, you should accept the challenge of these problems as an important learning experience. Longer assignments and projects are less well defined and are more representative of real life. There are no easy solutions to such assignments. Hard work and clear and original thinking are required. Of course, the ratio of original to cut-and-dried assignments will increase as you move up through the standard 4-year curriculum. You should exploit every opportunity to approach new problems and to develop your creative abilities. It is for this reason that engineering professors often respond to student complaints of "we've never had this before" by saying "of course not—that's why it's a good problem."

The professional engineer "exercises discretion and judgment" In engineering, the use of discretion and judgment in making responsible decisions is part of the job. Because so much engineering work is done in teams, however, the responsibility is not individual: engineers, particularly early in their careers, are typically responsible for only subtasks. Being responsible but not in total control often poses a dilemma for the working engineer: "how can I do my job when those I depend on are not doing theirs?" For example, "how can I produce a reliable brake when the forge is delivering warped brake disks?" In such a case the engineer must become a human-relations expert and use discretion, operating tactfully at first, but forcefully if necessary—being prepared to blow the whistle as a last resort if he or she determines that such an action is the only responsible solution.

In your school work you should also exercise discretion and judgment without making a nuisance of yourself. You should be sure that you know the educational purpose of your assignments, their relation to other subjects, and the constituents of a reasonable solution. When fellow students behave in what you consider an irresponsible way, you may want to try out what your later behavior in industry might be by using a person-to-person approach at first but blowing the whistle if necessary.

Professional work "cannot be standardized · · · to a given period of time" Engineers customarily receive monthly or yearly salaries. They are most often paid for the completion of a responsible task, not for the time spent on the job. In many cases, engineers might put in a reasonable amount of uncompensated overtime— time for which they are not directly paid. This contrasts with the nonprofessional who, according to Taft-Hartley, must be paid at premium rates for all work in excess of 40 hours per week.

In exchange, the professional engineer can expect greater freedom in working hours than the nonprofessional worker. Many employers have policies allowing compensatory time off in exchange for uncompensated overtime. Nevertheless, when the work at hand demands it, you should expect to work many evenings and weekends during your professional career. Frequently, the work itself is so challenging that you will truly enjoy the overtime work, particularly when it leads to the successful completion of an important project.

In school, too, you should measure your work by the results, not by the time put in. Frequently, when a subject is new to you, you will find yourself devoting more than the standard 9 hours per week that each three-credit course normally requires. Consider this as part of your initiation to the profession.

A profession "requires knowledge acquired by a prolonged course of instruction and study" It is certainly true that engineering requires prolonged study, typically a minimum of 4 years for the B.S. degree and longer for graduate degrees (see Sec. 1.3). Never-

A student working in a microprocessor computer laboratory. (*State University of New York at Buffalo.*)

theless, the *minimum* length of instruction required by the engineering profession is less than the minimum of 7 years beyond high school required of lawyers, and that of 8 years beyond high school (followed by a year of internship) required of physicians.

If you wish to pursue a career in research and development, you should plan to continue your education to at least the M. S. level. In happy contrast to the law student and medical student, however, you can frequently plan to have your tuition waived and some of your graduate school expenses paid through an assistantship or fellowship if your grades are good enough. Of course, you will be deferring some income, since assistantships and fellowships pay only a fraction of industrial salaries, but you will be compensated not only by higher salaries in your future work but also by more interesting and challenging assignments.

A professional has high standards of achievement and conduct The requirement for high standards of achievement and conduct is not spelled out in the Taft-Hartley definition, but it is included in every dictionary definition as well as in Kemper's book [2]. Certainly the public has every right to expect high standards of conduct from the engineer and to expect that the profession itself will define and enforce these standards. This is of particular importance in engineering, since so many topics that the engineer deals with are highly technical in nature, and engineers themselves are usually best qualified to define the standards.

In recent times, we have seen standards of conduct imposed on engineers from outside the profession (as in nuclear power plant regulation and automobile safety standards, for example). These standards have sometimes been far from perfect, but they have been imposed when the profession itself failed to grapple with the problem. In a later section, we will discuss the roles played by the engineering professional societies in setting both technical standards and standards for professional engineering conduct.

As a student, you can begin to act like a professional by personally striving for the best performance you are capable of and by scrupulous honesty in the classroom and laboratories.

SALARY AND OTHER MOTIVATIONS

Survey data consistently show that salary is important but not the most important thing to engineers. The top five motivating factors are: (1) opportunity for advancement, (2) challenge of work, (3) salary, (4) recognition of job well done, and (5) work at the leading edge of the field. The best engineers and engineering students are typically motivated by an interest in the subject itself and are later pleasantly surprised at the availability of jobs and the good starting salaries.

TABLE 1·1
WAGES OF NONPROFESSIONALS IN COMMERCE AND INDUSTRY (1980)

Industry	Average Hourly Wage, $	Yearly, $
Apparel and other textile manufacture	4.49	9,500
Wholesale and retail trade	5.40	11,400
Services	5.75	12,100
Printing and publishing	7.34	15,500
Automobile manufacture	9.04	19,000
Construction	9.68	20,000

Outside the profession, many people think that engineers are exceptionally well paid, while others tell tales of wealthy plumbers (or bricklayers or television repair persons). The true facts lie between these extremes and (as usual) are more complicated. The data shown in Table 1.1 were collected by the U.S. Bureau of Labor Statistics (USBLS) and are correct as of March 1980. Note that these figures are averages. For example, an individual construction worker might start at $4 per hour ($8400 per year) and work up to a top of $12 per hour ($25,000 per year). Overtime pay can increase these figures substantially, and layoffs caused by weather or economic conditions can severely reduce them.

Salaries for selected professions are shown in Table 1.2. Although professionals normally think of their salaries in monthly or yearly figures, we have computed hourly equivalents to allow easy comparison with nonprofessionals. The starting salaries are good estimates of current conditions, but the midcareer figures are subject to substantial variation.

Individuals in any profession can make less and—depending on individual skills, ambition, and luck—some can make a great

TABLE 1·2
EARNINGS OF PROFESSIONALS (1980)

Occupation	Minimum Education beyond High School, years	Starting Salary, $/year ($/h)*	Midcareer Salary, $/year ($/h)
Physician	9	$45,000 (21)	$82,000 (39)
Dentist	8	27,000 (13)	53,000 (25)
Lawyer	7	18,000 (9)	50,000 (24)
Engineer	4	19,000 (9)	35,000 (17)
Accountant	4	14,000 (7)	31,000 (15)

* Quantities in parentheses are for dollars per hour.

outlined above, mechanical engineers are deeply involved in such problems of the future as development of solar-energy power systems, wind turbines, and equipment for the control of environmental pollution.

CIVIL ENGINEERING

Four of the major areas within civil engineering are structures, transportation, sanitation, and soils. Civil engineers are involved in the design and construction of buildings, bridges, and towers of all types. Moreover, the use of civil engineering methods in the design of aircraft structures has played an important role in the development of the modern jet airplane.

Civil engineers are also designers and builders of roads and railroad beds. This has quite naturally led to the emergence of civil engineers as transportation planners. Mass transit systems always include civil engineers on the design team. Problems concerning waste disposal are also the province of civil engineers; construction of canals for transportation and dams for flood control is planned and supervised by civil engineers; water resources, municipal water supply systems, and sewage disposal plants are designed and managed by civil engineers. Aerial photography and mapping techniques are often used by civil engineers who

Civil engineers are often involved in large construction projects. The individuals shown here are directing the erection of part of a tall smokestack at the site of a new power plant. (*ASME.*)

must deal with water-resource and land-use problems on a large scale. Civil engineers also work with large-scale aspects of air and water pollution, making measurements in communities to identify problem areas and exploring the effects of currents and drainage on water-pollution problems.

Civil engineers deal with rock and soil as engineering materials, in some cases using geological methods. They know how to place building foundations in different types of soils. They may be responsible for tunneling through rock formations or may use their geological skills in the mining and petroleum industries.

Civil engineers are often employed by government agencies at the federal, state, and local level. They commonly work with consulting firms that specialize in large construction projects.

CHEMICAL ENGINEERING

Chemical engineers work on large-scale chemical processes. They are often involved in taking a laboratory procedure for producing a certain chemical product and scaling that procedure into a full-scale chemical plant. Oil refineries; paper mills; and manufacturers of paint, medicinal products, fertilizers, plastics, and rubber typically employ chemical engineers. They may be involved in designing new plants, in doing research in particular problems on existing plants, in maintaining the operation of an existing plant, and also in handling many other types of problems that may not be directly related to chemical plants.

Many chemical engineers work in combustion-related activities—studying how to obtain more efficient combustion in an

A chemical engineer developing coating materials for aerospace applications. Many chemical engineers work in laboratories that concentrate on such practical applications of chemistry. (*ASME*.)

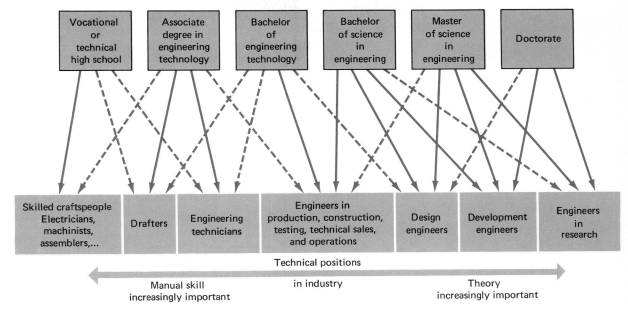

FIGURE 1·6
Relation of education to occupation in industry. Solid lines represent typical career paths; dashed lines represent less-typical career paths.

prototype models and experimental equipment, install measuring instruments for pressure, temperature, etc. They also take data and run many experiments. Technicians normally work closely with engineers, furnishing a combination of manual skills with practical as well as theoretical knowledge.

Normally, craftspeople, drafters, and engineering technicians are not graduates of 4-year B.S. engineering programs. The remaining technical positions in Fig. 1.6 cover a wide range of activities but are normally thought of as engineering functions and are normally staffed by engineering graduates. We will describe each of them below. Note that it is quite common for individuals to perform various ones during their career. In fact such is frequently the case in smaller companies. In larger companies, however, organization is more rigid. People usually stay focused on one area at a time and thus the educational patterns shown in Fig. 1.6 tend to be more strictly followed. For example, in some smaller companies, an experienced person with a 2-year technology degree may be promoted to a lower-level engineering position. In most larger companies, this is no longer common.

Production

Production engineers are responsible for producing the finished product of a company. They work closely with the craftspeople and laborers who actually perform the manufacturing operations.

They must also decide how to make new products being designed by their company by selecting the manufacturing methods and the machines which will be used. Production engineers play a major role in modern industry, where the costs of production must be carefully controlled and safety is an important concern.

Production engineers are involved in plant layout. They locate machines, aisles, and conveyer systems to facilitate a smooth flow of manufactured goods and people within a plant. They may also work with computers handling production and inventory records and, in fact, keeping actual control of various manufacturing operations. Production engineers often work closely with design engineers when production methods might have a serious impact on the design of a new product. They may also work with research and development engineers who are carefully studying specific operations in the production process or developing new production methods.

Production engineers normally have backgrounds in mechanical, industrial, or electrical engineering. However, chemical engineers and people from other engineering disciplines may also become involved in production engineering.

Construction

Building a bridge, a chemical plant, or a skyscraper is a construction project managed and supervised by construction engineers who convert the plans and drawings of the designing architect or engineer into reality. They select methods of construction and work with laborers and craftspeople during the construction process. Construction engineers continually face economic pressures, since the cost of construction is normally specified before construction begins and any cost overruns reduce the profit on the job. Construction projects must stay on schedule and can be seriously affected by poor weather.

Construction engineers most often have a background in civil or architectural engineering. However, electrical engineers can go into construction engineering because of the electric-power requirements of buildings; similarly mechanical engineers can become involved in heating and air conditioning systems as well as large mechanical projects such as electric-power plants. Chemical engineers are needed in the construction of refineries, chemical plants, and other similar facilities.

Test Engineering

Testing engineers are involved in many aspects of modern technology. In a typical manufacturing company testing facilities are needed for the evaluation of new designs and experimental proto-

A test engineer checking an electronic component on a missile guidance unit. (*Hughes Aircraft Company.*)

types. An automobile company, for example, has an engine-test facility for the study of newly proposed engine designs. The same company must also evaluate standard production engines as they are manufactured; that is, they must exercise what is called "quality control," which consists of a basic test of each engine plus extensive testing of typical engines periodically pulled from the production line. Similar testing is performed for all types of manufactured products, including electronic equipment and circuits.

In the production of raw materials, testing also plays an important role. For example, steel mills, chemical plants, and oil refineries must continually test the ingredients which go into their process and must carefully monitor the composition of the final product. Testing has also become more necessary as society's concern for the environment has grown. Industries and governmental agencies have begun environmental monitoring through regular air- and water-quality tests as well as noise tests.

Test engineers not only perform or supervise the actual test measurements but, more importantly, also plan the entire testing process. This begins with the choice of the test facilities and

1.3 EDUCATION AND EMPLOYMENT **29**

development of a test procedure. It also involves the selection and design of the instrumentation systems. Test engineering spans a wide range of technical fields and offers opportunities for individuals from nearly all the engineering disciplines.

Technical Sales

Sales engineers are often involved in the sale to businesses of products not usually sold to individual consumers—ones that are normally large, expensive, and technically oriented. Jet airliners, steam power plants, electric generators, mining and manufacturing equipment, radio transmitters, and scientific computers are only a few examples of these products. The people who handle the sales of such products are often engineers who can understand the technical details of the product and can communicate with the engineering staff of the company considering the purchase.

Sales engineers must be technically competent and personable and be able to relate to people. They must be able to bring technical concepts down to earth and explain in dollars and cents the engineering advantages of their product. Since millions of dollars could be involved in a single commercial sale, the work of sales engineers can be vital to the success of their company. Sales engineers are often called upon to propose solutions to particular customer problems. They may design solutions themselves or work with design engineers to define an appropriate solution and then present it to the customer. Since financial considerations are important in large investments, sales engineers must often develop a background in business and accounting to supplement their engineering knowledge.

Operations

The day-to-day operation of a manufacturing plant, a paper mill, nuclear power plant, or other large facility is controlled by operations and plant engineers. They are involved with every aspect of such facilities including turning on and controlling sophisticated production equipment (such as an automatic machine for assembling and testing electronic microcircuits), maintaining and repairing such machinery, installing new equipment, and coordinating with the designers and purchasers of new equipment. Operations engineers must also handle emergency equipment problems. In a manufacturing facility, operations and production engineers must work together to maintain continuous production. The shutdown of a large facility resulting in a loss of production is very costly to a company. Operations engineers make extreme efforts to promptly repair or replace a critical piece of production equipment.

Operations engineers are very practical people. They work with theoretical ideas occasionally but normally work directly with the machines, processes, and electric equipment that are a part of the production procedure. They also work closely with the craftspeople and laborers involved in construction, maintenance, and production. Mechanical, chemical, and electrical engineers are all required for the various problems that may be faced by an operations engineering staff.

Design Engineering

Design engineers have the responsibility for defining every operational detail of a company's product. They may be guided in their decisions by management and salespeople but must take full responsibility for the engineering aspects of the product. Design engineers make many detailed decisions such as how thick the concrete should be on a bridge, what type of electronic circuit should be used in a calculator, how the seat-adjustment mechanism should work in an automobile, etc. They make a large number of these types of decisions daily as they strive not only to create new products and improve older ones but also to ensure that the product they create is safe, reliable, and not overly expensive.

Design engineers work with drafting people who make the drawings required to define the object being designed; they work with technicians and test engineers who initially build and evalu-

An engineer using a computer graphics terminal in the design and analysis of a structure. Computers are playing an ever increasing role in modern engineering design. (*Information Displays, Inc.*)

ate their new designs; and they work with production engineers or construction engineers who build the final design, often in large quantities. Design engineers also become involved with research and development engineers who are developing new concepts for use in design or who often provide mathematical analysis expertise to aid the design engineer in making decisions. They also work frequently with sales engineers from other companies who would like their product incorporated into the design.

All the engineering disciplines are involved in design and take part in the design process. People who become design engineers must combine a practical knowledge of standard design ideas with a theoretical mind for use in the generation of new concepts and in the decision-making process. As our modern society strives to solve the problems of industrial productivity and energy-dependence, design engineers will be playing a vital role. Because of the importance of engineering design, Chap. 8 of this book is devoted to the process and includes examples of actual engineering designs.

Development

There is some overlap between the functions of design and development—the essential difference being that the design engineer is concerned with products or devices that are close to the production stage, whereas the development engineer focuses on how to build new products or devices which are still in the conceptual stage. The goal is to see if a new idea works—often by mathematically analyzing it first to see if it has merit and then by designing, building, and testing it. Successful concepts are then released to a design engineering team to be made suitable for actual production or incorporated into an existing product.

Engineers of all disciplines are involved in the many aspects of development engineering. They must be inventive and must understand theoretical concepts; they must also understand the many standard design ideas but not be so blinded by them that they can not generate new ideas. Development engineers very often have Master of Science or doctoral degrees, which enhance their ability to think analytically and increase their knowledge of theory. Development engineers often develop mathematical and computer methods which are used by design engineers and others in their day-to-day tasks.

Engineering Research

The function of the research engineer is to obtain an understanding of engineering phenomena and processes. This understanding increases the knowledge available to the profession and can then be used to guide development and design engineers in their daily work, helping to improve the things that they build.

Increased knowledge comes from both theory and experiment. Laboratory experiments lead to further knowledge in such areas as the fundamental behavior of materials, friction between materials, the interaction of electrons and matter in an electric arc, or the details of a biochemical reaction. These experiments may be explorations—the "let's see what happens" type—or carefully designed tests to come up with a new theory or to obtain very accurate measurements for use in engineering calculations. The people involved in research must have a special grounding in theory and mathematics as well as a talent and interest in careful investigations. Research can lead to new outlooks, approaches, and equations which describe engineering phenomena ranging from the operation of ball bearings to the performance of transistor materials and from the spread of air pollution to the behavior of the newest communication systems. Such understanding is often the first step in improving product characteristics or in reducing production difficulties. It should be mentioned, or course, that research is performed not only on a company's products but also on manufacturing and production methods and machinery as well as the by-products of production such as air and water pollution.

Research engineers are the most highly educated of engineering personnel, normally holding a Master of Science degree and very often the doctorate. A great deal of engineering research is carried on by universities, government laboratories, or companies devoted to engineering research performed under contract to the government or to private industry. All the engineering disciplines are involved in the many aspects of engineering research.

OTHER OPPORTUNITIES

In our discussion of the engineering functions, we concentrated on typical engineering roles in industry. There are many other opportunities for people with an engineering education. For example, government units of all types are major employers of engineers. Local and state governments hire engineers to handle road- and bridge-building tasks, to conduct pollution studies, and to deal with all types of urban planning. The federal government employs engineers for military programs which are highly technical, for energy, transportation and environmental programs, and for many other programs including the national space effort.

This section will describe briefly a few of the other tasks which engineers often perform. It should be remembered, of course, that many varied careers can begin with an engineering education. The fields of medicine and medical research are examples where the analytical abilities of an engineer can be valuable. An engineering background can also be helpful in the legal profession, where complex environmental problems and product liability cases are often engineering-oriented. Patent law has a similar engineering orientation.

Management

Many engineers become involved in management either as an original career goal or through growth and promotion in their field of engineering. Managers and supervisors are required in each of the technical fields shown in Fig. 1.6. The people most qualified to manage practicing engineers are, quite naturally, competent engineers who can deal with people. These engineering managers have the satisfaction not only of being promoted from within their company but also of having the opportunity and responsibility of directing a variety of engineering projects.

Engineers also fit well into the corporate management structure. Since so many large companies produce highly technical products and have large engineering staffs, an engineering background is a valuable asset to upper-level managers. In fact, the presidents of many major corporations are often engineers. For individuals who aspire to corporate management positions, a bachelor's degree in an engineering discipline combined with a master's degree in business administration (M.B.A.) can provide an appropriate background.

Consulting

There are a variety of opportunities for work as an engineering consultant. Engineering consultants may be hired by companies or governmental agencies for short-term projects that necessitate a particular expertise or for tasks that require a temporary increase in engineering staff size.

An engineering consultant may be an experienced engineer who specializes in a fairly narrow field such as vibration testing, electronic circuit design, or chemical reactor dynamics. Such individuals frequently work out of a small office aided only by a secretary and an engineering assistant. They work for a variety of industrial clients on design problems in their specialty. Or they solve emergency problems on a system already constructed. They may also furnish technical assistance to attorneys. Such consultants rely on their contacts with local and regional firms and their reputation to maintain their businesses.

Other engineering consulting work is performed by larger consulting firms—firms that specialize in broader areas such as the design of bridges and structures or electric power stations. Such firms are engaged by companies or government agencies to complete an engineering project in their specialty which may require 20 to 30 engineers for a short period. They prepare all the design details and may even supervise the actual construction phase. Some engineering consulting firms are actually quite unspecialized and thus are able to perform all types of engineering work, on order, for companies whose engineering staffs are either overloaded or inadequate. At times these consulting firms provide temporary engineering services for a client, with employees working directly in the client's office.

Engineering consultants see a variety of engineering problems and work on current and important projects. In many instances they have opportunities for extended travel and above-average compensation. Periods of general economic difficulty, however, often lead to substantial cutbacks in consulting engineering business.

Education

In Figs. 1.2 and 1.4 we presented the various options for education in engineering and engineering technology. The teachers in most of these programs are engineers. Instructors in associate degree programs generally have a B.S. or M.S. in an appropriate engineering discipline. People teaching in the bachelor of engineering technology programs usually have a doctorate or an M.S. degree and substantial experience. Nearly all faculty in B.S., M.S., or doctoral programs have doctoral degrees.

Engineering educators are involved in many other activities besides classroom teaching. Courses and programs of study are being frequently modified. New courses are developed to meet changing needs. Laboratories are organized and equipment purchased and maintained.

In addition to the above duties, engineering educators must keep up to date in their field through study, consulting, and research. Consulting opportunities are frequently available through local industries. Educators are involved in research, especially at the university level, to involve graduate students in actual engineering projects and to illustrate the methods of reserach. University faculties supervise graduate research projects and are normally expected to seek financial support for their research program. This support may come from industry or governmental agencies.

Engineering educators above all have the opportunity to watch students learn and mature under their guidance and find roles in today's technological society.

1·4 PROFESSIONAL DEVELOPMENT

HISTORICAL PERSPECTIVE: THE DEVELOPMENT OF THE ENGINEERING PROFESSION

Although the practice of engineering goes back to the early stages of civilization, the development of engineering as a well-defined profession is more recent, having occurred during the period 1747 to 1872. Nevertheless, some feats of engineering belong to ancient history. For example, the Babylonians practiced hydraulic engineering in 2000 B.C., the Greeks began applying geometry to architecture around 300 B.C., and the Roman Empire employed a corps of surveyors to plan its roads during the first century A.D.

The men who performed these feats were typically self taught or trained by apprenticeship. In contrast, a professional requires "advanced knowledge acquired through specialized intellectual instruction in an institution of higher learning." By this definition, the founding of the first college-level engineering school marks the beginning of engineering as a profession. This occurred when L'École des Ponts et Chaussées (The School of Bridges and Highways) was founded in Paris in 1747. Other schools soon followed in Europe, but by 1840 only two engineering schools existed in the United States: the U.S. Military Academy at West Point and the Rensselaer School (now Rensselaer Polytechnic Institute) at Troy, New York. Despite the small number of schools, there was considerable demand for technically trained persons to oversee the construction and operation of canals, roads, railroads, and mines. In response, a single act of Congress gave impetus to a manyfold increase in the supply of college-trained engineers. The Morill Land Grant Act of 1862 granted federal aid to the states for the establishment of schools of agriculture and the mechanic arts. A land grant act prompted the establishment of new colleges in many states and the creation of engineering departments in existing colleges. By the end of the next decade (1862–1872) the number of engineering schools in the United States had jumped from 6 to 70. By 1880 there were 85, and at present there are nearly 300.

The number of professional engineering societies has increased in parallel with the growth of engineering education. Five of these societies are known collectively as founder societies. They are:

American Society of Civil Engineers (ASCE), founded 1852
American Institute of Mining Engineers (AIME), founded 1871
American Society of Mechanical Engineers (ASME), founded 1880
Institute of Electrical and Electronic Engineers (IEEE), founded 1884

Roman engineers knew the properties of concrete. This section of the Eiffel-Cologne Aqueduct was constructed in 80 A.D. When dug up 18 centuries later, it still had its strength and waterproof qualities. (*Smithsonian Institution, National Museum of American History.*)

The Industrial Revolution in the United States brought a surge of construction of canals, roads, railroads, and mines. Shown here are five double locks at Lockport, New York, which were constructed in 1825 for the Erie Canal, one of the great engineering projects of its time. During the construction of the Erie Canal, American engineers rediscovered the long-lost secret of Roman concrete. (*Smithsonian Institution, National Museum of American History.*)

American Institute of Chemical Engineers (AIChE), founded 1908

At the present time there are over two dozen major engineering societies in the United States. Some are quite broad in scope, others are rather specialized. The following is a partial list of engineering societies and their addresses and founding dates.

Coordinating Societies

Accreditation Board for Engineering and Technology, Inc. (ABET), 345 East 47th Street, New York, NY 10017 (founded 1980)

American Association of Engineering Societies, Inc. (AAES), 345 East 47th Street, New York, NY 10017 (founded 1980) (successor to Engineers Joint Council, founded 1904)

National Council of Engineering Examiners, Inc. (NCEE), Box 5000, Seneca, SC 29678 (founded 1920)

National Society of Professional Engineers (NSPE), 2029 K Street, N.W., Washington, DC 20006 (founded 1934)

Technical Societies for Branches of Engineering

American Institute of Aeronautics and Astronautics, Inc. (AIAA), 1290 Avenue of the Americas, New York, NY 10019 (founded 1931)

American Institute of Chemical Engineers (AIChE), 345 East 47th Street, New York, NY 10017 (founded 1908)

American Institute of Industrial Engineers, Inc. (AIIE), 25 Technology Park/Atlanta, Norcross, GA 30071 (founded 1948)

American Institute of Mining, Metallurgical, and Petroleum Engineers (AIME), 345 East 47th Street, New York, NY 10017 (founded 1871)

Constituent Societies of AIME: Society of Mining Engineers; Society of Petroleum Engineers; The Metallurgical Society

The American Nuclear Society (ANS), 244 East Ogden Avenue, Hinsdale, IL 60521 (founded 1954)

American Society for Engineering Education (ASEE), One Dupont Circle, Washington, DC 20036 (founded 1893)

American Society for Testing and Materials, Inc. (ASTM), 1916 Race Street, Philadelphia, PA 19103 (founded 1898)

The American Society of Agricultural Engineers (ASAE), 2950 Niles Road, St. Joseph, MI 49085 (founded 1907)

American Society of Civil Engineers (ASCE), 345 East 47th Street, New York, NY 10017 (founded 1852)

American Society of Heating, Refrigerating and Air-Conditioning Engineers (ASHRAE), 345 East 47th Street, New York, NY 10017 (founded 1894)

The American Society of Mechanical Engineers (ASME), 345 East 47th Street, New York, NY 10017 (founded 1880)

The Institute of Electrical and Electronic Engineers (IEEE), 345 East 47th Street, New York, NY 10017 (founded 1884)

National Institute of Ceramic Engineers (NICE), 4055 North High Street, Columbus, OH 43214 (founded 1938)

Society of Automotive Engineers (SAE), 400 Commonwealth Drive, Warrendale, PA 15096 (founded 1905)

Society of Engineering Science (SES), c/o School of Engineering, VPI, Blacksburgh, VA 24061 (founded 1963)

Society of Manufacturing Engineers (SME), One SME Drive, Box 930, Dearborn, Michigan (founded 1932)

THE ROLE OF PROFESSIONAL ENGINEERING SOCIETIES

Professional engineering societies serve their members and the public in different ways. Traditionally, technical, educational and legal functions have been most important. Recently, however, so-called "professional" functions (related to ethics, employment, incomes and professional status) have gained added prominence.

Technical Functions

In their charters, engineering societies frequently list technical functions first. This reflects the importance of the technical functions to engineering societies.

Technical journals and technical meetings Engineering societies in the United States disseminate information at the leading edge of technology by publication of technical journals and sponsorship of technical meetings. A particular society usually publishes one or two journals of general interest to its members and may publish other journals of primary interest to specialists. The IEEE, for example, publishes two general-interest journals and over 30 special-purpose journals.

Technical meetings have as their principal purpose the rapid dissemination of information through formal lectures and informal discussion. In addition, the meetings give engineers the opportunity to meet their colleagues and to discuss common technical concerns.

Most technical societies pay special attention to the information needs of the student engineer. Many publish special journals directed to students and offer incentives for students to attend their meetings. The meetings frequently have special sessions to serve the student engineer. Many schools have student chapters of some of the national societies. These student chapters may be able to tell you about student-oriented services of their society, or you may write directly to the societies and inquire about student membership.

Technical standards In response to the need for standardized and interchangeable engineered products, engineering professional societies have become leaders in industrial standardization. A technical standard is a specialized form of technical writing. It is a carefully drafted set of written requirements, usually relating to:

At meetings and conferences sponsored by engineering societies, engineers are able to keep up to date on developments in their field. The group shown here are discussing recent advances in the field of mechanical engineering. (*ASME.*)

- Technical specifications for a material, process, device, or measurement method
- Technical regulations designed to promote the safety of workers or the general public.

You encounter standards and standardized products every day. For example:

- A cassette of ASA 400 color film made in Rochester, New York that fits perfectly into a camera made in Yokahama, Japan. If the ASA dial is set to 400, the prints will be correctly exposed.
- A 12-mm metric socket wrench made in Bridgeport, Connecticut that fits bolts on your bicycle made in Genoa, Italy. Because the fit is neither too loose nor too tight, the bolts can be adjusted on the bicycle without rounding the corners of the hexagonal bolt heads or skinning knuckles.
- A replacement light fixture that can be safely installed because it has color-coded wires—black for hot, white for neutral, and green for ground.

All three of these examples involve American National Standards [4 to 6] which are coordinated and published by the American National Standards Institute (ANSI). The history of this standards organization began in 1884 when the ASME formed a committee on measurement standards. Other technical societies set up similar standards committees and in 1918 the ASME, IEEE (then AIEE), AIME, AIChE, ASTM and SAE cooperatively founded the American Engineering Standards Committee. This organization eventually became the American Standards Association (remember the ASA number for film) and is now the American National Standards Institute.

ANSI is a clearing house for over 400 organizations which originate standards and is not a government agency. The National Bureau of Standards, however, is a government agency. It produces relatively few standards but supports the standards process by performing tests (most technical societies do not operate testing laboratories), conducting measurements, and establishing test procedures. New standards are drawn up by committees formed of people from industries related to the standard who are willing to write an initial draft. Service on such a committee is normally voluntary and the members are frequently engineers who are paid by their employers during the time they serve on the committee. The draft of the standard is submitted to the appropriate technical society for review and is circulated throughout the industry.

After revision, re-review, and further revision, possibly over a period of years, the standard may be accepted by the technical society and forwarded to ANSI. The standard is then publicized

An American National Standard pamphlet, which represents the work of many people over a period of years at a cost of several million dollars. (*American National Standards Institute, Inc.*)

and further reviewed and revised until approved by ANSI as a national standard. At present, ANSI supports over 10,000 national standards [7]. National standards are used "voluntarily" unless a local, state, or federal government adopts the standard, making its use compulsory. For example, the *National Electrical Code* [6] must be followed whereve⁻ it has been adopted by a city or town as part of the building code.

Your first use of standards as an engineering student will probably be in connection with nuts and bolts—machine screw specifications—or electric wire specifications, such as the American Wire Gauge. We have touched on these subjects in Chaps. 8 and 9. As a practicing engineer you will find that standards affect nearly everything you do. Indeed it is likely that sometime in your engineering career you will help add a new standard to the 10,000 which already exist.

Educational Functions

Many societies perform direct educational services, such as sponsoring short courses for the continuing education of their members. Sponsorship of student chapters is another educational activity. Societies carry out their principal educational function, however, by influencing the courses of study at schools of engineering. Thus they have a direct effect on *your* engineering education.

Accreditation of engineering schools   The Accreditation Board for Engineering and Technology, Incorporated (ABET)[3] accredits the curriculums of each engineering program or course of study. ABET's board of directors has representatives from nineteen professional societies, including all listed in Sec. 1.4. In practical terms, this gives the profession of engineering a direct input into the content of the course of study at every engineering school which seeks accreditation. Engineering resembles in this respect the professions of law, medicine, and dentistry. By way of contrast, the accreditation for other disciplines such as mathematics, natural sciences, and humanities is under the control of educational, rather than professional, organizations.

ABET coordinates the curricular requirements of the individual specialties of the profession. It also sets up broad, uniform requirements which all curriculums must satisfy. In individual engineering schools, these general requirements translate into specific required courses in mathematics, natural sciences (such as physics and chemistry), engineering sciences (such as statics, dynamics, thermodynamics), and electric circuits, which all cur-

---

[3] ABET was formerly known as Engineers' Council for Professional Development (ECPD).

riculums of the school must include in their engineering degree requirements. In addition, ABET requires that one-eighth of the courses taken by engineering students must be outside engineering—in the humanities and social sciences, for example.

Accreditation serves a number of purposes. For the individual student, completion of an ABET-accredited curriculum is a major step toward meeting the requirements for a professional license (see Legal Functions, below). On a broader level, accreditation is the means by which the profession oversees the relevance and quality of engineering education.

Legal Functions

The legal functions of engineering societies relate to the licensing of individuals to practice as engineers. Although the engineering societies do not license engineers directly, they play a key role in the licensing process through their participation in ABET accreditation of curriculums. To paraphrase Leo Young, president (1980) of the IEEE, engineering societies do not qualify individuals, they qualify universities. Graduates of the engineering schools of these universities then have a clear path toward a professional license.

Registration as a professional engineer (P.E.) At present, about 27 percent of all engineers are P.E.s. Must you become a P.E.? Should you become a P.E.? How does a graduate engineer become a P.E.?

How you become a P.E. is rather simple to answer. Whether you must become a P.E. depends on the licensing laws of the 50 states. Our advice on whether you should become a P.E. is discussed further below.

Any engineer who offers services directly to the public is required by the state in which he or she practices to register as a P.E. Engineers who are employees of manufacturing firms are not usually required to be P.E.s, but engineers who work for consulting firms or companies engaged in building or structural design, in construction, or in public-utility operation are almost always required to be P.E.s. Figure 1.7 shows the path from graduation to a P.E. license. This is the procedure followed in more than half the 50 states, and resembles the procedure in the others. The first step is to take and pass an 8-hour exam called the "fundamentals of engineering" examination. If you pass the "fundamentals" exam, you become an "engineer in training," or EIT. After 4 years of engineering experience as an EIT, you are eligible to take the "professional" exam. On successful completion of the "professional" exam, you become a P.E.

Our advice: every engineering student should make serious

FIGURE 1-7
Requirements for professional registration—a procedure followed in more than half the 50 states that resembles the procedure followed in the others.

P.E. status
↑
Take and pass 8-hour "professional examination"
↑
4 years of professional experience
↑ Engineer-in-training status
Take and pass 8-hour fundamentals of engineering exam
↑ B.S. degree
Complete an ABET-accredited engineering curriculum in 4 to 5 years

42 PATHS TO AN ENGINEERING CAREER

plans to take the fundamentals exam during his or her final year in engineering school (or at the earliest date permitted by the state licensing board). We advise this based on the following considerations:

1. The necessity for registration is continually increasing.
2. Your career path or your state of residence may change. Even if your first job does not require registration, you may later move from an area of optional registration into an area of required registration.
3. Many companies are requiring that all their management personnel be P.E.s. Hence, promotion to management could hinge on P.E. registration.

Indeed, engineer-in-training status may be a help in landing certain desirable entry-level jobs. For details of registration requirements in your state, contact the National Council of Engineering Examiners at the address given earlier in this section.

Professional Functions

In the past decade or two engineering societies have become increasingly concerned with some of the nontechnical aspects of engineering. Attention has been given to ethics in the practice of engineering and also to the conditions of engineering employment. But some society members believe that professional concerns should be the most important activities of their society while others believe that engineering societies should be purely technical in scope.

Ethical standards The word "ethics" has two subtly different meanings. The first, relatively narrow meaning relates to rules of professional conduct agreed upon by a particular profession or group, as in "medical ethics." The second and broader meaning concerns ethics in a philosophical and moral sense, with the assignment of values such as right or wrong to human actions. For over half a century, engineering societies have had codes of ethics based on the first, narrow meaning of the word. Ethical considerations in the second, broader sense are much newer to engineering.

The "Code of Ethics for Engineers" shown in Figure 1.8 is, at present, the most commonly cited summary of engineering ethics. It contains four fundamental principles and seven fundamental canons. The canons contain phrases such as "welfare of the public" and "honor, integrity, and dignity" which may require ethical interpretation in the broader sense. The lengthly guidelines which are published with the Code of Ethics (see the Appendix) help to provide such an interpretation. The guidelines are also useful

CODE OF ETHICS OF ENGINEERS

THE FUNDAMENTAL PRINCIPLES

Engineers uphold and advance the integrity, honor and dignity of the engineering profession by:

I. using their knowledge and skill for the enhancement of human welfare;

II. being honest and impartial, and serving with fidelity the public, their employers and clients;

III. striving to increase the competence and prestige of the engineering profession; and

IV. supporting the professional and technical societies of their disciplines.

THE FUNDAMENTAL CANONS

1. Engineers shall hold paramount the safety, health and welfare of the public in the performance of their professional duties.

2. Engineers shall perform services only in the areas of their competence.

3. Engineers shall issue public statements only in an objective and truthful manner.

4. Engineers shall act in professional matters for each employer or client as faithful agents or trustees, and shall avoid conflicts of interest.

5. Engineers shall build their professional reputation on the merit of their services and shall not compete unfairly with others.

6. Engineers shall associate only with reputable persons or organizations.

7. Engineers shall continue their professional development throughout their careers and shall provide opportunities for the professional development of those engineers under their supervision.

Approved by the Board of Directors, October 1, 1974

FIGURE 1·8
This code was developed by the Engineers Council for Professional Development in 1977. It has been recommended for use by the American Association of Engineering Societies. (*Accreditation Board for Engineering and Technology, Inc.*)

in resolving possible conflict between the canons. For example, engineers are expected to "hold paramount the safety, health and welfare of the public" and also to act as "faithful agents" for an employer. There may be situations where these canons conflict. The guidelines state that

> Should the Engineers' professional judgement be overruled in circumstances where the safety, health and welfare of the public are endangered, the Engineers shall inform their clients or employers of the possible consequences and notify other proper authority of the situation, as may be appropriate.

The common term for notifying "other proper authority" is "whistle blowing." There is no more controversial engineering topic in these days of environmental pollution, automobile recalls, etc. Perhaps the most disputed issue is defining the role that engineering professional societies should play in ensuring due process for their members. Many people believe that engineering societies should provide a hearing and appeal process to solve ethical disputes between engineers and employers. The discussion is a complex one. It will continue for years to come, indicating the expanding role of engineering societies.

Engineering employment The salaries of practicing engineers are almost always a matter of individual negotiation between engineer and employer. Engineering societies are seldom involved directly in matters of salary. They do play an important informational role, however, and they have become increasingly active in the area of fringe benefits such as insurance and pension programs.

At present the question of vesting is receiving considerable attention. If an employee has vested rights in a company's pension plan, he will receive pension payments when he retires even though he was not working for that company at retirement. Benefits frequently become vested only after 5 or more years of continuous employment. The case of an engineer who has worked 26 years under five different employers and never became vested in a pension plan is not at all unusual. For this reason, engineering societies are devoting considerable effort toward the development of portable pensions which will automatically follow an engineer who changes jobs.

Engineering societies also play an important role in aiding the free market of engineering employment. Membership lists are provided to employers recruiting engineers; facilities are available at national society meetings to facilitate contacts between engineers and prospective employers; and professional journals are used for employment advertising. Advertisements in professional journals can give you valuable insight into available job opportunities and the engineering skills in current demand.

OTHER READING

For those interested in further reading related to the engineering profession see Refs. 8 to 23. The first group [8 to 10] includes introductory engineering books which provide a broad coverage of the profession. References 2 and 11 give a more detailed coverage of professional engineering registration and 12 is an annual survey of the technical job market. Ralph Nader's well-known book [13] is an important critique of the automotive industry which has had substantial impact on both society and engineering. In reading it you should remember that the author is an attorney, practiced at presenting only one side of an argument. Similar treatments of other industries may be found in Refs. 14 and 15.

Ethics in engineering is discussed in Refs. 16 to 19. You may have particular interest in Ref. 19, which considers ethics in the classroom. The problems of engineers torn between conscience and employer are described in Refs. 20 to 23, which deal with whistle blowing. Readings of this type should help you to understand the difficult decisions sometimes faced by practicing engineers.

REFERENCES

1. Harmon, W. W., J. B. Franzini, W. G. Ireson, and S. J. Kline, "Abstract Report of the Stanford University Committee on Evaluation of Engineering Education," *Journal of Engineering Education*, vol. 44, p. 258, December 1953.
2. Kemper, J. D., *The Engineer and His Profession*, 2d ed. New York, Holt, 1975.
3. Sheridan, P. J., "Engineering and Engineering Technology Enrollments, Fall 1979," *Engineering Education*, vol. 71, no. 1, October 1980, pp. 46–54.
4. American National Standards Institute, *Method of Determining the Speed of Color Negative Films for Still Photography*, ANSI PH 2.27–1979.
5. American National Standards Institute, *Metric Hex Cap Screws*, ANSI B18.2.3.1M-1979.
6. National Fire Protection Association, *National Electrical Code*, ANSI/NFPA 70-1978.
7. American National Standards Institute, *Catalog of American National Standards*, ANSI (1430 Broadway, New York, NY 10018).
8. Beakley, G. C. and H. W. Leach, *Engineering*, 3d ed. New York, Macmillan, 1977.

9. Eide, A. R., et al., *Engineering Fundamentals and Problem Solving,* New York, McGraw-Hill, 1979.
10. Glorioso, R. M., and F. S. Hill, Jr., *Introduction to Engineering,* Englewood Cliffs, N.J., Prentice-Hall, 1975.
11. Constance, J. D.: *How to Become a Professional Engineer,* New York, McGraw-Hill, 1958.
12. "Salaries of Scientists, Engineers and Technicians—A Summary of Salary Surveys," Washington, D.C. The Scientific Manpower Commission (1776 Massachusetts Avenue, NW 20036)
13. Nader, Ralph, *Unsafe at Any Speed; The Designed-In Dangers of the American Automobile,* New York, Grossman, 1972.
14. Olson, M. C., *Unacceptable Risk: The Nuclear Power Controversy,* New York, Bantam, 1976.
15. Godson, J., *The Rise and Fall of the DC-10,* New York, David McKay, 1975.
16. American Society of Civil Engineers, *Proceedings of the Conference on Engineering Ethics,* May 18–19, 1975, New York, ASCE, 1976.
17. Flores, A. (ed., vol. I) and R. J. Baum, (ed., vol. II), *Ethical Problems in Engineering,* 2d ed., Troy, N.Y., Center for the Study of Human Dimensions of Science and Technology, 1980.
18. Hughson, R. V., and P. M. Kohn, "Ethics," *Chemical Engineering,* vol. 87, no. 19, September 22, 1980, pp. 132–147.
19. Christiansen, D., "On Falsifying Data," *IEEE Spectrum,* vol. 17, no. 12, December 1980, p. 19.
20. Nader, R., P. J. Petkas, and K. Blackwell, *Whistle Blowing; The Report of the Congress on Professional Responsibility,* New York, Grossmann, 1972.
21. Chalk, R., and F. von Hippel, "Due Process for Dissenting 'Whistle-Blowers'," *Technology Review,* vol. 81, no. 7, June/July 1979, pp. 48–55.
22. Raven-Hansen, P., "Dos and Don'ts for Whistleblowers; Planning for Trouble," *Technology Review,* vol. 82, no. 6, May 1980, pp. 34–44.
23. Anderson, R. M., et al., *Divided Loyalties—Whistle Blowing at BART,* Lafayette, Ind., Purdue Research Foundation, 1980.

PART TWO

CALCULATIONS AND COMPUTING

The world of engineering is quantitative and depends heavily on numbers, data, and calculations. In Chap. 2 we discuss the proper methods of performing calculations and presenting results in ways that are convenient and easily understood. These techniques are common to the many branches of engineering.

Of course, calculations today are very often carried out with electronic calculators and computers. These machines are used to handle simple arithmetic and to perform operations that an engineer could never carry out unassisted. Computers have become a dominant part of technology not only for engineers but for society as a whole. Chapter 3 is devoted to the history and principles of these electronic marvels.

2 INTRODUCTION TO ENGINEERING CALCULATIONS

The work of an engineer is very often quantitative, so the use of numbers, arithmetic, graphs, and other basic tools becomes second nature to the practicing engineer. The special terms used to describe standard procedures and techniques become a part of an engineer's working vocabulary. In this chapter, you will find an introduction to the basic tools for calculations as used by all types of scientists and engineers. We begin with the simplest ideas about writing numbers and performing arithmetic and end with discussion of the proper amount of precision necessary in calculations as well as a consideration of units of measurement, graphs, and methods of handling statistics.

2·1 NUMBERS AND ARITHMETIC

In performing engineering calculations and presenting engineering data, it should be recognized that the numbers being used have only a limited amount of precision and accuracy. When you present the results of an engineering calculation, you do not simply copy a number containing eight or more digits from the face of a pocket calculator. By doing this, you imply that your results are accurate to eight digits, a type of accuracy that is only rarely possible in engineering. The number of digits that should be presented is usually less than eight, because it depends on the particular problem at hand and involves much more than the accuracy of the arithmetic performed by a calculator or computer. Material which follows furnishes a guide to presenting results with a proper number of digits, or significant figures.

SIGNIFICANT FIGURES

Engineering data and the results of engineering computations should be presented to an appropriate number of significant figures to avoid giving a misleading impression of their precision and accuracy. For example, the numbers

$$3.11$$
$$42.3$$
$$602$$
$$0.0123$$

each contain three significant figures. The numbers

$$1.2$$
$$678.1$$
$$0.34056$$

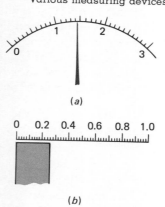

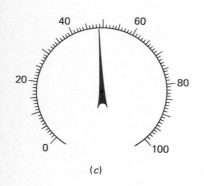

FIGURE 2·1
Various measuring devices.

have two, four, and five significant figures, respectively. For any number, the significant figures are those which indicate the quality of the measurement or computation being presented. Zeros are not always significant figures when included in a number, since they may be necessary only to locate the decimal point, as in the last examples in each of the two groups above. There may be confusion with larger numbers because the trailing zeros in larger numbers may be meant to be significant or to be simply an indication of the magnitude of the number. For example, the number 4500 may have two, three, or four significant figures depending on interpretation, but this is not clear from simply looking at the number. The confusion can be corrected by the use of scientific notation, which is discussed in the next section. There is no confusion, however, when zeros are added after a decimal point at the right of a number, since they are always then understood to be significant. That is, the number 4.50 is said to have three significant figures, and the number 45.00, four significant figures. The zeros are obviously not necessary to define the magnitudes of these numbers, so they are not added in these cases unless they are intended to be significant.

We can see how actual measurements are recorded to a proper number of significant figures by looking at Fig. 2.1. Three different types of measuring devices are shown: first, a voltmeter measuring the voltage from a flashlight battery; second, a machinist's scale measuring a thickness; and third, a pressure gage measuring home water pressure. In each case, the reading obtained can not be expected to be more accurate than one-half the smallest subdivision of the scale used. It is customary, however, to estimate the instrument reading within the smallest subdivision, subdividing it by eye and recording the reading to the proper number of significant figures. Possible error in the last significant figure is introduced by the estimation and should, of course, be recognized. Thus, the readings of the three scales in Fig. 2.1 can be recorded as 1.43 volts, 0.260 inch, and 48.6 pounds per square inch. Each measurement is to three significant figures, but the last figure is somewhat in doubt. Measurement quality of this sort can be obtained in engineering with a modest amount of care. Improvements in measurement quality can be made when necessary by using more sophisticated instruments and readout devices that are more precise than a simple scale and pointer. A device called a "vernier scale," for example, is often used to improve on the eyeball estimation within a subdivision and help one obtain an extra significant figure. Digital readout displays actually present the digits of interest in the measurement and further increase the number of significant figures which can be read. An automobile odometer uses a digital display which normally presents as many as six significant figures.

The rules for using an appropriate number of significant figures are followed closely in practice for presenting and applying engineering data and performing calculations. However, engineers and scientists are not continually preoccupied with significant figures; there are many instances where the rules are somewhat informal. One such case is when a given number is quite certainly intended to be a simple integer. In this situation the significant figures in the number can be increased by adding an appropriate number of zeros. For example, when we write that 1 foot = 0.3048 meter or that 1 pound = 454 grams, it should be clear that exactly 1 is intended. On the left sides of these equal signs, zeros which increase the significant figures of the ones as desired may be understood. The same is true when integers occur naturally in algebraic equations. An example of this might be that the radius of a circle is one-half the diameter, or $r = d/2$. In the equation, we need not write 2.0000; it is simply understood that the 2 is exact. Similarly in problem definitions (especially in textbooks), where it might be stated that "given a voltage supply of 6 volts," or "for a mass of 1 kilogram," or "with a velocity of 10 feet/second," if no additional information is available, it is reasonable to add one or two zeros to the numbers presented and perform any required calculations using three significant figures. It would not normally be appropriate, however, to go beyond three significant figures without further justification.

Precision and Accuracy

At this point we should discuss the terms precision and accuracy, which we have mentioned earlier. They are used throughout engineering in describing calculations and instrument measurements. *Precision* refers to the number of significant figures representing a quantity or to the spread or variation in repeated readings of an instrument measuring a constant physical quantity. *Accuracy* refers to the nearness of a given number or measurement to the actual physical quantity it represents.

Digital displays are normally precise, since they can present several significant figures. However, the accuracy of an instrument using such a display is not demonstrated unless it has been carefully calibrated so that its readings properly indicate the quantity being measured. In some cases, you may find a digital display to be very precise but the number presented inaccurate. The automobile odometer, mentioned earlier, may read to six significant figures; but it is attempting to measure the distance of automobile travel by counting the revolutions of the drive shaft. Its accuracy depends not only on how well it counts those revolutions but also on how well the revolutions reflect the distance traveled. Wheel slip and wheel diameter are two things among others which can influence its performance. In fact, an accuracy of 1 percent would be very good for an automotive odometer. Even at this level of accuracy, however, errors of about 100 miles in the odometer reading would be expected after 10,000 miles of travel, despite the fact that the odometer reads very precisely to a tenth of a mile.

Similar situations occur in the use of digital computers and pocket calculators. These are very precise devices that can easily perform calculations to eight and more significant figures. However, computer results can represent physical quantities only as accurately as are the data and equations used in the calculations. In most normal engineering applications, these do not represent physical quantities more accurately than 1 percent. Three significant figures are adequate for presenting numbers to this accuracy; four significant figures can represent an accuracy of one-tenth of 1 percent. In practice, one should consider carefully the actual accuracy of an engineering calculation before presenting the results to more than three significant figures. It is seldom necessary for a number to be more precise than it is accurate.

SCIENTIFIC NOTATION

In engineering and scientific calculations, a variety of very large and very small numbers are frequently found. For example, the elastic modulus of steel, which indicates its stiffness, is 30,000,000 pounds per square inch and a typical electronic capacitor may have a capacitance of 0.00000012 farads. In order to work conveniently with various-sized numbers such as these, scientific notation is used to eliminate unnecessary zeros. In scientific notation, numbers are written in a standard form as the product of a number between 1 and 10, and the number 10 raised to the appropriate exponent. The numbers mentioned above become 3.00×10^7 pounds per square inch and 1.2×10^{-7} farads, respectively. This compact form removes the strings of zeros and also preserves the intended number of significant figures. In this form there need not be confusion in larger numbers about whether zeros are significant or not. It is clear now that the elastic modulus is defined to three significant figures. Our capacitor, of course, is defined to only two significant figures.

The original form of numbers expressed in scientific notation can always be reconstructed by carrying out the multiplication indicated in the notation. For the numbers above,

$$3.00 \times 10^7 = 3.00 \times 10{,}000{,}000 = 30{,}000{,}000$$

and

$$1.2 \times 10^{-7} = 1.2 \times 0.0000001 = 0.00000012$$

Another way of interpreting scientific notation is to visualize the exponent of the 10 as a decimal-point shifter. For example, the $+7$ means to move the decimal point seven places to the right, the -7 means to move it seven places to the left. Thus:

$$3.00 \times 10^7 = 3\underbrace{0\,0\,0\,0\,0\,0\,0}_{1\;2\;3\;4\;5\;6\;7}. = 30{,}000{,}000$$

and

$$1.2 \times 10^{-7} = .\underbrace{0\,0\,0\,0\,0\,0\,1}_{7\;6\;5\;4\;3\;2\;1}2 = 0.00000012$$

In carrying out calculations, particularly when using pencil and paper, it is frequently useful to manipulate numbers into nonstandard forms of scientific notation so that arithmetic operations like addition and subtraction can be more easily performed. These forms can be obtained by simply adjusting the decimal-point position and exponent simultaneously until the necessary form is achieved. If desired, our numbers above could be written in any of the following forms:

$$3.00 \times 10^7 = 30.0 \times 10^6 = 0.0300 \times 10^9 = 3000 \times 10^4$$

and

$$1.2 \times 10^{-7} = 0.12 \times 10^{-6} = 120 \times 10^{-9} = 0.0012 \times 10^{-4}$$

During the process of this type of manipulation, the clear indication of the number of significant figures present may be lost. This is shown in two places above, where it has been necessary to add trailing zeros.

EXPRESSING RESULTS OF CALCULATIONS

In performing a series of arithmetic operations, each number in the calculation is known only to a certain number of significant figures, so the results of the calculation should be expressed with an appropriate number of significant figures. As a general guide to performing computations and presenting results with the proper precision,

1. Carry out the computation so that an excess number of digits is available.
2. Round off to the proper number of significant figures. In rounding off, increase the last digit retained by 1 if the first digit discarded is larger than 5. If the discarded digits represent 5 exactly, the last digit retained should be increased by 1 only if it is odd.
3. For multiplication and division, round off so that the number of significant figures in the result equals the smallest number of significant figures contained by a quantity in the operation.
4. For addition and subtraction, round off so that the least-significant digit appearing in the result corresponds to the most-significant doubtful digit contained in the addition or subtraction.
5. For combinations of arithmetic operations, perform multiplications and divisions in quantities to be added or subtracted and round off as required before performing the addition or subtraction. If additions or subtractions are involved in quantities to be multiplied or divided, perform them and round off before multiplying or dividing.
6. In more complicated calculations, such as the solution of simultaneous algebraic equations, when it is necessary to obtain intermediate results with extra significant figures to be sure that the final results will be reasonably accurate, use common sense and set aside the rules above.
7. When extensive computer calculations are being performed, also treat the preceding rules less formally. It is not customary to interrupt the computation at each stage to establish the number of significant figures. However, upon completing the computation, consider the overall accuracy and round off the results appropriately.

Examples

Rounding off

$$1.7543 \to 1.75$$
$$6.342 \to 6.3$$
$$6.857 \to 6.9$$
$$10.660 \to 10.7$$
$$4.365 \to 4.36$$
$$0.6750 \to 0.68$$

Multiplication and division

$$0.743 \times 6.815 = 5.06$$

$$\frac{853}{0.71} = 1.2 \times 10^3$$

$$\frac{(7.4 \times 10^{-4})(9.95 \times 10^6)}{1.273 \times 10^4} = 0.58$$

Addition and subtraction The most-significant doubtful digits are underlined below and the results are rounded off to that point. Note that a digit in the tenths column, for example, is more significant than is a digit in the hundredths column.

Problem and proper result	Work
$0.27 + 17.1 = 17.4$	$.27$ $17.\underline{1}$ $\overline{17.37}$
$6.3 \times 10^4 - 1.564 \times 10^3 = 6.1 \times 10^4$	$6.\underline{3} \times 10^4$ -0.1564×10^4 $\overline{6.1436 \times 10^4}$
$7.386 \times 10^{-4} + 6.2 \times 10^{-6}$ $+ 4.33 \times 10^{-7} = 7.452 \times 10^{-4}$	$7.38\underline{6} \times 10^{-4}$ $0.06\underline{2} \times 10^{-4}$ 0.00433×10^{-4} $\overline{7.45233 \times 10^{-4}}$

Combinations

$$\frac{0.645(3.14)}{0.173(221)} - \frac{7.63(4.8)}{7543} = 0.0530 - 0.0049 = 0.0481$$

$$\frac{(1.74 - 0.02)(274 + 1.36)}{2.030[(9.3 \times 10^4) - (1.57 \times 10^5)]} = \frac{1.72(275)}{2.030(-6.4 \times 10^4)} = -3.6 \times 10^{-3}$$

TABLE 2·1
EXPRESSING THE RESULTS OF CALCULATIONS

Arithmetic Operation	Calculated Result	Calculated Upper Limit	Calculated Lower Limit	Properly Expressed Result
$x + y$	218.1	219.2	217.0	218
$x - y$	211.9	213.0	210.8	212
$x \cdot y$	666.5	691.2	642.0	6.7×10^2
x/y	69.3548	72.0000	66.8750	69
y/x	0.01442	0.01495	0.01389	0.014

We will now consider a further series of examples to help illustrate rules 3 and 4 above in more detail. Let $x = 215$ and $y = 3.1$. The quantity x is defined to three significant figures, so there is some doubt associated with the 5, which is the least-significant figure. It is therefore a doubtful digit. If it is incorrect by ± 1, then the upper limit of x is $x_u = 216$ and the lower limit $x_l = 214$. Similarly, the quantity y defined to two significant figures could range from $y_u = 3.2$ to $y_l = 3.0$. The first column of Table 2.1 shows the basic operations that might be performed with x and y. The second column shows the calculated results for the values of 215 and 3.1. However, we do not actually know x and y exactly, and since they may vary, the results of the arithmetic operations will also vary. With our values for x_u, x_l, y_u, and y_l, we can find the upper and lower limits for each arithmetic operation, shown in columns 3 and 4 of Table 2.1. For example, the upper limit of the sum is $x_u + y_u = 219.2$ and the lower limit on the x/y quotient is $x_l/y_u = 66.8750$. Columns 3 and 4 in the table, give the range that the sum, the product, etc., may take on. It is clearly not reasonable to represent the operations by the calculated results in the second column. Rules 3 and 4 have been applied in the last column to properly express the calculated results. The number of significant figures in this column is appropriate compared with the expected range of the results. Ideally, the least-significant digit presented in the result should be the only doubtful digit.

PROBLEMS

2.1 How many significant figures are in each of the following numbers?
  (a) 700
  (b) 7.00×10^2
  (c) 0.00560
  (d) 0.05060
  (e) 7.0056
  (f) 50
  (g) 5
  (h) 5.000×10^3
  (i) 6.03
  (j) 6.30

2.2 Carry out the following multiplications or divisions and show the results to the proper number of significant figures. Indicate clearly how many significant figures are in your answers.
(a) 1.57×191
(b) $(6.54 \times 10^3) \times (5.4 \times 10^{-7})$
(c) $\dfrac{7.534 \times 10^{-3}}{6.54 \times 10^5}$
(d) $51.6/8.536 \times 10^{-3}$
(e) $\dfrac{(7.5 \times 10^2)(6.43 \times 10^{-4})}{2.411 \times 10^{-4}}$
(f) $\dfrac{(6.523 \times 10^5)(4.38 \times 10^{-4})}{9274(6.29 \times 10^{-8})}$
(g) $\dfrac{(6.894 \times 10^{-7})(8.593 \times 10^6)}{(1.4 \times 10^8)(4.231 \times 10^{-8})}$
(h) $\dfrac{5425}{0.0052 \times 7005}$

2.3 For the additions and subtractions below, write the results to the proper number of significant figures. Your answers should show the number of significant figures unambiguously.
(a) $757 + 7.3$
(b) $1075 + 4.8$
(c) $7.57 \times 10^5 + 6.3 \times 10^2$
(d) $9.532 \times 10^6 + 6.5 \times 10^4$
(e) $8.9 \times 10^{-7} - 7.543 \times 10^{-6}$
(f) $8.1 \times 10^{-4} - 6.243 \times 10^{-5}$

2.4 For the combined operations below, determine the results to the proper number of significant figures.
(a) $7.53(6.2 \times 10^{-3}) + (7.23 \times 10^6)(8.5 \times 10^{-8})$
(b) $(7.53 + 8.0 \times 10^2)(3.53 \times 10^{-4} - 7.543 \times 10^{-3})$
(c) $\dfrac{2.853 \times 10^{-5} - 3.2 \times 10^{-7}}{5.671 \times 10^3 + 7.4}$
(d) $\dfrac{6.54 \times 10^{-3}}{1.756 \times 10^3 + 27.4} - 3.15 \times 10^{-6}$

2.5 At 5 P.M. on September 15, 1972, the population of Pittsburgh was 4.62×10^5. In the next 24 hours, there were 36 births and 23 deaths in the city. What was the new population expressed to the proper number of significant figures?

2.6 The formula for kinetic energy is $KE = \frac{1}{2}mv^2$. If a particle has a mass $m = 6.82 \times 10^5$ kilograms and a velocity $v = 9.5 \times 10^{-1}$ meters per second, what is its kinetic energy in joules? Show the proper number of significant figures in your answer. Note that the $\frac{1}{2}$ is known exactly.

2.7 A ruler has been used to measure the dimensions of a quarter. The diameter is found to be 2.40 centimeters and the thickness is measured at 0.17 centimeters. What is the volume of material used in the coin, expressed in cubic centimeters, to the proper number of significant figures?

2·2 DIMENSIONS AND UNITS

The preceding sections dealt with ways of properly expressing numbers to indicate precision and accuracy. Since numbers are used by engineers to represent physical quantities, attention must be given to the units of measurement relating the numbers to the physical quantities of interest. Units of measurement have varied from country to country since the ancient civilizations of Babylon and Egypt and have often been barriers to communication and international trade. Today, nearly all the nations of the world are committed to the metric system, so it seems that the difficulties will finally end. The United States remains the only major nation which has not yet made complete commitment to the metric system. However, metric measurements are being used more and more in American science and technology and even in daily life.

The modern metric system is known by the French term as Le Système International d'Unités (International System of Units), normally abbreviated as SI (for Système International). In what follows, we first discuss the concept of dimensions, which forms the basis of measurement systems, and then describe both the SI units and the U.S. Customary System (USCS) of units. Until SI units are completely accepted in the United States, it will be necessary for engineers to work comfortably in both. And throughout this book we will present examples using both systems of units.

DIMENSIONS

A dimension is a characteristic of an object, condition, or event which can be defined quantitatively. In order to satisfactorily describe some particular object, condition, or event for a given purpose, the dimensions of interest must be identified and a numerical representation must be established for each. Every dimension is measured in units. It is in the relative size and subdivision of the units that the conflicts between measurement systems occur.

Consider water in a container. Two dimensions which are of obvious interest in describing the water are the dimension of volume, measured perhaps in units of cubic meters or cubic feet, and the dimension of temperature, which can be measured in units of Celsius degrees (formerly centigrade) or Fahrenheit degrees. Other dimensions describing the water which may be more significant for some purposes are its mass (units of kilograms or pounds), its density (units of kilograms per cubic meter or pounds per cubic foot), or its height in a container, which is a length dimension measured simply in units of meters or feet. There are more dimensions which could be listed to describe the properties or geometry of the water—and, of course, we have not yet begun to consider the container holding the water.

TABLE 2·2
SYMBOLS FOR THE FUNDAMENTAL DIMENSIONS

Fundamental Dimension	Length	Time	Mass	Force	Temperature	Electric charge	Luminous intensity
Symbol	L	T	M	F	θ	Q	I

Thus, many different types of dimensions may be easily identified. In fact, when we consider all possible physical quantities, we find that the number of dimensions is essentially limitless. Fortunately, certain fundamental dimensions (also called primary or basic dimensions) may be defined. These can be combined to obtain all other dimensions, which are then referred to as derived, or secondary, dimensions. Fundamental dimensions are shown in Table 2.2 along with their symbols. Derived dimensions are established from these fundamental dimensions: velocity, which is length per unit time (L/T); area, which is the product of two lengths (L^2); density, which is mass per unit volume (M/L^3), and volume flowrate, which is volume per unit time (L^3/T). Other examples are shown in Table 2.3.

We use the symbols L, T, M, etc., to represent fundamental dimensions so that we can speak of dimensions in general without having to assign particular units. The units of the fundamental dimensions are called fundamental, or basic, units. Any derived dimension can be easily given specific units by substituting basic units for each of the fundamental dimensions found in the expression for the derived dimension. Thus, velocity can be measured in units of meters per second, area in square feet, pressure in pounds per square inch, and rate of temperature change in degrees Celsius per hour.

Work, Energy, and Power

Two important and widely used dimensions which are often not completely understood describe the concepts of work, energy, and power. Technically, work is the process of applying a force through a distance. If a force is applied to some object and there is no motion (as when one leans on a shovel), the work is zero. Also, if there is motion without force (a coasting spacecraft), the work is

TABLE 2·3
TYPICAL DERIVED DIMENSIONS

Dimension	Area	Volume	Velocity	Acceleration	Pressure	Density	Volume flowrate	Mass flowrate	Work, energy	Power	Electric current
Symbol	L^2	L^3	$\dfrac{L}{T}$	$\dfrac{L}{T^2}$	$\dfrac{F}{L^2}$	$\dfrac{M}{L^3}$	$\dfrac{L^3}{T}$	$\dfrac{M}{T}$	FL	$\dfrac{FL}{T}$	$\dfrac{Q}{T}$

zero. For a constant applied force, work is the product of force and distance. It has the dimension of force times length (FL). Thus, if 100 lbf is required to lift a weight and the weight is simply raised a height of 10 ft, then 1000 ft·lbf of work has been done.[1] Energy is the ability to do work. It therefore has the same dimension (FL) and is measured in the same units as is work. By dropping the raised weight back through the 10-ft height in a controlled way, work is done—and thus the raised weight has energy (in this case potential energy). The work which can be recovered is equal to the work done in raising the the weight for this simple case, so the potential energy of the raised weight is 1000 ft·lbf.

Power is the rate at which work is done or the rate at which energy is dissipated. Its dimension is work or energy per unit time (FL/T). If our raised weight is used in a bell tower to drive a large clock and is able to drive it for a full day, we can say that the clock requires a power of $1000/(24)(3600) = 0.012$ ft·lbf/s for its operation. Since one horsepower is defined to be 550 ft·lbf/s, this is a power level of 2.1×10^{-5} hp. The raised weight might be dropped without control and not do any useful work except deforming the ground where it lands. The entire energy of the weight is then dissipated in the impact. If the impact occurs in, say, one-hundredth of a second, the average power is $1000/0.01 = 1.0 \times 10^5$ ft·lbf/s, or 180 hp during the impact.

LE SYSTÈME INTERNATIONAL D'UNITÉS (SI)

An international treaty establishing the metric system resulted from meetings held in France in 1872. In addition to providing standards for the metric system, this treaty formed the General Conference of Weights and Measures, an international body which meets every 6 years. The United States signed the original treaty and is a member of this international body. At the 1960 meeting of the General Conference of Weights and Measures, the representatives of 40 nations modernized the metric system and created Le Système International d'Unités. Legislation was enacted by Congress in 1974 for the adoption of SI. Even though no commitment has been made with respect to a compulsory changeover date, changeover is beginning in the United States.

The fundamental, or base, units of SI and their symbols are shown in Table 2.4. Symbols of units that derive from proper names are capitalized, other symbols are not. Periods are not used after the symbols and the symbols do not take on plural forms. Each unit has been defined with extreme care, as outlined below.

---

[1] lbf is used to designate pounds-force. The significance of this is described in our discussion of USCS units.

TABLE 2·4
BASIC UNITS IN SI*

	Unit Name	Symbol
Length	meter	m
Mass	kilogram	kg
Thermodynamic temperature	kelvin	K
Time	second	s
Electric current	ampere	A
Amount of matter	mole	mol
Luminous intensity	candela	cd

* K used without a degree symbol represents temperature in kelvins. For temperatures on the Celsius (°C), Fahrenheit (°F), or Rankine (°R) scales, the degree symbol is used.

The meter At the French National Convention of 1795, the meter was intended to be one ten-millionth of the circumferential distance from the equator to the North Pole. It is now formally defined as 1,650,763.73 wavelengths of an orange-red light radiated by the krypton 86 atom.

The kilogram Established originally as the mass of one-thousandth of a cubic meter of water, the kilogram is now the mass of a platinum specimen located at the headquarters of the International Bureau of Weights and Measures in France.

The second Originally defined so that one mean solar day contains 86,400 seconds, the second is now defined as 9,192,631,770 times the period of radiation emitted by cesium 133.

Thermodynamic temperature The kelvin is defined as the fraction 1/273.16 of the thermodynamic temperature at the triple point of water. This makes the kelvin equal in size to the Celsius degree. Thermodynamic temperature differs from the Celsius scale, however, since zero corresponds to absolute zero (-273.15 °C). Of course, on the Celsius scale, zero is the freezing point of water at standard atmospheric pressure.

The ampere Inasmuch as current represents the flow of electric charge, the ampere is intended to designate a flow of one coulomb per second, which corresponds to a flow of 6.25×10^{18} electrons per second. The ampere is now formally defined as the current in two parallel wires 1 meter apart which produces a specific force between them (2×10^{-7} N).

The mole Representing an amount of matter containing Avagadro's number of atoms or molecules (6.03×10^{23}), the mole is now defined as an amount of matter containing the same number of elementary entities as there are atoms in 0.012 kg of carbon 12.

TABLE 2·5
DERIVED UNITS IN SI

	Unit Name	Formula	Symbol
Acceleration—linear	meter per second per second	m/s²	
Velocity—linear	meter per second	m/s	
Frequency	hertz	s⁻¹	Hz
Force	newton	kg·m/s²	N
Pressure or stress	pascal	N/m²	Pa
Density	kilogram per cubic meter	kg/m³	
Energy or work	joule	N·m	J
Power	watt	J/s	W
Electric charge	coulomb	A·s	C
Electric potential	volt	W/A	V
Electric resistance	ohm	V/A	Ω
Electric capacitance	farad	C/V	F
Magnetic flux	weber	V·s	Wb
Inductance	henry	Wb/A	H

The candela Originally one candle power, the candela is now defined as the luminous intensity of one six-hundred-thousandth of a square meter of a blackbody at the freezing temperature of platinum at standard atmospheric pressure.

The derived units for other physical quantities are obtained by combining the fundamental units. Table 2.5 shows the more common derived units along with their formulas in terms of the fundamental units and/or preceding derived units. Symbols are shown for those units which have them.

Perhaps the most difficult of the derived units to understand is the unit for force, named after Isaac Newton. Although force is frequently considered to be a fundamental dimension, force and mass cannot be considered independently to be fundamental dimensions or fundamental units. Newton's law ($f = ma$) is used as a basis for relating force, mass, and acceleration, so that we can express the derived dimension of force in terms of the fundamental mass, length, and time dimensions as ML/T^2. Substituting the SI units for these dimensions, we see that force can be measured in units of kg·m/s². One newton is defined to be equal to one kilogram-meter per second squared. Thus, a mass of 1 kg at sea level and 45° latitude, where the acceleration of gravity is 9.81 m/s², will require an upward force of $f = (1.00 \text{ kg})(9.81 \text{ m/s}^2) = 9.81$ N to support it. One might be tempted to say that this is the weight of 1 kg of mass on earth, but the term weight is not used in SI.

Quantities we are interested in are frequently relatively small or large compared with the standard SI units. For example, a computer operation may occur in 1.0×10^{-9} s, a sheet of metal may be

TABLE 2·6
SI PREFIXES

Multiple	Prefix	Symbol
10^{12}	tera	T
10^9	giga	G
10^6	mega	M
10^3	kilo	k
10^2	hecto	h
10	deka	da
10^{-1}	deci	d
10^{-2}	centi	c
10^{-3}	milli	m
10^{-6}	micro	μ
10^{-9}	nano	n
10^{-12}	pico	p
10^{-15}	femto	f
10^{-18}	atto	a

0.0005 m thick, and atmospheric pressure may be 1.01×10^5 Pa. To avoid having to write such small and large numbers, the prefixes shown in Table 2.6 may be used with the SI units. It is normally desirable to hold the number being expressed to a magnitude between 0.1 and 1000 and to use those prefixes for quantities that have an exponent for the number 10 that is a multiple of 3. The quantities mentioned above would thus be conveniently written as

$$1.0 \times 10^{-9} \text{ s} = 1.0 \text{ ns}$$
$$0.0005 \text{ m} = 0.5 \text{ mm}$$
$$1.01 \times 10^5 \text{ Pa} = 101 \text{ kPa}$$

CUSTOMARY AMERICAN UNITS—THE U.S. CUSTOMARY SYSTEM (USCS)

The SI units represent an orderly and scientifically based system of measurement. Unfortunately, remains of the customary American system of units will be with us for many years to come. It is just as necessary for us to discuss this system as it will be necessary for you to use it on many occasions.

In the discussion below and in Table 2.7, we focus on units which differ from the SI designations—popularly used units and

TABLE 2·7
U.S. CUSTOMARY UNITS

	Dimensional Representation	Base Units	Other Units
Fundamental:			
Length	L	ft	1 ft = 12 in = $\frac{1}{3}$ yd = 1.89×10^{-4} mi
Time	T	s	
Force	F	lbf	1 lbf = 16 oz = 5.00×10^{-4} ton
Temperature	θ	°R	°R = °F + 459.69
Derived:			
Mass	$\dfrac{FT^2}{L}$	(lbf)(s²)/ft	1 slug = 1 (lbf)(s²)/ft = 32.17 lbm
Acceleration	$\dfrac{L}{T^2}$	ft/s²	
Velocity	$\dfrac{L}{T}$	ft/s	1 ft/s = 0.682 mi/h
Energy or work	FL	ft·lbf	1 ft·lbf = 1.29×10^{-3} Btu
Power	$\dfrac{FL}{T}$	ft·lbf/s	1 ft·lbf/s = 1.818×10^{-3} hp
Volume	L^3	ft³	1 ft³ = 7.48 gal
Area	L^2	ft²	1 ft² = 2.296×10^{-5} acre
Pressure	$\dfrac{F}{L^2}$	lbf/ft²	1 lbf/ft² = 0.00694 lbf/in²

units associated with mechanical phenomena. Other units used in the United States, particularly electrical and science-oriented units, agree with the SI units shown in Tables 2.4 and 2.5.

The most obvious difficulty with the customary American system is that the various units created by kings, merchants, and scientists over the centuries are still used to measure particular dimensions (e.g., inches, feet, yards, and miles for length) and these lead to inconvenient multiples for smaller or larger units (e.g., 12 in/ft, 16 oz/lb). Another difficulty for engineers and especially engineering students is the confusing treatment of force and mass. A force is the push or pull which can cause motion or deflect a spring. Mass is the property of a body which requires a force to move it or to support it in a gravitational field. Mass and force are related by Newton's law:

$$f = ma \tag{2.1}$$

which states that a force applied to a body equals the mass of the body times the resulting acceleration. We can substitute the dimensional representation for each of the quantities in this equation so that

$$F = \frac{ML}{T^2} \tag{2.2}$$

relates the dimensions of force, mass, length, and time. As stated earlier, force and mass cannot both be primary dimensions. In SI, mass is taken to be primary and Eq. (2.2) defines the derived force dimension. In customary American usage, force is selected as the primary dimension and Eq. (2.2) is rearranged to derive the mass dimension as

$$M = \frac{FT^2}{L} \tag{2.3}$$

The basic units used are feet for the length dimension, seconds for time, and pounds-force for force. Substituting these into Eq. (2.3) we obtain a derived mass unit of pounds-force times second squared per foot. This mass unit is given the name slug but the smaller unit of pounds-mass is commonly used, where

$$1 \text{ slug} = 1 \frac{(\text{lbf})(s^2)}{\text{ft}} = 32.17 \text{ lbm} \tag{2.4}$$

To conveniently apply Newton's law with customary American units, we must have force in units of pounds-force, acceleration in units of feet per second squared, and mass in the derived units of slugs; otherwise results from Eq. (2.1) may require substantial interpretation.

At sea level and 45° latitude, the acceleration of gravity is 32.17 ft/s². Consider an object having a mass of 1 lbm. The upward force required to support this mass in the standard graviational field is obtained from Eq. (2.1), realizing that 1 lbm = 1/32.17 slugs or 1/32.17 (lbf)(s²)/ft, so that

$$f = ma = \left[\frac{1}{32.17} \frac{(lbf)(s^2)}{ft}\right]\left(32.17 \frac{ft}{s^2}\right) = 1 \text{ lbf}$$

Thus, our mass of 1 lbm has a "weight" of 1 lbf. If we consider a mass of 1 slug, since 1 slug = 1 (lbf)(s²)/ft, we substitute directly into Eq. (2.1) and

$$f = \left[1 \frac{(lbf)(s^2)}{ft}\right]\left(32.17 \frac{ft}{s^2}\right) = 32.17 \text{ lbf}$$

which means that the 1 slug mass will weigh 32.17 lbf. The advantage of using the mass unit lbm is that an object's mass in pounds-mass equals its weight in pounds-force in a standard gravitational field. The advantage of using mass units of slugs is that 1 slug = 1 (lbf)(s²)/ft and no conversion is necessary before applying the $f = ma$ equation of Newton's law provided that the units pounds-force and foot per second squared are used for force and acceleration. Similarly, to use American units in any basic equation, mass should be in slugs or a conversion should be made from pounds-mass. The use of the two units pounds-mass and slug and the continual conversion between mass units create a source of confusion which plagues even experienced engineers. Confusion becomes more likely as equations become more complex. The SI approach reduces this confusion by focusing on mass, which is a property of an object and not dependent on local acceleration. The concept of weight is very much deemphasized and the fact that a 1-kg mass does *not* require a 1-N supporting force is ignored. A metric force unit called a kilogram-force may occasionally be seen (1 kgf = 9.81 N), but it is *not* an acceptable SI unit.

For many of the units in Table 2.7, the development was very disorganized, and initially rather arbitrary definitions were often used. For example, King Henry I of England defined the yard as being the distance from the tip of his nose to the fingertips of his outstretched arm. A sixteenth-century German regulation defined the rod as the distance spanned by the left feet of 16 men including "tall ones and short ones." The acre was originally the amount of land that could be plowed by a team of oxen in one day.

The Fahrenheit temperature scale, popularly used in the United States, was based on G.D. Fahrenheit's setting 0°F to match what he thought to be the lowest possible temperature and setting 100°F to match the "standard temperature" of human blood. The Rankine temperature scale uses degrees of the same size as the degree Fahrenheit, but it is an absolute scale with a reference of absolute zero, so that 459.69°R = 0°F. A popular unit of energy, the British thermal unit (Btu), is the energy needed to raise the temperature of one pound-mass of water one degree Fahrenheit. One horsepower is naturally based on the rate at which a horse can comfortably do work.

CONVERTING UNITS

It is frequently necessary to make conversions between the various units in which a given dimension can be measured. The most convenient way to do this and avoid errors is to begin with an equality relating the two units of interest; for example,

$$1 \text{ ft} = 0.3048 \text{ m} \tag{2.5}$$

The equation can then be rearranged, dividing by one side (including the units) to obtain either

$$\frac{1 \text{ ft}}{0.3048 \text{ m}} = 1 \tag{2.6}$$

or the inverse. The expression is now unity and can be multiplied or divided into any quantity without changing its actual value. To convert any number of meters to the corresponding number of feet, we can thus multiply by this version of unity, treating the units as algebraic quantities. We find that

$$4478 \text{ m} = \left(\frac{1 \text{ ft}}{0.3048 \text{ m}}\right) 4478 \text{ m} = 14{,}690 \text{ ft}$$

converts the height of the Matterhorn from meters to feet. The inverse of our relation is also unity and can similarly be used in conversions, so

$$6288 \text{ ft} = \left(\frac{0.3048 \text{ m}}{1 \text{ ft}}\right) 6288 \text{ ft} = 1917 \text{ m}$$

converts the height of New Hampshire's Mt. Washington from feet to meters. Conversions of all types of units can be handled in the same way. Table 2.8 presents the equivalents for many of the more common units.

TABLE 2·8
EQUIVALENCE STATEMENTS FOR VARIOUS UNITS

Lengths:
 1 m = 3.281 ft = 39.37 in = 6.214 × 10^{-4} mi
 1 ft = 12.00 in = 1.894 × 10^{-4} mi = 0.3048 m
 1 in = 1.578 × 10^{-5} mi = 0.02540 m = 0.08333 ft
 1 mi = 1609 m = 5280 ft = 63360 in

Mass:
 1 kg = 0.06854 slug = 2.205 lbm
 1 slug = 32.17 lbm = 14.59 kg
 1 lbm = 0.4535 kg = 0.03108 slug

Force:
 1 N = 0.2248 lbf
 1 lbf = 4.448 N

Pressure:
 1 Pa = 0.02089 lbf/ft² = 1.450 × 10^{-4} lbf/in² = 9.869 × 10^{-6} atm = 2.953 × 10^{-4} inHg
 1 lbf/ft² = 6.944 × 10^{-3} lbf/in² = 4.725 × 10^{-4} atm = 0.01414 inHg = 47.88 Pa
 1 lbf/in² = 0.06805 atm = 2.036 inHg = 6895 Pa = 144 lbf/ft²
 1 atm = 29.92 inHg = 1.013 × 10^5 Pa = 2116 lbf/ft² = 14.69 lbf/in²
 1 inHg = 3386 Pa = 70.73 lbf/ft² = 0.4912 lbf/in² = 0.0334 atm

Work and energy:
 1 J = 0.7376 ft·lbf = 9.478 × 10^{-4} Btu
 1 ft·lbf = 1.285 × 10^{-3} Btu = 1.356 J
 1 Btu = 1055 J = 778.2 ft·lbf

Power:
 1 W = 0.7376 ft·lbf/s = 1.341 × 10^{-3} hp = 3.412 Btu/h
 1 ft·lbf/s = 1.818 × 10^{-3} hp = 4.626 Btu/h = 1.356 W
 1 hp = 2545 Btu/h = 745.7 W = 550.0 ft·lbf/s
 1 Btu/h = 0.2931 W = 0.2162 ft·lbf/s = 3.930 × 10^{-4} hp

Temperature:
 $\Delta 1$ K = $\Delta 1$°C = $\Delta 1.8$°F = $\Delta 1.8$°R
 K = 273.15 + °C
 °R = 459.67 + °F
 °F = 32 + 1.8°C

Quantities with more complex units can be converted by applying repeatedly the simple equivalents as required algebraically. Converting area from square feet to square meters, for example, is performed by inverting and squaring the specially written version of unity shown in Eq. (2.6):

$$100 \text{ ft}^2 = \left(\frac{0.3048 \text{ m}}{1 \text{ ft}}\right)^2 100 \text{ ft}^2 = 9.29 \text{ m}^2$$

Multiple conversions can be performed simultaneously, so using Eq. (2.6) and an equivalent from the force entries of Table 2.8,

$$1 \text{ J} = 1 \text{ N·m} = \left(\frac{0.2248 \text{ lbf}}{1 \cancel{\text{N}}}\right)\left(\frac{1 \text{ ft}}{0.3048 \cancel{\text{m}}}\right) 1 \cancel{\text{N·m}} = 0.7375 \text{ ft·lbf}$$

This can be checked against the work and energy entries in Table 2.8. Conversion operations can become quite lengthy, but they are not difficult if the proper simple equivalents are set up and the basic algebraic operations are kept in mind, as shown in the various conversions below:

Velocity in miles per hour to feet per second:

$$60 \frac{\text{mi}}{\text{h}} = \left(\frac{5280 \text{ ft}}{1 \text{ mi}}\right)\left(\frac{1 \text{ h}}{3600 \text{ s}}\right) \quad 60 \frac{\text{mi}}{\text{h}} = 88.00 \frac{\text{ft}}{\text{s}}$$

Density in kilograms per cubic meter to pounds-mass per cubic inch:

$$1 \frac{\text{kg}}{\text{m}^3} = \left(\frac{0.3048 \text{ m}}{1 \text{ ft}}\right)^3 \left(\frac{1 \text{ ft}}{12 \text{ in}}\right)^3 \left(\frac{2.205 \text{ lbm}}{1 \text{ kg}}\right) \quad 1 \frac{\text{kg}}{\text{m}^3} = 3.61 \times 10^{-5} \frac{\text{lbm}}{\text{in}^3}$$

Rotational speed in revolutions per minute to radians per second:

$$1000 \frac{\text{r}}{\text{min}} = \left(\frac{2\pi \text{ rad}}{1 \text{ r}}\right)\left(\frac{1 \text{ min}}{60 \text{ s}}\right) \quad 1000 \frac{\text{r}}{\text{min}} = 104.7 \frac{\text{rad}}{\text{s}}$$

Converting kilogram-meters per second squared to slug-feet per second squared should confirm the relation shown in Table 2.8 for the newton and the pound-force:

$$1 \text{ N} = 1 \frac{\text{kg} \cdot \text{m}}{\text{s}^2} = \left(\frac{0.06854 \text{ slug}}{1 \text{ kg}}\right)\left(\frac{1 \text{ ft}}{0.3048 \text{ m}}\right) 1 \frac{\text{kg} \cdot \text{m}}{\text{s}^2}$$
$$= 0.2249 \frac{\text{slug} \cdot \text{ft}}{\text{s}^2} = 0.2249 \text{ lbf}$$

Thermal conductivity in Btu per hour per foot per degree Fahrenheit to watts per meter per Kelvin:

$$1 \frac{\text{Btu}}{(\text{h})(\text{ft})(°\text{F})} = \left(\frac{1 \text{ W}}{3.412 \text{ Btu/h}}\right)\left(\frac{1 \text{ ft}}{0.3048 \text{ m}}\right)\left(\frac{1.8°\text{F}}{1 \text{ K}}\right) 1 \frac{\text{Btu}}{(\text{h})(\text{ft})(°\text{F})}$$
$$= 1.731 \text{ W/m} \cdot \text{K}$$

Conversion of Units within Equations

Equations used in engineering and science are valid only when the quantities in the equations are expressed in consistent units. The equation for the volume of a rectangular solid

$$V = lwh \qquad (2.7)$$

works properly only if l, w, and h are in the same units (say, meters) and the volume computed is in the cube of those units (cubic meters). If for some reason the length l were measured as 2.30 ft, w as 7.10 in, and h as 1.00 cm, substituting these directly into Eq. (2.7) would not produce a meaningful result. It would be necessary to convert each into the same units before computing the volume. Thus,

$$l = \frac{30.48 \text{ cm}}{1 \text{ ft}} \, 2.30 \text{ ft} = 70.1 \text{ cm}$$

$$w = \frac{2.540 \text{ cm}}{1 \text{ in}} \, 7.10 \text{ in} = 18.0 \text{ cm}$$

$$h = 1.00 \text{ cm}$$

results in

$$V = lwh = (70.1)(18.0)(1.00) = 1260 \text{ cm}^3$$

An alternative to requiring conversions before applying equations is to build the conversion into the equation. If we would like to frequently compute volume in cubic centimeters given l in feet, w in inches, and h in centimeters, the equation can be modified. Thus,

$$V = \left(\frac{30.48 \text{ cm}}{1 \text{ ft}}\right)(l \text{ ft}) \left(\frac{2.540 \text{ cm}}{1 \text{ in}}\right)(w \text{ in})(h \text{ cm})$$

$$= 77.42 \, lwh \quad \text{cm}^3 \tag{2.8}$$

produces an equation that can be useful but is no longer a general equation. Now, l must be in feet, w in inches, and h in centimeters to get the correct result and the number 77.42 has units of centimeters squared per foot per inch. For our example above, Eq. (2.8) results directly in

$$V = (77.42)(2.30)(7.10)(1.00) = 1260 \text{ cm}^3$$

which checks with the earlier calculation.

Perhaps the most commonly seen converted equation is Newton's law, as it has been used in the United States so that the mass unit of pounds-mass can be applied. This requires that a conversion be inserted to change m from pounds-mass to slugs:

$$f = \frac{1 \text{ slug}}{32.17 \text{ lbm}} (m \text{ lbm}) \left(a \, \frac{\text{ft}}{\text{s}^2}\right)$$

$$f = \frac{1}{32.17} ma \tag{2.9}$$

This equation now requires m in pounds-mass and a in foot per second squared. The result is f in pounds-force. The conversion factor here is frequently written with the 32.17 replaced by g_c and a common form for Newton's law is

$$f = \frac{1}{g_c} ma \qquad (2.10)$$

The conversion factor g_c appears in many other equations and can be confusing. It is set to $g_c = 1$ when consistent units such as newtons, kilograms, meters, and seconds are used in SI or when the customary American units of pounds-force, slugs, feet, and seconds are used. It is set to $g_c = 32.17$ when slugs are replaced by pounds-mass. After many years of use, the popularity of g_c is finally fading. It is best in all general equations to simply use fundamental units but not intermix SI and American fundamental units. In special equations where mixed units occur, they should be clearly identified with the equation and the equation should contain the numerical values of any conversion factors. It is a good habit to check the units in equations by carrying them along algebraically, as we have done in this section. Examples 2.1 and 2.2 illustrate these concepts.

Example 2·1

The mass m of a body is density ρ times volume v:

$$m = \rho v$$

where v has dimension L^3
ρ has dimension M/L^3
m has dimension M

What are suitable SI and American units in this equation?

Solution

In this general equation, fundamental units can be substituted directly. Thus, in SI we have volume in cubic meters, density in kilograms per cubic meter, and mass in kilograms. To check the consistency of these we can simply write the equation $m = \rho v$ in terms of units alone and treat the units algebraically:

$$\text{kg} = \frac{\text{kg}}{\cancel{\text{m}^3}} \cancel{\text{m}^3} \quad \text{or} \quad \text{kg} = \text{kg}$$

Both sides have the same units; the equation checks.

Using fundamental American units, we would have volume in cubic feet, density in slugs per cubic foot, and mass in slugs. The consistency of these units can also be checked

$$\text{slugs} = \frac{\text{slugs}}{\cancel{\text{ft}^3}} \cancel{\text{ft}^3} \quad \text{or} \quad \text{slugs} = \text{slugs}$$

This is satisfactory.

Example 2·2

In a certain laboratory, the mass of a liquid must be determined in units of pounds-mass. This is done by measuring the liquid volume in a graduated cylinder labeled in cubic centimeters and looking up the density in a table with units of kilograms per cubic meter. Modify the general equation $m = \rho v$ so that it will give mass directly in pounds-mass when ρ and v are in kilograms per cubic meter and cubic centimeters, respectively.

Solution

First consider the conversion of volume from cubic centimeters to cubic meters

$$v \text{ [in cubic meters]} = v \text{ [in cubic centimeters]} \left(\frac{1 \text{ m}}{100 \text{ cm}}\right)^3$$
$$= 10^{-6} \, v \text{ [in cubic centimeters]}$$

The equation

$$m = 10^{-6} \, \rho v$$

will thus compute m in kilograms given ρ in kilograms per cubic meter and v in cubic centimeters. Now we convert mass from kilograms to pounds-mass, referring to Table 2.8:

$$m \text{ [in pounds-mass]}$$
$$= m \text{ [in kilograms]} \frac{2.205 \text{ lbm}}{1 \text{ kg}}$$
$$= 2.205 \, m \text{ [in kilograms]}$$

We now include this in the previous equation to obtain

$$m = 2.205 \times 10^{-6} \, \rho v$$

This special equation is now valid for m in pounds-mass, ρ in kilograms per cubic meter, and v in cubic centimeters. It is only valid for this set of units. Note that the included conversion factor of 2.205×10^{-6} has units of $\text{lbm} \cdot \text{m}^3/\text{kg} \cdot \text{cm}^3$.

DIMENSIONAL ANALYSIS

Dimensional analysis is used to help generalize the results of experiments and to provide organization and insight into the solution of practical engineering problems. The results of an experiment are "generalized" when we apply them to a different but similar application involving a change in size or material from the original experiment. This is done frequently in engineering, an important example of which is the use of wind tunnels, where scale models of aircraft are often tested to predict the behavior of full-sized aircraft.

Real engineering problems frequently involve several variables, so it is necessary to find an equation or at least a graphical relationship between the variables. In applying dimensional analysis to such a problem, the first step is to list all the variables involved in the problem. Dimensional analysis can then be used to examine the relationships between the variables or to form combinations of variables. It can often reduce the total number of variables that must be considered in performing experiments to study the problem.

The methods of dimensional analysis are related to the techniques we applied in performing the conversion of units in the previous section. We deal more generally now with dimensions instead of units, but we still work with similar types of algebraic manipulations. Two of the basic principles involved are that the dimension of each side of an equation must be the same and that only quantities of like dimension can be added or subtracted. These principles in themselves are important for checking the validity of equations and can also be an aid in deriving simple equations. As an example of this, imagine that you have been told that the stress at the inside of the wall of a cylindrical pressure vessel is given as below and you would like to check the plausibility of the equation.

$$\sigma = \frac{Pr}{R^2 - r^2}\left(1 + \frac{R^2}{r^2}\right) \qquad (2.11)$$

where σ = interior tangential stress, dimension F/L^2
P = cylinder pressure, dimension F/L^2
r = inner radius, dimension L
R = outer radius, dimension L

Although this equation does not have any real meaning to you at this time, it can be quickly checked for its plausibility by seeing if the dimensions of both sides of the equation are the same and if the additions and subtractions are allowable. Examine first the terms involving addition and subtraction to see if they are legitimate:

$$\left(1 + \frac{R^2}{r^2}\right)$$
$$(R^2 - r^2)$$

In the first of these terms, 1 is dimensionless and the dimension of R^2/r^2 is $L^2/L^2 = 1$, which is also dimensionless. Since both quantities are dimensionless, this addition is permissible, so the result will be dimensionless. In the second term, R^2 and r^2 both have dimension L^2, so this subtraction may be performed. The result will also have dimension L^2. Now substituting these dimensions into Eq. (2.11) to compare the two sides,

$$\frac{F}{L^2} \stackrel{?}{=} \frac{\frac{F}{L_2}L}{L^2} \quad (1)$$

where the 1 represents the dimensionless sum. This reduces algebraically to

$$\frac{F}{L^2} \stackrel{?}{=} \frac{F}{L^3}$$

Since these dimensions are not equivalent (because their exponents are not the same), Eq. (2.11) cannot be valid as presented. We cannot identify exactly the error here by dimensional analysis, but it is clear that an extra length dimension is needed in the numerator of the right-hand side. (Actually the r in the numerator should be r^2.)

The derivation of simple equations can also be aided by dimensional considerations. As an example of this, if one knows that the average velocity of fluid in a pipe is directly related to the fluid flow rate in the pipe and its cross-sectional area, an equation can be written by considering these variables:

V = average velocity, dimension L/T
Q = volume flow rate, dimension L^3/T
A = area, dimension L^2

The form of an equation for velocity cannot involve an addition or subtraction of Q and A since the dimensions are not equivalent. However, an equation of the form

$$V = Q^a A^b \quad (2.12)$$

is possible where a and b can be obtained dimensionally. Substituting dimensions into Eq. (2.12), we have

$$\frac{L}{T} = \left(\frac{L^3}{T}\right)^a (L^2)^b$$

or, rearranging,

$$LT^{-1} = L^{3a}T^{-a}L^{2b}$$

For the dimensions on both sides of this equation to be equivalent, the exponents of L and T must also be equivalent. For L

$$1 = 3a + 2b \quad (2.13)$$

and for T,

$$-1 = -\alpha \qquad (2.14)$$

From this last equation, $\alpha = 1$. Then, from Eq. (2.13),

$$1 = 3 + 2b$$

or

$$b = -1$$

Returning to Eq. (2.12), we find that

$$V = QA^{-1} = \frac{Q}{A}$$

This equation now relates average velocity to flow rate and area.

The Buckingham Pi Theorem

The pi theorem published by Buckingham in 1915 provides the basis of applying dimensional analysis for the combination and reduction of variables in an engineering problem. If n variables are involved in a problem and among these variables there are k fundamental dimensions (counting the dimension of force or mass but not both), then only $n - k$ variables are actually important. These are formed by combining the original n variables into $n - k$ dimensionless groups which are each given the symbol π with a subscript (hence the name of the theorem). The functional relationships for the problem can then be expressed by defining π_1 as a function of the other π's; that is,

$$\pi_1 = f(\pi_2, \pi_3, \ldots, \pi_{n-k}) \qquad (2.15)$$

FIGURE 2·2
A simple pendulum.

The details of the function may be obtained by experiment or theoretical analysis. In practice, the experimental result is usually available first, but any theoretical equation which is obtained must also satisfy the form of Eq. (2.15).

As a first example of dimensional analysis, consider the simple pendulum formed by suspending a mass on a string, as indicated in Fig. 2.2. The time required for the pendulum to complete one full oscillation is known as its period. We would like to determine the relationship between the period and the pendulum characteristics. It is believed that the suspended mass m, the string length l, and the acceleration of gravity g may influence the period. These $n = 4$ variables and corresponding dimensions are

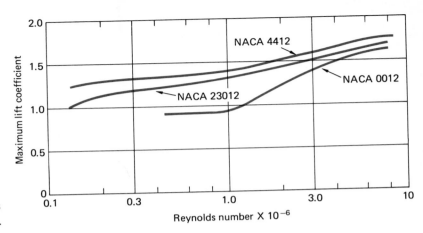

FIGURE 2·3
Actual airfoil performance.

tant variable is not included, results of experiments based on the dimensional analysis cannot be expected to be accurate. However, care must be used in selecting the variables. If insignificant variables are included and this does not become obvious in the dimensional analysis, a great deal of extra experimentation may be required to determine functional relationships.

PROBLEMS

2.8 Each of the quantities below is shown with its dimension expressed in terms of fundamental dimensions. Determine the most appropriate derived units for each of these quantities in SI and also in the customary American system. For example, area (L^2) has units of square meters in SI and square feet in customary American units.

(a) Volume (L^3)
(b) Pressure (F/L^2)
(c) Velocity (L/T)
(d) Acceleration (L/T^2)
(e) Power (FL/T)
(f) Heating value (FL/M)
(g) Specific heat ($FL/M\theta$)
(h) Viscous friction (FT/L)
(i) Viscosity (FT/L^2)
(j) Viscosity (M/LT)

2.9 Make the following conversions from the units given to the units indicated. Begin only with the equivalents between units of length, time, mass, and force.

(a) 15 ft² = m²
(b) 17.3 N = lbf
(c) 75.3 mi/h = m/s
(d) 1.94 slugs/ft³ = kg/m³
(e) 62.4 lbm/ft³ = kg/m³
(f) 135 m³/s = ft³/h
(g) 0.635 lbf/in² = Pa

(h) 16.4 W   =   ft·lbf/s
(i) 0.54 ft·lbf   =   J
(j) 10 m/s²   =   ft/s²

2.10 Make the following conversions:
(a) 10 kW   =   hp
(b) 73.5 W   =   Btu/h
(c) 25 W/(m²)(K)   =   Btu/(h)(ft²)(°F)
(d) 1000 J/(kg)(K) =   Btu/(lbm)(°R)
(e) 2.5 Pa·s   =   kg/(m)(s)
(f) 2.5 Pa·s   =   lbf·s/ft²
(g) 10 Btu/lbm   =   J/kg

2.11 The equation for the kinetic energy of a moving body is $KE = \frac{1}{2} mv^2$, where m is the mass of the body and v is the velocity of the body. Show that the dimensions on both sides of the equation are the same. Note that the number $\frac{1}{2}$ is dimensionless. What is a consistent set of SI units for each of the three variables in this equation? What is a consistent set of customary American units?

2.12 A person has a mass of 150 lbm. On the moon the gravitational acceleration is 5.4 ft/s². What force would this person exert while standing on the surface of the moon?

2.13 The pressure at the bottom of a tank of liquid is given by $P = \rho g h$, where P is the pressure, ρ is the mass density of the liquid, g is gravitational acceleration, and h is the depth of the liquid. For $\rho = 1000$ kg/m³, $g = 32.2$ ft/s², and $h = 10$ yd, determine the resulting pressure in kilopascals and in pounds per square inch.

2.14 The equation $P = \rho g h$ is to be used repeatedly to calculate P in kilopascals when given the mixture of units used in Prob. 2.13. Determine the conversion factor C_f which should be inserted in the equation $P = C_f \rho g h$ so that units of kilograms per cubic meter, feet per second squared, and yards can be used directly for ρ, g, and h.

2.15 The potential energy PE of an elevated body is known to depend on the mass m of the body, the acceleration of gravity g, and the height of the body h. Using dimensional analysis, combine these variables into a single dimensionless group. From your experience in physics, what is the constant value of your dimensionless group?

2.16 The tube shown in Fig. P2.16 could be a home-made boat speedometer. When it is moved through a liquid, the force of the liquid acting on the open lower end causes the liquid level in the tube to rise above the surrounding surface level. With proper calibration, the liquid level in the tube could be used to measure the velocity. The height h of the liquid level is thought to depend on mass density ρ and gravitational

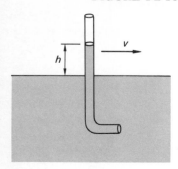

FIGURE P2·16

82   INTRODUCTION TO ENGINEERING CALCULATIONS

acceleration g as well as the velocity v. Determine a dimensionless group which represents this problem.

2.17 It has been suggested that the behavior of the boat speedometer of Prob. 2.16 should also depend on the diameter d of the tube in addition to the variables mentioned. This means that the height h might be a function of ρ, g, v, and d. How many dimensionless groups are now required to represent this problem? Define suitable groups.

2·3 GRAPHS

Graphs are an effective means of summarizing large amounts of quantitative information. In most cases, presenting an equivalent amount of information with tables of numbers is a confusing and impossible task. Graphs allow a reader to observe trends in data easily. A physical sense of functional relationships can be put forth and comparisons between various conditions can be made conveniently. In almost every situation in which more than a few pieces of quantitative information are to be discussed, even in an informal way, a graph is an important aid. Today, with modern computer and instrumentation technology, companies and careers have developed based on convenient and automated techniques for simply plotting graphs.

Figure 2.4 shows a few of the more common types of graphs. The circle graphs in Fig. 2.4a depict fractions of some whole entity as "slices of a pie," a convenient method for presenting budget costs. The particular data in Fig. 2.4a show world consumption and production of energy, clearly indicating that some industrial nations, which are major consumers of energy, produce only small amounts, whereas countries in the Middle East and Africa produce far more energy than they consume. The bar graph in Fig. 2.4b presents data on the surface area of the major lakes in the world, for convenient comparisons of sizes. Bar graphs are often used for such simple comparisons and are thus a common advertising tool. By such means, for example, snow tire brand x may be compared with brands y and z in a bar-graph display of stopping distance to proclaim the desirable feature of one brand as opposed to the others.

In Fig. 2.4c, several curves have been drawn on the same set of axes: curves of starting salaries vs. time for students receiving degrees at different levels of engineering education. On this graph, the general increase of starting salaries is obvious, since the dollar values have not been corrected for inflation. The difference in starting salaries among the various education levels is also shown. Note that these differences are consistent and nearly un-

FIGURE 2·4
Example graphs. (a) A circle graph. Data from *The World Energy Book*, Nichols Publishing Company, New York, 1978. (b) A bar graph. Data from *Water Resources of the World*, Water Information Center, Inc., Port Washington, NY, 1975. (c) Graph of curves to compare salaries. Data from The College Placement Council, Inc., and the Engineering Manpower Commission. (d) Curves drawn by a recording instrument. (e) Graph of straight line drawn through scattered data points. Data from "Sensing and Communication between Vehicles," NCHRP Rept. 51, Highway Research Board, 1968.

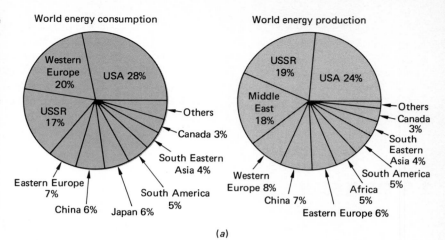

(a)

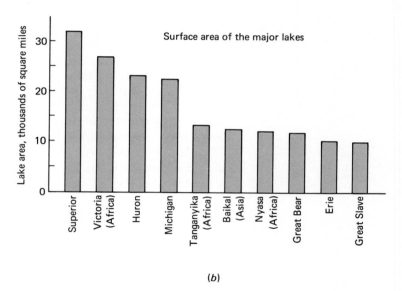

(b)

varying over the 15-year span of the graph. The next family of curves, shown in Fig. 2.4d, was made by an automatic plotting instrument called a recorder. In the report from which this was taken, a rule of graphical presentation is violated in that the vertical and horizontal axes are not labeled. This can be a problem with respect to many recorders, since labeling is not usually automatic so it must be added at a later time. In this case, the missing labels are tolerable, since the purpose of presenting these curves is to show the extreme differences between them. The vertical axis here represents gas-pedal position. It is shown as a function of time, represented by the horizontal axis. The curves, for six different subjects (numbered 1 to 6) driving over the same stretch of road, indi-

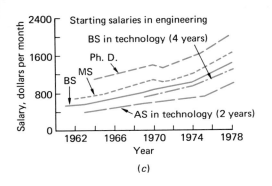

(c)

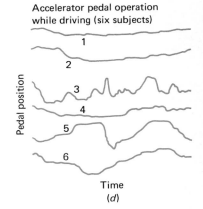

(d)

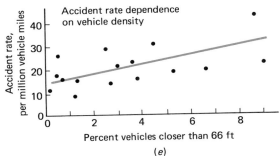

(e)

cate substantial individual differences in driver behavior. The zero position on the vertical axis is different for each of the curves.

The final graph, Fig. 2.4e, presents results of an automotive accident study based on data from a variety of highways. The data are scattered, but despite the scatter they show a clear trend of increased accident rate per million vehicle miles as a function of the percentage of vehicles traveling at spacings less than 66 feet apart. The straight line drawn through these points indicates the trend.

The two graphs in Fig. 2.5 illustrate how more complex relationships can be depicted graphically. Figure 2.5a shows how various new energy sources and conservation strategies could be initiated and then developed in order to meet the increased energy demands of the year 2000. The energy unit quad represents one quadrillion (10^{15}) Btu. Figure 2.5b is a three-dimensional plot that presents the results of a survey of practicing engineers who are machine designers. The question posed was, "What percentage of your time do you spend doing mechanical drawing?" The graph exhibits the percentage of respondents as a function of education level and percentage of working time spent drafting. Note that most engineering drafting in this situation is done by people at the high school and associate degree levels.

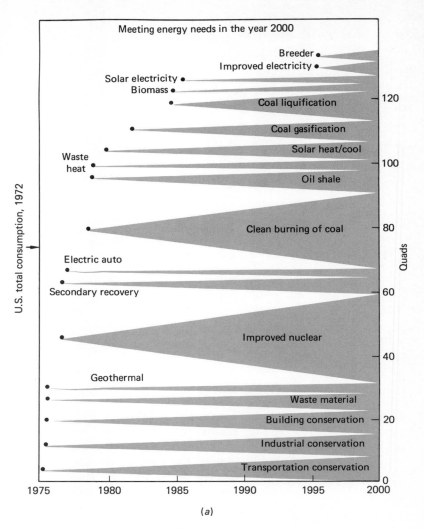
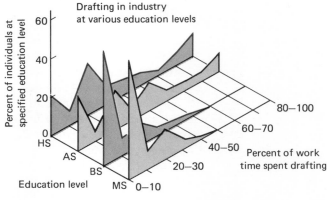

FIGURE 2·5
Some uniquely designed graphs. (a) From G. L. Decker, "The Impact of the New Energy Technologies, *Mechanical Engineering*, May, 1977. (b) From A. H. Soni and L. Torafson, "State of the Art in Machine Design at the Practice Level," ASME Design Engineering Technical Conference, 1974.

Reading graphs and drawing graphs play an important role in engineering; in fact, many engineering studies are pursued by someone visualizing the sorts of graphs that might be desired and then gathering the data necessary to plot them. The use of graphs to present complex technical information in a clear form is limited only by an individual's patience and creative ability.

BASIC RULES IN PLOTTING GRAPHS

Most graphs are initially plotted on special paper containing a prepared grid. For insertion in a report or publication, the graph may later be traced and all or part of the grid may be removed (as in some of the graphs of Figs. 2.4 and 2.5). Work done by engineering students is normally on graph paper (similar to that in Fig. 2.6). A few simple rules which should be followed in plotting graphs are

1. Select scales on the two axes such that the smallest subdivision on the grid represents a convenient number. It is usually convenient to use subdivisions of 1, 2, or 5 times 10 raised to an appropriate power.
2. Make the horizontal axis, or abscissa, the independent variable and the vertical axis, or ordinate, the dependent variable. The dependent variable is thus said to be plotted versus (or as a function of) the independent variable.
3. Label both axes clearly and indicate the units being used. Insert 10 raised to a power in the units, if necessary, to keep the numbers on the axes to reasonable magnitudes.
4. If the relation between the variables being plotted is given directly by an equation, plot points lightly and cover them with the curve when it is drawn.
5. If the relation between the variables is obtained from experimental data, plot small dots and then circle them. (Squares, triangles, etc., may also be used around the dots to identify different conditions or curves being plotted on the same graph.) Draw a curve smoothly through the points which averages or summarizes them but does not necessarily pass through each point. Do not draw the curve into the circles or symbols surrounding the data points.
6. Title the graph, label the curves, and add a key if necessary for any symbols used in plotting. This lettering should not interfere with the curves.

The graphs in Fig. 2.6 have been drawn according to these rules: that in Fig. 2.6a has the smallest subdivision equivalent to 1°. Figure 2.6b shows 0.01 N·m per subdivision on the abscissa and 10 r/min per subdivision on the ordinate. The first graph is simply a plot of the conversion equation

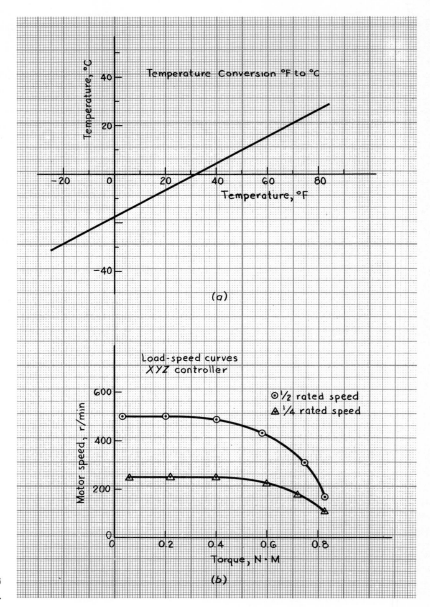

FIGURE 2·6
Typical graphs.

$$°C = \tfrac{5}{9}(°F - 32) \qquad (2.27)$$

It does not involve experimental data, so no data points are shown. Because degrees Celsius on the vertical axis is plotted versus degrees Fahrenheit on the horizontal axis, the graph is intended to convert Fahrenheit temperature (the independent variable) to Celsius temperature (the dependent variable) according to Eq. (2.27). Of course, the graph can also convert degrees Celsius to degrees Fahrenheit, but if this were the intended primary use, the axes would be reversed, just as Eq. (2.27) would be rewritten with degrees Fahrenheit on the left to make frequent conversions from degrees Celsius to degrees Fahrenheit. The graph in Fig. 2.6b depicts the results of two separate tests to determine the load-speed characteristics of a small dc motor operated by a special electronic controller. Two sets of data were obtained, one with the controller speed setting at one-half the rated motor speed and the other at one-quarter the rated speed. In obtaining these data, the load torque was applied and then the speed was observed. Thus speed, plotted vertically on the ordinate, is shown as a function of torque, the independent variable plotted on the abscissa.

LOGARITHMIC SCALES

Labeling an axis on a graph actually scales lengths along that axis in terms of the variable which the axis represents. For axes using linear divisions, as in Fig. 2.6, the scale factor, in units of length per unit of variable, is

$$S = \frac{L}{v_f - v_i} \qquad (2.28)$$

where L = length of axis
v_f = variable's final value on axis
v_i = variable's initial value on axis

A distance d along the axis then represents a change Δv in the value of the variable v according to

$$d = S\, \Delta v \qquad (2.29)$$

The change $\Delta v = v_2 - v_1$, where v_2 and v_1 are values of the variable v at two points of interest. Consider the abscissa in Fig. 2.6a, where the grid is based on 10 subdivisions to the centimeter. The scale factor is

$$S = \frac{L}{v_f - v_i} = \frac{11}{90 - (-20)} = 0.10 \text{ cm/°F}$$

and Eq. (2.29) can compute the distance between points corresponding to particular temperatures. For example, between -20 and $40°F$,

$$\Delta v = v_2 - v_1 = 40 - (-20) = 60$$
and
$$d = S \, \Delta v = 0.10(60) = 6.0 \text{ cm}$$

There are 60 subdivisions, as can be seen on the graph.

We can handle the linear subdivision of axes quite adequately without consciously thinking of Eqs. (2.28) and (2.29); however, similar equations can be important for preparing logarithmic scales. With log scales, the scale factor becomes

$$S = \frac{L}{\log v_f - \log v_i} \tag{2.30}$$

and distances can be laid out proportional to increments in the log of the represented variable as

$$d = S \, \Delta \log v \tag{2.31}$$

where $\Delta \log v = \log v_2 - \log v_1$ is the difference between logarithms at two points of interest. When we observe values of v which differ by a factor of 10 so that $v_2 = 10 \, v_1$, we say that v_2 and v_1 are one decade apart. The difference between the logarithms of two numbers one decade apart is

$$\begin{aligned}\Delta \log v &= \log 10v_1 - \log v_1 \\ &= \log\left(\frac{10v_1}{v_1}\right) \\ &= \log 10 \\ &= 1\end{aligned}$$

Inserting this into Eq. (2.31), we find $d = S$, so that the distance spanned by one decade on a log scale is equal to the scale factor. The logarithmic scale factor may thus be referred to in units of length per decade. Commercial graph paper having log scales on both axes is frequently made with $S = 3.33$ in per decade, so a graph two decades by three decades fits well onto a standard $8\frac{1}{2} \times 11$ in sheet of paper. Many other scales, however, are also available. Of course, by properly applying Eqs. (2.30) and (2.31), one can generate any logarithmic grid. In fact, ordinary linear graph paper can be labeled logarithmically if desired.

TABLE 2·9
DATA FOR TIME AND DISTANCE

Time, s	0.35	0.50	0.71	1.12	1.58	1.93	2.23
Distance, ft	2.00	4.00	8.00	20.00	40.00	60.00	80.00

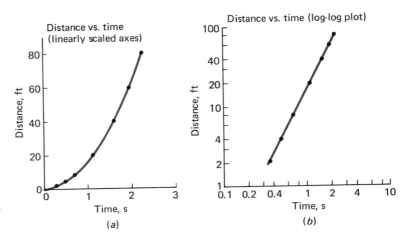

FIGURE 2·7
Plots of distance vs. time.

The data in Table 2.9 represent the distances fallen by a freely falling body at various times following its release. These data are plotted on a graph with linearly scaled axes in Fig. 2.7a and on logarithmic axes in Fig. 2.7b, so two of the reasons for using logarithmic axes become clear. First, the scale is expanded for the smaller quantities. On the vertical axis, the decade from a distance of 1 ft to a distance of 10 ft becomes equal in length to the decade from a distance of 10 ft to a distance of 100 ft. On the linear scale, the first decade is one-tenth the length of the second decade and the data points presented in that range are very close together. The expansion capability of a logarithmic scale becomes particularly important when data are presented spanning several decades, as shown in the microphone response curves of Fig. 2.8, where frequencies from 1.0 to 200,000 Hz are indicated. The second reason for using logarithmic scales in graphs has to do with the possibility of producing a linear, or straight-line, plot when the ordinary linear scale produces a nonlinear curve. This is the case with respect to the data in Table 2.9, where the relation between distance and time plotted in Fig. 2.7a is not a linear one, but a straight-line plot has been obtained on the log-log graph of Fig. 2.7b. A linear plot makes it more convenient to obtain an equation describing experimental data. This will be discussed further in the next section.

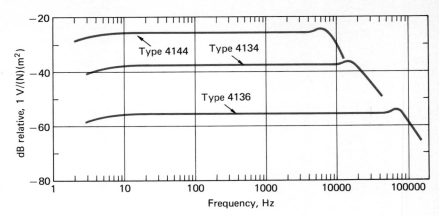

FIGURE 2·8
Microphone response plots. (*B & K Instruments, Inc.*)

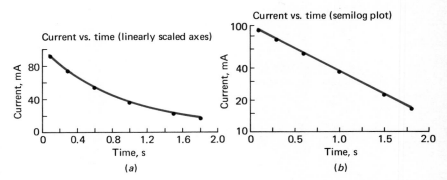

FIGURE 2·9
Plots of current vs. time.

TABLE 2·10
DATA FOR TIME AND CURRENT

Time, s	Current mA
0.100	90.5
0.300	74.1
0.600	54.9
1.00	36.8
1.50	22.3
1.80	16.5

It is not necessary to scale both axes on a graph logarithmically. One axis can remain linear while the other is logarithmic. Such a graph is called semilogarithmic. It can be used to preserve the proper sense of proportionality that one gets on the linear scale of one axis while gaining the expansion capabilities of a log scale on the other axis. It can also be used to convert nonlinear relationships into linear plots. An example is shown in Table 2.10 and Fig. 2.9. These data represent the current through a certain electric resistance when power is removed from the circuit in which it is located.

DEVELOPING EQUATIONS FROM GRAPHS

In engineering and science, useful equations can be developed by theory or by experiment and, in some cases, by a combination of both techniques. Theoretical equations are developed by one's understanding the physical phenomena involved in a relationship well enough to combine basic mathematical expressions into an overall equation describing the relationship. Empirical equations are obtained by performing an experiment to observe a physical relationship and then analyzing the resulting data to establish a useful equation based only on the data. Since relatively few things are understood so well that theoretical equations can be readily developed, experiments and empirical equations are very important to engineers. A graph may frequently be an adequate description of experimental results. However, if computations are to be regularly performed using the results, an empirical equation is desirable. A large number of techniques have been developed for fitting equations to experimental data. Many of these are sophisticated techniques and beyond the scope of this book. We will discuss a few of the basic approaches which are used quite commonly.

The equation for a straight line on ordinary rectangular coordinate graph paper is

$$y = mx + b \qquad (2.32)$$

which shows y as a linear function of x where $m =$ the slope and $b =$ the y intercept (value of y when $x = 0$). To define this equation for any particular straight line, the values of m and b must be determined. As an example, return to Fig. 2.6a. Observing this graph alone, we define the equation for the straight line which it shows. The coefficient

$$m = \frac{\Delta y}{\Delta x} = \frac{y_2 - y_1}{x_2 - x_1}$$

where x_2, y_2 and x_1, y_1 represent any two points on the line. Choosing on the horizontal axis $x_1 = 10.0$ and $x_2 = 60.0$, we find that the corresponding values on the vertical axis are $y_1 = -12.5$ and $y_2 = 15.5$. Thus, we get

$$m = \frac{15.5 - (-12.5)}{60.0 - 10.0} = 0.560$$

According to the value of y shown when $x = 0$ in Fig. 2.6a, $b = -18.0$ and the resulting linear equation is then

$$y = 0.560x - 18.0 \qquad (2.33)$$

Recall that Fig. 2.6a is a plot of the conversion in Eq. (2.27). It can be seen that our representation in Eq. (2.33), obtained by simply inspecting the graph, agrees well with the original Eq. (2.27). Similarly, if our straight line had been drawn through a set of data points, Eq. (2.33) would describe those data as closely as the straight line did.

The simple straight-line equation can also be applied to obtain nonlinear relations by use of logarithmic and semilogarithmic graphs. These graphs, as mentioned earlier, may be used to obtain linear plots of certain nonlinear functions. Consider first the log-log graph of a straight line, as shown in Fig. 2.7b; we will determine the equation which that line represents, considering distance to be on the y axis and time on the x axis. A linear equation in the form of Eq. (2.32) is still valid, but to consider the log scales, x and y must be replaced by their logarithms, so that the equation becomes

$$\log y = m \log x + b$$

or, equivalently,

$$\log y = \log x^m + b \tag{2.34}$$

The slope m is now defined by

$$m = \frac{\Delta \log y}{\Delta \log x} = \frac{\log y_2 - \log y_1}{\log x_2 - \log x_1}$$

where x_1, y_1 and x_2, y_2 represent two points on the line and b is the value of $\log y$ when $\log x = 0$. If we pick two points along the line in Fig. 2.7b as $y_1 = 4.0$ and $y_2 = 40.0$ where $x_1 = 0.50$ and $x_2 = 1.58$, we find that

$$m = \frac{\log 40.0 - \log 4.0}{\log 1.58 - \log 0.50} = 2.00$$

Note that at the value $x = 1$, where $\log x = 0$, the value of $b = \log 16$. Taking the antilog of Eq. (2.34) gives us

$$y = \text{antilog} (\log x^m + b)$$
$$y = 10^b x^m$$

or
$$y = a x^m \tag{2.35}$$

with $a = 10^b$. Substituting the values of m and b produces the equation $y = 16 x^2$ or $D = 16 t^2$ if D is distance and t is time. This equation describes the data shown in Table 2.9. Relationships in the form of Eq. (2.35) will always produce linear plots on log-log graphs.

Consider the semilogarithmic graph in Fig. 2.9b. We now develop an equation which represents the straight line plotted on this graph, where current is on the y axis and time is on the x axis.

The standard linear equation again applies, but this time the y is replaced by log y to compensate for the logarithmic scale on the ordinate while the x is retained, so that

$$\log y = mx + b \qquad (2.36)$$

The slope

$$m = \frac{\Delta \log y}{\Delta x} = \frac{\log y_2 - \log y_1}{x_2 - x_1}$$

for any two points on the line. The intercept b is the value of log y when $x = 0$. Taking the antilog of Eq. (2.36), we find that

$$y = \text{antilog }(mx + b)$$
$$y = 10^{mx} 10^b$$

or
$$y = cd^x \qquad (2.37)$$

where $c = 10^b$ and $d = 10^m$. Equations in the form of Eq. (2.37) will plot linearly on semilog coordinates if y is on the log axis and x is on the linearly scaled axis.

Picking two points in Fig. 2.9b, where $x_1 = 0.4000$ and $x_2 = 1.60$, and reading $y_1 = 67.0$ and $y_2 = 20.0$ yields

$$m = \frac{\log 20.0 - \log 67.0}{1.60 - 0.40} = -0.438$$

Extending the straight line to $x = 0$, we find that $b = \log 100 = 2.0$, so Eq. (2.37) becomes $y = 100(0.365)^x$. With i for current and t for time, the equation is then $i = 100(0.365)^t$, which describes the linear plot in Fig. 2.9b and the data of Table 2.10.

PROBLEMS

2.18 The power supply in a certain stereo amplifier contains an electric capacitor which becomes charged to a voltage of 10 V when the amplifier is operating. When it is turned off, the voltage gradually decays. The data in Table P2.18 show the capacitor voltage at various times after turn off. Make a neat, well-labeled graph of capacitor voltage vs. time which

TABLE P2·18

	0	1.0	2.0	3.0	4.0	5.0	6.0
Time, s							
Capacitor voltage, V	10	6.1	3.7	2.2	1.4	0.81	0.5

shows these data points. Plot a curve which passes through the points.

2.19 A pump is being used to pump water through a pipe 2000 ft long with a 6-in diameter. The pump produces the required pressure P necessary to force the water through the pipe. This pressure depends on the water flow rate Q desired. The data relating P to Q has been obtained in a series of tests, as shown in Table P2.19. Make a neat, well-labeled graph of pressure vs. flow rate which shows these data points. Plot a curve passing through the points.

TABLE P2·19

Flow rate, ft^3/s	0.50	1.0	2.0	3.0	4.0	6.0	8.0
Pressure, lbf/in^2	3.15	11.2	39.2	81.9	140	302	515

2.20 The Acme Rod Company produces steel rods in lengths of 2, 3, and 4 m and in various diameters up to 50 mm. Plot a single graph which shows rod mass in kilograms as a function of diameter in millimeters. Your graph should contain a curve for each of the Acme rod lengths and be neat and completely labeled. The density of steel is 7850 kg/m^3.

2.21 In Fig. P2.21 logarithmic axes have been drawn on commer-

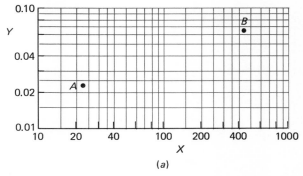

(a)

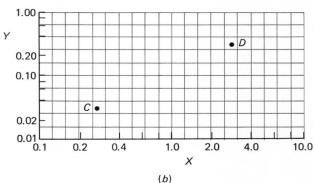

(b)

FIGURE P2·21
(a) Log paper. (b) Log scale on standard graph paper.

cial log paper and on standard graph paper. Determine the x and y coordinates of points A, B, C, and D on the graphs.

2.22 A "poor-man's" boat speedometer consisting of a piece of bent glass tubing has been calibrated as shown in Fig. P2.22. The height of fluid rise in the tube is a function of the velocity at which the tube moves through the water. Plot the data points shown on a well-labeled log-log graph. Obtain the equation relating height H to velocity V.

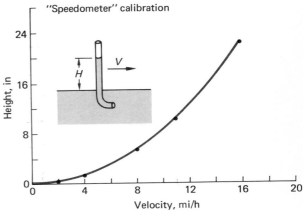

FIGURE P2·22

2.23 Plot the data of Prob. 2.19 on a log-log graph. Obtain the equation relating pressure P to flow rate Q.

2.24 A pot of water has been brought to a boil and taken outdoors where the temperature is just at freezing. A thermometer placed in the pot is read every half hour. The results are given in Table P2.24. Draw a straight-line plot of these results on semilogarithmic graph paper. Determine the equation relating temperature T to time t during this transient.

TABLE P2·24

Time, h	0	0.5	1.0	1.5	2.0	2.5	3.0
Temperature, °C	100	61	37	22	14	8	5

2.25 Draw a semilogarithmic graph for the data of Prob. 2.18. You should find a straight-line plot. Obtain the equation relating voltage v to time t.

2·4 STATISTICAL CONCEPTS

There is randomness, or chance, involved in many aspects of daily life. Engineers commonly encounter randomness when analyzing the results of measurements or experiments. The data ob-

TABLE 2·11
WIDTH MEASUREMENTS

Measurement Number	Width, cm	Measurement Number	Width, cm	Measurement Number	Width, cm	Measurement Number	Width, cm
1	89.70	11	89.83	21	89.86	31	89.98
2	89.75	12	89.92	22	89.78	32	89.80
3	89.93	13	89.80	23	89.84	33	99.85
4	89.78	14	89.94	24	89.88	34	89.89
5	89.83	15	89.81	25	89.89	35	89.82
6	89.74	16	89.93	26	89.88	36	89.87
7	89.86	17	89.85	27	89.81	37	89.88
8	89.90	18	89.86	28	89.87	38	89.87
9	89.84	19	89.90	29	89.86	39	89.95
10	89.84	20	89.91	30	89.93	40	89.88

tained may be scattered, repeated experiments may not yield the same numerical results, and plotted data points may not produce a smooth curve. Random variations may be a natural consequence of the experiment or an unavoidable shortcoming from the difficulty in measuring forces, temperatures, distances, or other quantities. In any case, engineers need statistical methods to deal with problems involving random behavior. Statistics is the area of mathematics which furnishes an organized approach for handling measurements and data that show random variations.

An example of a measurement which has been repeated many times and yields varying results is presented in Table 2.11. The data here were obtained while measuring the width of a conference table in centimeters. The 40 measurements shown were all made by the same person in about $\frac{1}{2}$ hour using a machinist's scale 30 cm in length. The variations reflect only one person's ability to make a consistent measurement with the available equipment. In statistical terminology, this group of measurements is viewed as a sample taken from a population of all the table width measurements that could be made under these conditions. The data as presented in Table 2.11 are referred to as *raw data*, that is, a list of quantities in the order in which they were obtained. By simply looking at this table, we cannot make any statement about the measurement of interest or its accuracy. A first step in analyzing this data is to rewrite the table, ignoring the order in which they were taken and putting them in numerical order. We have done this in Table 2.12, where numbers which are found more than once in the sample are also written repeatedly.

At this point we can begin to make some sense out of the data. We can easily identify two common statistical measures—the median and the mode. The *median* is at the midpoint of the data. With the data ordered as in Table 2.12, the median may be ob-

TABLE 2·12
ORDERED WIDTH MEASUREMENTS, cm

89.98	89.89	89.86	89.82
89.95	89.89	89.86	89.81
89.94	89.88	89.86	89.81
89.93	89.88	89.85	89.80
89.93	89.88	89.85	89.80
89.93	89.88	89.84	89.78
89.92	89.87	89.84	89.78
89.91	89.87	89.84	89.75
89.90	89.87	89.83	89.74
89.90	89.86	89.83	89.70

tained by counting down $(n + 1)/2$ from the top of the data where n is the sample size or total number of data points. When n is an even number, as is the case here, the median is one-half the sum of the numbers in the $n/2$ and $n/2 + 1$ positions. For example,

For 7, 10, 11, 13, 15, $n = 5$ and the median $= 11$
For 1.0, 1.0, 1.1, 1.2, 1.4, 1.5, 1.5, $n = 7$ and the median $= 1.2$
For 75, 76, 77, 78, 79, 80, $n = 6$ and the median $= 77.5$

For our measurement data in Table 2.12, the median is 89.86 cm. The median is frequently used as a single number to summarize all the data points, since it lies at the center of the ordered data and should therefore be a representative point.

The *mode* is the number which appears most frequently in the data. In situations with a small sample size, it may not be a suitable number to summarize the data. However, with a large amount of data, the tendency is for the most representative numbers in the range to be repeated most frequently. There may be one mode in a given set of data (unimodal data), two modes (bimodal data), or more (multimodal). In some instances, there may not be a mode. For example,

For 1.0, 1.1, 1.2, 1.3, 1.5, there is no mode
For 6, 7, 7, 8, 9, 10, the mode $= 7$
For 16, 17, 17, 18, 19, 19, 30, there are two modes, 17 and 19, so the list is bimodal

Our data in Table 2.12 is bimodal, showing modes at 89.88 and 89.86 cm.

THE MEAN

At this point, we must introduce the concept of the *mean*—the most common number used to represent a sample or set of data. It is simply the arithmetic average of all the numbers in the sample. There is, however, a very appealing mathematical reason for its being a useful number to represent the sample. Consider each piece of data in the sample to be identified by x_i, where i can be any integer from 1 to n, the number of data points. With our data in Table 2.11 as an example, $n = 40$ and $x_1 = 39.70$, $x_2 = 89.75$, . . . , $x_{13} = 89.80$, . . . , $x_{40} = 89.88$. The mean value of the data is given the symbol $\bar{x}$. If we use $\bar{x}$ as a number representative of the overall sample, the error involved between $\bar{x}$ and any particular piece of data x_i in the sample is $E_i = x_i - \bar{x}$. Some of the errors will be positive and some negative; however, the square of the error $E_i^2 = (x_i - \bar{x})^2$ must always be a positive number. If we add the square errors for each of the points $E_1^2 + E_2^2 + \cdots + E_n^2$, we have an indication of the overall error in letting $\bar{x}$ represent the sample. We call this the sum of the squared errors, represented by

$$\sum_{i=1}^{n} E_i^2 = E_1^2 + E_2^2 + \cdots + E_n^2 \qquad (2.38)$$

where the summation sign Σ means that we are to add the quantity to the right of the sign with subscript values from $i = 1$ to $i = n$ as shown in the equation. If we choose $\bar{x}$ so that we minimize the sum of squared errors, we have a good value for the mean. Thus, we want to minimize

$$\sum_{i=1}^{n} E_i^2$$

or, remembering the definition for E_i,

$$\text{minimize } \sum_{i=1}^{n} (x_i - \bar{x})^2 \qquad (2.39)$$

In calculus, a simple minimum results from setting the first derivative equal to zero. Differentiating Eq. (2.39) with respect to $\bar{x}$, we find

$$-2 \sum_{i=1}^{n} (x_i - \bar{x}) = 0$$

to be the condition for minimum square error. Simplifying, we get

$$\sum_{i=1}^{n} x_i - n\bar{x} = 0$$

$$n\bar{x} = \sum_{i=1}^{n} x_i$$

Finally,

$$\bar{x} = \frac{\sum_{i=1}^{n} x_i}{n} \qquad (2.40)$$

is the solution for $\bar{x}$ which minimizes the square error. This is, as mentioned earlier, simply the arithmetic average of the data, but you should know mathematically why it is a useful number. Thus,

For 5.0, 6.0, 7.0, 8.0, $\bar{x} = 6.50$
For 1.3, 1.6, 1.6, 1.7, 1.9, $\bar{x} = 1.62$
For 16, 17, 19, 20, 23, $\bar{x} = 19.0$

For the data given in Tables 2.11 and 2.12, we find that with the sample size $n = 40$, our mean $\bar{x} = 89.858$ cm.

TABLE 2·13
SUBDIVIDING THE RANGE

Subdivisions, cm	Frequency	Relative Frequency f/n
89.695–89.725	1	0.025
89.725–89.755	2	0.050
89.755–89.785	2	0.050
89.785–89.815	4	0.100
89.815–89.845	6	0.150
89.845–89.875	9	0.225
89.875–89.905	8	0.200
89.905–89.935	5	0.125
89.935–89.965	2	0.050
89.965–89.995	1	0.025
Total	40	1.000

HISTOGRAMS

We have discussed the mean, median, and mode, all statistical measures that tend to locate the middle of a sample or set of data. However, none gives significant information about the distribution of the data. To find the nature of the actual distribution of a sample we must first subdivide the sample into suitable groups or class intervals. We then break up the range which the data covers into more narrow intervals of equal size and we identify the appropriate interval for each data point. To be sure that none of the data points falls on a boundary between intervals, we define the boundary location with one more significant figure than the data itself contains. Our class intervals for the data of Table 2.11 are shown in Table 2.13. We have used 10 different subdivisions here, each 0.030 cm in width. If more data points were available, a larger number of more narrow subdivisions could be useful. We have counted the number of data points which fall into each class interval by inspecting Table 2.12. This number is the frequency of observation, shown in the second column of Table 2.13. Relative frequency is obtained by dividing each frequency by the sample size. For our data with $n = 40$, relative frequency is shown in the third column of Table 2.13. To visualize the distribution of this data, we now plot a *histogram*—a bar graph showing the frequency of observation in each subdivision over the range of the sample. It is the primary tool for examining statistical distributions. The histogram in Fig. 2.10 uses the data of Table 2.13. Histograms can also be plotted based on relative frequency; they then have the advantage of being independent of sample size. Comparisons of statistical distributionss involving different sample sizes can be made easily by graphing their relative-frequency histograms on the same coordinates.

We have shown some additional examples of histograms in Fig. 2.11: plots are presented for two human statistics in Fig. 2.11a and b and for two engineering-type statistics in Fig. 2.11c and d.

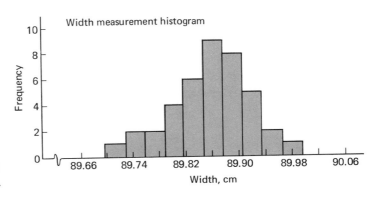

FIGURE 2·10
Histogram for the measurements.

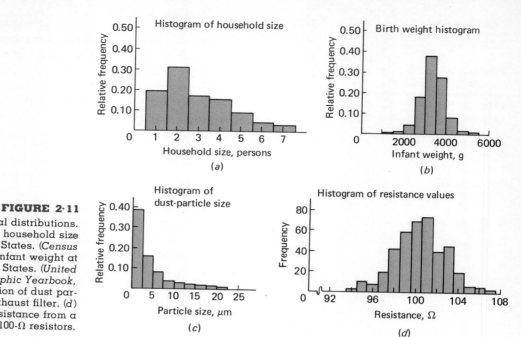

FIGURE 2·11
Typical statistical distributions. (a) Distribution of household size in the United States. (*Census Bureau*, 1975.) (b) Infant weight at birth in the United States. (*United Nations Demographic Yearbook*, 1975.) (c) Distribution of dust particle size in an exhaust filter. (d) Distribution of resistance from a batch of 100-Ω resistors.

Note that the data are derived from natural situations which have an inherently random aspect, not from an exploration of errors in measurement, as we have been discussing. The distributions in Fig. 2.11*a* and *c* are called *skewed* because they are not symmetrically distributed, and the largest number of data points is found to one side of the center of the range. This often occurs when a definite limitation exists on one side of the range, as in Fig. 2.11*a*, for example, where it is not possible to have a negative household size. However, as more often happens, the statistical distributions of interest to engineers are symmetric and unskewed, resembling in form Figs. 2.10 and 2.11*b* and *d*. These can usually be described by the so-called normal distribution, which can be defined by a convenient mathematical representation.

THE NORMAL DISTRIBUTION

A normal distribution for a sample can be established by knowing its mean and standard deviation. The mean $\bar{x}$ has been defined in Eq. (2.40) for a sample of size n, and the standard deviation is defined as

$$s = \left[\frac{\sum_{i=1}^{n} (x_i - \bar{x})^2}{n - 1} \right]^{1/2} \tag{2.41}$$

Note that the standard deviation has the same units as the x's. The standard deviation essentially measures the average distance from the mean $\bar{x}$ to the points x_i and thus indicates the spread of the data. If s is small, the data are all close to the mean. If s is large, the data are more spread out. For example,

For 5.0, 6.0, 7.0, 8.0, $\bar{x} = 6.5$ and $s = 1.29$
For 1.3, 1.6, 1.6, 1.7, 1.9, $\bar{x} = 1.62$ and $s = 0.217$
For 16, 17, 19, 20, 23, $\bar{x} = 19.0$ and $s = 2.74$

We customarily present the standard deviation to at least three significant figures, as shown above. If we return to the original data for our table-width measurements presented in Table 2.11, we compute a standard deviation of $s = 0.0595$ cm.

After values for $\bar{x}$ and s have been obtained, we are able to describe the normal distribution by mathematical equations. The probability of finding a data point in a sample between two particular values of x is

$$P = \frac{1}{s\sqrt{2\pi}} \int_{x_l}^{x_h} e^{-(x-\bar{x})^2/2s^2} \, dx \qquad (2.42)$$

where x_l is the lower value of x and x_h is the higher value. The e represents the number 2.718, which is the base of natural logarithms. The probability P is a number between 0 and 1, which is the fraction of data points we expect to find between x_l and x_h. For example, if $P = 0.1$, we expect to fine 1 out of 10 data points in this range of x. Also, if we have a total of n data points, we can expect $f = nP$ to be the number of points between x_l and x_h. The probability P may be considered to equal the relative frequency mentioned earlier so that

$$P = \frac{f}{n} \qquad (2.43)$$

The form of the normal distribution is symmetrical and bell-shaped, with its center at the mean. The general shape is shown in Fig. 2.12.

There are two ways to conveniently integrate Eq. (2.42). First, by keeping the distance $\Delta = x_h - x_l$ between the upper and lower values of x small, we can use the center value of $x = (x_h + x_l)/2$ in the expression

$$P = \frac{\Delta}{s\sqrt{2\pi}} e^{-(x-\bar{x})^2/2s^2} \qquad (2.44)$$

as an approximate equation and avoid an actual integration. The class intervals used in histograms are normally small enough for

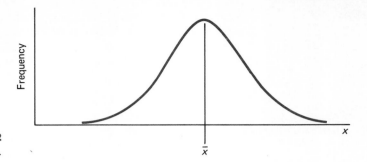

FIGURE 2·12
Shape of the normal distribution.

this to be helpful. Second, we can make use of tables which are readily available and include the integration as performed very accurately. For values of $x_l = \bar{x}$ and $x_h = X$, Table 2.14 can be used for this purpose, where you will need to compute

$$z = \frac{|X - \bar{x}|}{s} \tag{2.45}$$

before using the table. This variable z is the number of standard deviations between $\bar{x}$ and X. The result obtained from the table is the probability of finding data between $\bar{x}$ and X, but it is also simply the area contained under the normal distribution curve between $\bar{x}$ and X. We have indicated this as A in the small figure shown with the table. Since the normal distribution curve is symmetric, Table 2.14 may also be applied in the case where $x_l = X$ and $x_h = \bar{x}$ by using Eq. (2.45) for z and recognizing the absolute-value signs.

There are also a variety of other ways for utilizing Table 2.14. For example, to determine the probability of data occurring in the shaded region of Fig. 2.13, we substitute the value corresponding to X in this figure into Eq. (2.45) and then with the resulting z go to Table 2.14 to find A. We know that each half of the distribution has an area of 0.5000, as indicated in Fig. 2.13 from $x = \bar{x}$ to $x = +\infty$.

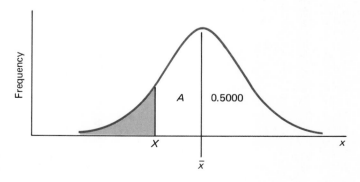

FIGURE 2·13
Areas under the normal distribution curve.

TABLE 2·14
INTEGRATION TABLE

$$A = \int_0^z \frac{1}{\sqrt{2\pi}} e^{-z^2/2} \, dz$$

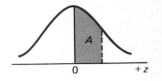

z	0	1	2	3	4	5	6	7	8	9
0.0	0.0000	0.0040	0.0080	0.0120	0.0160	0.0199	0.0239	0.0279	0.0319	0.0359
0.1	0.0398	0.0438	0.0478	0.0517	0.0557	0.0596	0.0636	0.0675	0.0714	0.0754
0.2	0.0793	0.0832	0.0871	0.0910	0.0948	0.0987	0.1026	0.1064	0.1103	0.1141
0.3	0.1179	0.1217	0.1255	0.1293	0.1331	0.1368	0.1406	0.1443	0.1430	0.1517
0.4	0.1554	0.1591	0.1628	0.1664	0.1700	0.1736	0.1772	0.1808	0.1844	0.1879
0.5	0.1915	0.1950	0.1985	0.2019	0.2054	0.2088	0.2123	0.2157	0.2190	0.2224
0.6	0.2258	0.2291	0.2324	0.2357	0.2389	0.2422	0.2454	0.2486	0.2518	0.2549
0.7	0.2580	0.2612	0.2642	0.2673	0.2704	0.2734	0.2764	0.2794	0.2823	0.2852
0.8	0.2881	0.2910	0.2939	0.2967	0.2996	0.3023	0.3051	0.3078	0.3106	0.3133
0.9	0.3159	0.3186	0.3212	0.3238	0.3264	0.3289	0.3315	0.3340	0.3365	0.3389
1.0	0.3413	0.3438	0.3461	0.3485	0.3508	0.3531	0.3554	0.3577	0.3599	0.3621
1.1	0.3643	0.3665	0.3686	0.3708	0.3729	0.3749	0.3770	0.3790	0.3810	0.3830
1.2	0.3849	0.3869	0.3888	0.3907	0.3925	0.3944	0.3962	0.3980	0.3997	0.4015
1.3	0.4032	0.4049	0.4066	0.4082	0.4099	0.4115	0.4131	0.4147	0.4162	0.4177
1.4	0.4192	0.4207	0.4222	0.4236	0.4251	0.4265	0.4279	0.4292	0.4306	0.4319
1.5	0.4332	0.4345	0.4357	0.4370	0.4382	0.4394	0.4406	0.4418	0.4429	0.4441
1.6	0.4452	0.4463	0.4474	0.4484	0.4495	0.4506	0.4515	0.4525	0.4535	0.4545
1.7	0.4554	0.4564	0.4573	0.4582	0.4591	0.4599	0.4608	0.4616	0.4625	0.4633
1.8	0.4641	0.4649	0.4656	0.4664	0.4671	0.4678	0.4686	0.4693	0.4699	0.4706
1.9	0.4713	0.4719	0.4726	0.4732	0.4738	0.4744	0.4750	0.4756	0.4761	0.4767
2.0	0.4772	0.4778	0.4783	0.4788	0.4793	0.4798	0.4803	0.4808	0.4812	0.4817
2.1	0.4821	0.4826	0.4830	0.4834	0.4838	0.4842	0.4846	0.4850	0.4854	0.4857
2.2	0.4861	0.4864	0.4868	0.4871	0.4875	0.4878	0.4881	0.4884	0.4887	0.4890
2.3	0.4893	0.4896	0.4898	0.4901	0.4904	0.4906	0.4909	0.4911	0.4913	0.4916
2.4	0.4918	0.4920	0.4922	0.4925	0.4927	0.4929	0.4931	0.4932	0.4934	0.4936
2.5	0.4938	0.4940	0.4941	0.4943	0.4945	0.4946	0.4948	0.4949	0.4951	0.4952
2.6	0.4953	0.4955	0.4956	0.4957	0.4959	0.4960	0.4961	0.4962	0.4963	0.4964
2.7	0.4965	0.4966	0.4967	0.4968	0.4969	0.4970	0.4971	0.4972	0.4973	0.4974
2.8	0.4974	0.4975	0.4976	0.4977	0.4977	0.4978	0.4979	0.4979	0.4980	0.4981
2.9	0.4981	0.4982	0.4982	0.4983	0.4984	0.4984	0.4985	0.4985	0.4986	0.4986
3.0	0.4987	0.4987	0.4987	0.4988	0.4988	0.4989	0.4989	0.4989	0.4990	0.4990
3.1	0.4990	0.4991	0.4991	0.4991	0.4992	0.4992	0.4992	0.4992	0.4993	0.4993
3.2	0.4993	0.4993	0.4994	0.4994	0.4994	0.4994	0.4994	0.4995	0.4995	0.4995
3.3	0.4995	0.4995	0.4995	0.4996	0.4996	0.4996	0.4996	0.4966	0.4996	0.4997
3.4	0.4997	0.4997	0.4997	0.4997	0.4997	0.4997	0.4997	0.4997	0.4997	0.4998
3.5	0.4998	0.4998	0.4998	0.4998	0.4998	0.4998	0.4998	0.4998	0.4998	0.4998
3.6	0.4998	0.4998	0.4999	0.4999	0.4999	0.4999	0.4999	0.4999	0.4999	0.4999
3.7	0.4999	0.4999	0.4999	0.4999	0.4999	0.4999	0.4999	0.4999	0.4999	0.4999
3.8	0.4999	0.4999	0.4999	0.4999	0.4999	0.4999	0.4999	0.4999	0.4999	0.4999
3.9	0.5000	0.5000	0.5000	0.5000	0.5000	0.5000	0.5000	0.5000	0.5000	0.5000

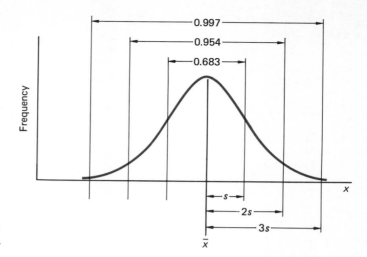

FIGURE 2·14
Areas within 1s, 2s, and 3s.

Thus $0.5000 + A$ is the probability of data occurring in the unshaded portion of Fig. 2.13 and $0.5000 - A$ is the probability of its occurring in the shaded portion. Using further manipulations of this type, we can establish some important properties of the normal distribution. It can be shown that within one standard deviation from the mean ($\pm s$), 68.3 percent of the data is found. Within $\pm 2s$, 95.4 percent of the data is found, and within $\pm 3s$, 99.7 percent is found. This is indicated in Fig. 2.14. Note that it is rare to find data points more than three standard deviations from the mean when a sample is normally distributed.

Examples 2.3 to 2.5 illustrate uses of the normal distribution. Example 2.3 shows how actual distribution data can be compared against a normal distribution curve. A utilization of the normal distribution for predicting failures is shown in Example 2.4. The concept of reliability is introduced in Example 2.5. Reliability is a popularly used term—often by the news media to describe spacecraft performance or nuclear reactor safety—but it is not widely understood by lay people. *Reliability* in a technical sense is the probability of successful performance under prescribed conditions. It must be based on a statistical distribution, as indicated in Example 2.5.

THE LEAST-SQUARES FIT

When we obtain real data and plot the points on a graph, we are frequently not sure exactly where to draw a line or curve to represent the data points, because real data may scatter. In Figs. 2.7b and 2.9b we found it easy to draw straight lines through the

Example 2·3

Using the data originally presented in Table 2.11 for the table-width measurements, plot the normal distribution curve and compare it with the actual histogram.

Solution

Here, we can make use of the approximation in Eq. (2.44). We have found the mean, $\bar{x} = 89.858$ cm, and the standard deviation, $s = 0.0595$ cm, for this data. In the frequency presentation of Table 2.13, we used subdivisions 0.0300 cm in width, so we will use $\Delta = 0.0300$ cm in obtaining the normal distribution from Eq. (2.44). We will also use x corresponding to the center of each subdivision. For example, in the first interval of Table 2.13, $x = 89.710$ cm, so we can compute

$$P = \frac{\Delta}{s\sqrt{2\pi}} \exp \frac{-(x-\bar{x})^2}{2s^2}$$
$$= \frac{0.0300}{0.0595\sqrt{2\pi}} \exp \frac{-(89.710 - 89.858)^2}{2(0.0595)^2}$$
$$= 0.00912$$

(To make the equation easier to write we have used the notation $\exp B = e^B$.) This number can be compared with the relative frequency experienced for the first interval. Repeating the calculation using the values of x in the other intervals, we can establish Table 2.15. These numbers can be plotted in a relative-frequency histogram to compare the actual distribution and the theoretical one. This has been done in Fig. 2.15 using the data in Tables 2.13 and 2.15. It can be seen that our measurement data follows a normal distribution reasonably well.

TABLE 2·15

x, cm	89.68	89.71	89.74	89.77	89.80	89.83	89.86	89.89	89.92	89.95	89.98	90.01
P	0.00229	0.00912	0.0281	0.0674	0.125	0.180	0.201	0.174	0.117	0.0609	0.0246	0.0070

FIGURE 2·15

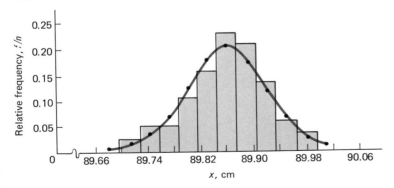

idealized data presented. But Fig. 2.4e shows some real data which are very scattered, so it is not obvious where a straight line should be drawn to represent these data points. The mathematical procedure known as least-squares fitting, or linear regression, can be used to determine the best fit to scattered data. The straight line shown in Fig. 2.4e was obtained by use of the least-squares method.

Example 2·4

In Table 2.16 we have the data obtained for the lifetime in hours of 20 different light bulbs. Assuming that this data is adequate and that it is normally distributed, determine the portion of refunds that would be expected if these light bulbs were guaranteed for 750 h.

TABLE 2·16

Bulb Number	Life, h	Bulb Number	Life, h
1	870	11	1140
2	955	12	820
3	1030	13	1030
4	770	14	970
5	1025	15	1065
6	940	16	925
7	1120	17	1170
8	960	18	975
9	865	19	1040
10	1070	20	920

Solution

First we must calculate $\bar{x}$ and s for this data. We find that

$$\bar{x} = \frac{\sum_{i=1}^{20} x_i}{20}$$

$$= \frac{1}{20}(870 + 955 + \cdots + 920)$$

$$= 983 \text{ h}$$

and

$$s = \left[\frac{1}{19}\sum_{i=1}^{20}(x_i - \bar{x})^2\right]^{1/2}$$

$$= \left\{\frac{1}{19}[(870 - 983)^2 + (955 - 983)^2 + \cdots + (920 - 983)^2]\right\}^{1/2}$$

$$= 105 \text{ h}$$

Calculate now that

$$z = \frac{|X - \bar{x}|}{s}$$

$$= \frac{|750 - 983|}{105}$$

$$= 2.22$$

Enter Table 2.14 to find $A = 0.4868$. The probability of failure is thus

$$P = 0.500 - 0.4868$$
$$= 0.0132$$

This means that approximately 13 early failures per thousand are to be expected.

Example 2·5

When a very large amount of data is taken, it is frequently presented only by class interval and frequency. This is the case for a large number of cast-iron specimens whose strength has been tested and for which the results are shown in Table 2.17. If this type of cast iron is loaded to 30,000 lbf/in² in an actual application, find its reliability assuming that a normal distribution is valid.

TABLE 2·17

Subdivision, lbf/in²	Frequency
29,950–34,950	4
34,950–39,950	8
39,950–44,950	12
44,950–49,950	15
49,950–54,950	19
54,950–59,950	13
59,950–64,950	11
64,950–69,950	8
69,950–74,950	2
Total	92

Solution

We must first calculate the mean and the standard deviation. With the data in the form of Table 2.17, we use the center of the class interval to represent all data within the class interval. Thus,

$$\bar{x} = \frac{\sum_{j=1}^{m} f_j x_j}{n} \qquad (2.46)$$

is an equivalent statement of the arithmetic mean computation in Eq. (2.40). Here m is the total number of class intervals. Now for Table 2.17, we have

$$\bar{x} = \frac{4(32,450) + 8(37,450) + \cdots + 2(72,450)}{92}$$

$$= 51,800 \text{ lbf/in}^2$$

We can apply the following equation for standard deviation:

$$s = \left[\frac{\sum_{j=1}^{m} f_j (x_j - \bar{x})^2}{n - 1} \right]^{1/2} \qquad (2.47)$$

as an equivalent form of Eq. (2.41). For our data, we find that

$$s = \left[\frac{\begin{array}{c} 4(32,450 - 51,800)^2 \\ + 8(37,450 - 51,800)^2 \\ + \cdots + 2(72,450 - 51,800)^2 \end{array}}{91} \right]^{1/2}$$

$$= 9920 \text{ lbf/in}^2$$

Now we compute

$$z = \frac{|X - \bar{x}|}{s}$$

$$= \frac{|30,000 - 51,800|}{9920}$$

$$= 2.20$$

Looking at Table 2.14, we find $A = 0.4861$. Thus, the *probability of failure* of an arbitrary piece of this cast iron when a load of 30,000 lbf/in² is first applied is $P = 0.5000 - 0.4861 = 0.0139$. In other words, the *probability of no failure* is $1.0000 - 0.0139 = 0.986$. We call this the reliability and represent it by the symbol R. In this case, $R = 0.986$.

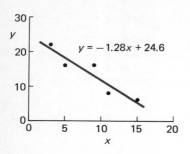

FIGURE 2·16
The least-square fit.

We will now study the least-squares procedure for obtaining a straight-line fit by considering the five data points plotted in Fig. 2.16. A straight line drawn here must be defined by

$$y = mx + b \tag{2.48}$$

and the least-squares curve-fitting process defines m and b so that the line fits the data as well as possible in a mathematical sense. Let each data point be represented by its coordinates x_i, y_i, where i is a subscript which can take on integer values from 1 to n, and n equals the number of data points. In our example of Fig. 2.16, $n = 5$ and a tabulation of the data shows

$$x_1 = 3, \ y_1 = 22$$
$$x_2 = 5, \ y_2 = 16$$
$$x_3 = 9, \ y_3 = 16$$
$$x_4 = 11, \ y_4 = 8$$
$$x_5 = 15, \ y_5 = 6$$

If Eq. (2.48) were used to estimate the value of y at any of the locations $x = x_i$ above, we would calculate

$$y = mx_i + b$$

The difference between the actual value at the data point which is y_i and the value of y estimated by this equation is an error which we can express as

$$E_i = y_i - (mx_i + b) \tag{2.49}$$

By letting $i = 1$ through n, the error at each data point can be given by this equation. To indicate how well the line fits the data points, the value of E_i can be squared at each data point and we sum the squares to get

$$\sum_{i=1}^{n} E_i^2 = \sum_{i=1}^{n} (y_i - mx_i - b)^2 \tag{2.50}$$

where, as mentioned earlier, the summation sign Σ means to add the quantity to the right of the sign using subscripts from $i = 1$ to $i = n$. In this way,

$$\sum_{i=1}^{n} E_i^2 = E_1^2 + E_2^2 + E_3^2 + \cdots + E_n^2$$

We work with squared errors so that positive and negative values of error contribute equally to the summation.

The least-squares fit selects m and b in the straight-line equation to find the smallest possible value for the sum of the squares. This is obtained by differentiating Eq. (2.50) with respect to m and b and finding the values of m and b which make the derivatives zero. The equations resulting from equating the derivatives to zero are

$$\sum_{i=1}^{n} x_i(y_i - mx_i - b) = 0 \qquad (2.51)$$

and

$$\sum_{i=1}^{n} (y_i - mx_i - b) = 0 \qquad (2.52)$$

with the first equation resulting from differentiation with respect to m and the second with respect to b. Since all the x_i and y_i are known from data, Eqs. (2.51) and (2.52) are simultaneous equations in m and b which may be solved to define the desired straight line.

For our case, with $n = 5$ and the given data, Eqs. (2.51) and (2.52) become

$$3(22 - 3m - b) + 5(16 - 5m - b) + 9(16 - 9m - b)$$
$$+ 11(8 - 11m - b) + 15(6 - 15m - b) = 0$$
$$(22 - 3m - b) + (16 - 5m - b) + (16 - 9m - b)$$
$$+ (8 - 11m - b) + (6 - 15m - b) = 0$$

These two equations simplify to

$$461m + 43b = 468$$
$$43m + 5b = 68$$

which show the simultaneous equations in more normal form. Solving these we find $m = -1.28$ and $b = 24.6$ and the resulting linear equation for our straight line is

$$y = -1.28x + 24.6 \qquad (2.53)$$

This line, plotted in Fig. 2.16, fits appropriately into the scattered data.

The least-squares curve-fitting technique described here can be applied directly to logarithmic and semilogarithmic plots by working with $\log x$ and $\log y$ in the equations instead of x and y. The technique can also be extended to fit higher-order curves (e.g., quadratic) through data which does not show a linear trend.

PROBLEMS

2.26 Find the median, the mean, and any modes for each set of data shown below.
(a) 8, 9, 10, 11, 12, 13, 14
(b) 10.1, 10.5, 10.4, 10.3, 10.4, 10.7, 10.3, 10.4
(c) 0.104, 0.114, 0.101, 0.123, 0.115, 0.115, 0.117
(d) 153, 152, 154, 155, 151, 156, 150, 149
(e) 6.54, 6.53, 6.52, 6.53, 6.54, 6.52, 6.51, 6.55, 6.53

2.27 A certain manufacturer produces machined shafts of about $\frac{1}{2}$-in diameter. In a selection of 20 samples from the production line, the following shaft diameters were measured in inches:

0.5001	0.5007	0.5008	0.4999
0.5013	0.5004	0.4987	0.5002
0.4986	0.5011	0.5006	0.4997
0.4982	0.5018	0.5004	0.4993
0.4991	0.4992	0.4995	0.4998

Subdivide this data into reasonable class intervals and tabulate the frequency and the relative frequency of observation. Plot a neat, well-labeled histogram for the relative-frequency distribution.

2.28 For the data in Prob. 2.27, determine the median, the mean, and the standard deviation.

2.29 A series of 1-min counts has been taken by a well-shielded Geiger counter in a 20-min period. The Geiger counter was placed in an open field to measure background radiation and the following data were obtained.

3	18	5	17
5	6	11	12
7	14	15	15
8	11	6	8
10	9	11	4

Subdivide this data into reasonable class intervals and tabulate the frequency of observation. Plot a neat, well-labeled histogram of the frequency distribution.

2.30 For the data in Prob. 2.29, determine the median, the mean, and the standard deviation.

2.31 Mass production of shafts in a certain factory results in a mean shaft diameter of 12.00 mm with a standard deviation of 0.045 mm. Assume that the diameters are normally distributed.
(a) What percentage of shafts are expected to fit without interference into a hole 12.10 mm in diameter?
(b) A diametral clearance of more than 0.20 mm will result in a fit between shaft and hole which is unacceptably loose. What percentage of the shafts is expected to fit acceptably into the 12.10-mm hole without interference?

and gases; trace the fate of smokes and other pollutants in the human lung; calculate the behavior of flows of charged particles in a magnetic field; study the behavior of tsunamis (large harbor waves) and, in collaboration with civil engineers, evaluate their effects on buildings on shore.

Industrial engineers design computer controls to be comfortable for human operators; calculate the number of machines needed in a plant, balancing these numbers so that machines are seldom idle yet jobs are infrequently kept waiting; simulate the performance of a person faced with a complex control task; and balance the risk of inattention and boredom against the risk of mistakes caused by having to make too many decisions too fast.

Mechanical engineers work on design and production of mechanical and electromechanical components of computers such as printers, plotters, memory-disk and tape drives, keypunch machines, and terminal keyboards; solve problems in stress analysis, fluid flow, and heat transfer; apply computers to automatic control, particularly machine control and control of consumer products.

Nuclear engineers calculate the temperatures inside reactor fuel elements; assess the probable consequences of postulated accidents and evaluate their effect on the environment; study fuel cycles, determining how to get the most energy out of a given mass of uranium; use computers to process information to help operators manage nuclear power plant operation.

PROBLEMS

3.1 Review your own past and future experience with digital computers. Where did you first encounter computers? What computer courses are required of all engineering students at your school? If you are contemplating an engineering specialty, what computer courses are required of, or optional for, students in that specialty?

3·2 HISTORICAL PERSPECTIVE: FROM EVOLUTION TO REVOLUTION

Although a study of the history of computers could be justified on cultural grounds alone, our reasons for looking at the history of computing are very practical ones. By studying the history of past projects—both successful and unsuccessful—we can identify the factors essential to the design and production of computing machinery that works. We will see that a successful project requires

Reliable components available in requisite numbers
An organization capable of coordinating the efforts of the many people involved in the project
Funding sufficient to permit orderly and uninterrupted progress toward one project goal
Realistic anticipation of the needs and abilities of potential users in coordination with the machine's development

ORIGINS OF MECHANICAL COMPUTATION

Ancient Babylonians, Greeks, and Romans had clumsy nonpositional number systems, using V for 5, L for 50, and D for 500, for example. They never tried to compute by using their number systems directly; they used a simple mechanical device—the abacus—to do the computations. It incorporates a positional number system and, often, sophisticated alternates to decimal notation. The Chinese abacus obtains ten-decimal-digit precision from its ten horizontal columns and uses biquinary (base 2 and base 5) notation, but the operator's fingers and brains do the addition, subtraction, and carrying.

The abacus had been in use for thousands of years when the French philosopher and mathematician Blaise Pascal (1623–1662) built a mechanical device capable of adding, subtracting, and carrying. The model shown can count to 999,999 French francs.

The work of the German philosopher, mathematician, and writer, Gottfried Wilhelm Leibniz (1646–1716) has had a profound effect on both the practice and theory of computation.

His experience with computing mechanisms convinced Leibniz that computing *ought* to be automatic. His theoretical investigations led him to systematic development of the *binary* system. He showed that the universe of real numbers (starting with zero) could be expressed by some combination of the two digits 0 and 1. The binary system found no immediate practical application, but it remained in the system of formal thought taught by mathemati-

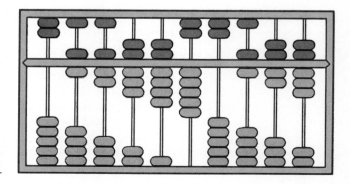

A Chinese abacus.

Pascal's adding machine. (Crown Copyright. Science Museum, London.)

cians at universities. When $2\frac{1}{2}$ centuries later von Neumann applied the system to computers, all the required mathematical techniques were available.

BABBAGE, LOVELACE, AND THEIR ENGINES

The period from 1700 to 1850 was an epoch of rapid growth in mathematical analysis. During this time, Sir Isaac Newton, Leon-

Gottfried Wilhelm Leibniz. "For it is unworthy of excellent men to lose hours like slaves in the labor of calculation which could safely be relegated to anyone else if machines were used." (Science Museum, London.)

Charles Babbage at age 49 (1841). (*Science Museum, London.*)

hard Euler, Joseph Louis Lagrange, Pierre Simon de Laplace, and Karl Friedrich Gauss developed many of the mathematical ideas used in engineering analysis today. Their ideas found almost immediate application in astronomy, surveying, and navigation, so there was a pressing need for extensive and accurate tables of logarithms, trigonometric functions, and astronomical positions. Because no automatic means of calculation were available, the tables calculated by overworked clerks were full of errors.

The English mathematician Charles Babbage (1792–1871) developed two types of automatic calculating machines:—*the difference engine* and *the analytical engine*—with the intention of eliminating the errors in mathematical tables.

The difference engine was a specialized automatic calculator, suitable for calculating polynomials by the difference method. Babbage's 1833 working model handled second differences and six-digit decimal numbers.

A simple extension of this machine would have been of immediate practical value for calculating and checking tables. [Indeed, two practical difference engines (which automatically printed 14-digit decimal results) were built in Sweden from Babbage's designs and used in computing mathematical tables.] Babbage left the development of the difference engine to others and turned to a more ambitious but less fruitful plan—the analytical engine. It

Hollerith's "statistical piano" tabulator and sorter. (*U.S. Bureau of the Census.*)

around the cylinder in the background. With her eyes focused on the data, she finds by touch the proper row and column on the template below her hand and punches one hole in the card clamped in the middle of the machine. In addition to the pantograph punch, Hollerith developed a tabulator and card sorter, known informally as the "statistical piano."

In operation, census clerks inserted cards into a press. The press handle lowered an array of spring-loaded metal fingers onto the punched card, which was precisely placed on an insulating plate. Wherever the card had a hole, the metal fingers passed through the card and made contact with mercury in cups below the insulating plate—thus generating an electric signal. These signals were sent to the array of 40 electromechanical clockwork counters located on an upright. Applied to the census, the machine operated as follows: if a citizen was a married female from Maine, holes were punched in the appropriate card columns, metal fingers made contact, relays clicked, and the counters registered one more married female inhabitant for Maine.

Hollerith was the first person to demonstrate that automatic computing machinery could do useful work and earn its keep. Much of his work was done less than 30 years after the death of Charles Babbage. Ironically, the first substantial applications of Babbage's difference-engine method for the routine tabulation of functions were carried out in England in 1929 by L. J. Comrie of His Majesty's Nautical Almanac Office using Hollerith-type American equipment.

BOOLE, SHANNON, AND LOGIC

The elements of computer development were not strictly chronological. Two developments, one in the field of pure mathematics in 1854 and one in electrical engineering in 1938, have had a profound effect on computer design, but were not generally applied until after 1946. The English mathematician George Boole (1815–1864) published *An Investigation of the Laws of Thought, on which are founded the Mathematical Theories of Logic and Probabilities* in 1854, an attempt to link arithmetic and logic in order to reduce complicated logical proofs to simple exercises in arithmetic. To do this, he found he had to restrict his variable quantities to the two values 0 and 1. Thus, his work connected logic, arithmetic, and the binary system.

Nearly a century later, Claude Shannon (born 1916), an American, used binary logic to do arithmetic.

In his master's thesis, "A Symbolic Analysis of Relay and Switching Circuits," Shannon associated closed and open switches with Boole's mathematical 0s and 1s. He also associated series and parallel circuits with logical AND and logical OR. He then showed that any function in Boole's logical calculus could be

George Boole, mathematician. (*Science Museum, London.*)

realized by suitable series and parallel combinations of open and closed switches. At the end of his paper, he showed how binary addition could be done using electric switches. Although the value of Shannon's work was recognized immediately, its broad application had to wait for von Neumann's advocacy of the binary system in his EDVAC report of 1946.

ENIAC, A PIONEER ELECTRONIC MACHINE

About 20 years elapsed between the formation of International Business Machines Corporation (IBM) in 1924 and the first demonstration of an electronic computer in the mid-1940s. During this period, electromechanical tabulating machines pioneered by Hollerith were further improved. In the late 1930s and early 1940s, large-scale, fully automatic (but still electromechanical) machines were developed by Howard Aiken at Harvard (with support from IBM), by George Stibitz and collaborators at Bell Telephone Laboratories, and by Konrad Zuse in Germany. Although these electromechanical machines performed reliably, the needs of the scientists serving in the United States military services during the Second World War required order of magnitude increases in speed. These needs led to the electronic (vacuum-tube) computer.

During the Second World War, the U.S. Army Ordnance Department faced an unprecedented need for tables of the trajectories of artillery shells. Even when all available computing equipment was pressed into service, the production of tables fell far behind the requirements. In this atmosphere, J. Presper Eckert (born 1919) and John W. Mauchly (1907–1980) persuaded the U.S. Army to support development of an Electronic Numerical Integrator and Calculator, the ENIAC. The ENIAC proved to be 1000 times as fast as the then-existing electromagnetic calculators. However, their preliminary design called for the use of 18,000 vacuum tubes at a time when very few people believed that a machine using hundreds, let alone thousands, of vacuum tubes would ever be reliable.

The problem of reliability was solved by using two key ideas:

- Vacuum tubes would be operated in only two states—nominally "on" and "off"—corresponding to 1 and 0.
- Worst-case design—which was invented in the ENIAC project—would be used to allow the circuits to function despite the worst anticipated variations in the performance of vacuum tubes and other components.

The organization of the ENIAC's design team still serves as a model for large-scale computer design projects involving the coordinated efforts of dozens or even hundreds of professionals. Of

The ENIAC was the result of the coordinated efforts of University of Pennsylvia researchers and military users. This photograph, taken at the ENIAC dedication in 1946, shows J. Presper Eckert, chief engineer, first at the left; John W. Mauchly, consulting engineer, fifth from the left; as well as (from left to right) J. G. Brainerd, University of Pennsylvania; S. Feltman, U.S. Army Ordnance; H. H. Goldstine, U.S. Army Ordnance; Mauchly; H. Pender, University of Pennsylvania; G. M. Barnes, U.S. Army Ordnance; P. M. Gillon, U.S. Army Ordnance. (*University of Pennsylvania Archives.*)

these, perhaps the most important was the designers' concentration on meeting contract specifications.

In the special case of ENIAC, this required concentration on building a *special-purpose* machine for artillery trajectory calculations. As the work proceeded, the designers saw many ways in which the circuits might have been improved and the overall design might have been modified to give the machine more of a *general-purpose* character. They resisted the temptation to deviate from the contractual goal, however, reasoning that improvements could be incorporated into later machines. They were designing a one-of-a-kind prototype, and their attitude was to get it done on time and make it work. The final product was huge (3000 ft^3 of machinery consuming 80 kw of electric power), but it worked! It satisfied the needs and suited the abilities of its intended users. Subsequently, the ENIAC designers built general-purpose machines for the Census Bureau under the name UNIVAC.

VON NEUMANN AND THE STORED PROGRAM

At the time that the ENIAC was completed, it shared a common design feature with the few electromechanical computers then in operation: the program and the data were both stored, but the

The ENIAC in military use, calculating ballistic trajectories (1946). (*University of Pennsylvania Archives.*)

designers regarded the storage of programs and data as separate functions. In the ENIAC, for example, the program was wired into the machinery, and to change the program, the operators had to change the wiring, which could take days. Data, on the other hand, was stored electronically and could be quickly retrieved and altered.

The mathematical physicist, John von Neumann (1903–1957), who had been a consultant on the ENIAC project, recognized the need to speed up and generally improve the programming process. He saw that if the binary system were adopted, both program instructions and numerical data could be expressed as strings of 1s and 0s, which could be stored by identical devices and recalled, if necessary, at electronic speed (in microseconds). EDVAC was to be ENIAC's successor. In a series of reports written in 1945 and 1946, von Neumann and his associates laid out the principles for designing EDVAC as a stored program machine. These reports circulated widely within the computing community and spread von Neumann's stored program concept through the United States and Europe.

By 1952, a whole generation of machines incorporating the

Dr. John von Neumann standing next to his MANIAC (mathematical analyzer, numerical integrator and computer) at the Institute for Advanced Study, Princeton, New Jersey.

stored program/binary arithmetic idea (now known as the "von Neumann architecture") burst on the scene with such names as SEAC, EDSAC, JOHNIAC (in successive models), and even MANIAC. This basic design principle has remained as the standard for all computers, and today is embodied in the million-dollar super computers as well as in today's inexpensive, hand-held programmable calculators.

FORRESTER AND MAGNETIC MEMORY

Program storage equipment at the time of von Neumann and his contemporaries was complicated, expensive, and unreliable. For example, the ENIAC used 2000 vacuum tubes to store only 20 ten-digit decimal numbers. Von Neumann's own machine used 40 hand-made television tubes to store 1024 binary numbers, each containing 40 binary digits (1s or 0s). This memory (5 kilobytes in today's terms) was the largest available to von Neumann; today, a better memory of the same capacity can be built with solid-state technology for a few hundred dollars. As often happens, progress toward larger, cheaper, more reliable memories required that entirely new principles be exploited.

In 1951, J. W. Forrester (born 1918), an electrical engineer, provided the needed breakthrough with his invention of the coincident-current magnetic-core memory. In its very first application—the Whirlwind I computer—the magnetic-core memory increased memory capacity thirty-two times over the pre-

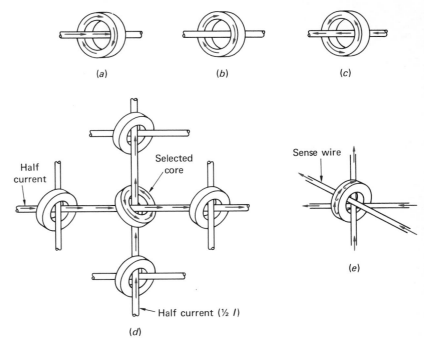

The von Neumann architecture demanded a large-scale, inexpensive, and reliable memory. The coincident current, magnetic-core memory (invented by J. W. Forrester) met these needs. (a) Electric current in wire produces circular magnetic field (Ampere's law), which induces a magnetic field in doughnut. (b) When current stops, doughnut remains magnetized. (c) When current is reversed, direction of magnetization is reversed. (d) A specific doughnut in a large array is magnetized by stringing the doughnuts on two wires and sending half the necessary current down two of the wires. Where the currents meet, their magnetic fields will add and magnetize the desired doughnut. (e) A third wire, called a "sense" wire, is used to sense the direction when the magnetic doughnut is "interrogated."

viously available technology. Memory designers began to measure memory in multiples of K, where K is 1024 units of memory. As methods of mass production of magnetic-core memory improved, memory sizes could be increased to any desired capacity. In 1978, for example, memories of 256K were quite common, and the cost for storing one binary digit had dropped to only one cent. Data could be retrieved from memory in a fraction of a microsecond. At the present time, magnetic memories are gradually being supplanted by solid-state integrated-circuit devices (which are even cheaper), but many magnetic-core memories remain in use. Alternate magnetic technologies, particularly discs and magnetic bubbles, are finding increasing application.

THE MASS-PRODUCED COMPUTER

By 1955, all the elements for the mass production of computers were in place: the techniques for electronic arithmetic had been perfected, the von Neumann architecture had been adopted as standard, and Forrester had solved the problem of memory capacity. Thus, the stage was set for *mass production*.

After three years of development, IBM built several hundred 704s, machines using vacuum tubes for arithmetic and logic and equipped with 32K of magnetic-core memory. The high-speed and

Engineers and technicians checking the performance of production models of the IBM 701. These large machines, built on production lines, served as prototypes for the mass-produced IBM 704. (*Smithsonian Institution.*)

large memory of the 704 made possible the development of the FORTRAN language, which made computer programming much easier for engineers and scientists and in turn created a still larger market for mass-produced machines.

Many companies began to compete with IBM and UNIVAC in the commercial market. During the decade 1946 to 1956, new computer designs were introduced at the rate of two or three per year. In succeeding years, the rate of introduction of new computer models jumped to twelve to fifteen per year. Development proceeded still faster in the late 1950s and early 1960s as transistors replaced vacuum tubes and integrated circuits replaced transistors, first for arithmetic and logic and then for memory.

In October 1964 the Control Data Corporation (CDC) produced the CDC 6600, making unprecedented computational speed and accuracy available for scientific and engineering computing. The 6600 uses a central logic unit to control arithmetic operations performed by ten specialized functional units and has ten separate peripheral processors to channel data to and from input terminals and output devices. It is not one computer but a group of coordinated computers. Its speed, efficiency, and innovative design were not surpassed for nearly a decade. Some are still in use. Other well-known mass-produced supercomputers include the

Designer Seymour Cray stands next to the Cray model 1 supercomputer. (*Cray Research*.)

larger computers in the IBM 360 and 370 series and the UNIVAC 1100 series. The offspring of the 6600 are the CDC CYBER series computers. Meanwhile, the designer of the 6600, Seymour Cray, founded his own corporation to build computers whose speed now approaches the inviolable limits set by the speed of light.

The CDC and Cray machines have million-dollar price tags. Beginning in 1961, the Digital Equipment Corporation (DEC) recognized that there was a market for machines with more modest performance specifications and a lower cost. By 1968, the price for their mass-produced PDP 8 had dropped below $10,000. Shortly thereafter (1972), Hewlett Packard and Texas Instruments introduced the hand-held scientific calculator. By 1975, retail sales of microcomputers had begun and the computer revolution of supercomputers, minicomputers, microcomputers, and hand-held calculators was here.

THE FUTURE OF COMPUTERS

Throughout their history computers have become faster and faster, have packed more computing power into smaller and smaller packages, and have become cheaper and cheaper. These trends will probably continue into the future, but further prog-

Production line for PDP-8 mini-computers. (*Digital Equipment Corporation.*)

ress along these lines will eventually be slowed by the approach of two barriers, both of which are associated with the speed of light.

When computer speeds (on a logarithmic scale) are plotted versus time (in years, on a linear scale) we can see a tendency for the curve to flatten as speed approaches one billion (10^9) operations per second.

This happens because at 10^9 operations per second, light (and electric signals) travel 30 cm (about 1 ft) during an operation. Thus, the entire computer must be packed into a very small volume or else the time spent for data to get from one place to another in the machine becomes large compared with the time spent doing something useful with the data. Thus, increases in speed beyond 10^9 operations per second require dramatic decreases in computer size.

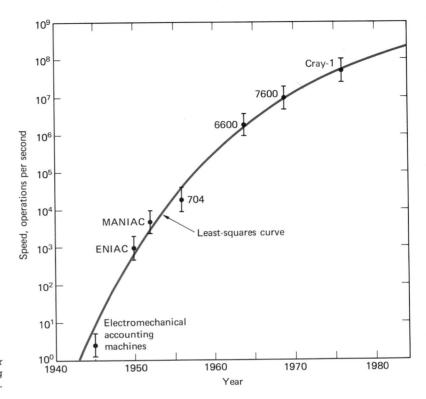

The trend of increasing computer speeds shows signs of leveling off.

Scanning electron microscope photograph of LSI (large-scale integration) computer circuit. (*Magnified 1000 times actual size.*) (*GCA Corporation, Burlington Division.*)

3.2 HISTORICAL PERSPECTIVE: FROM EVOLUTION TO REVOLUTION

Present-day integrated circuits are so densely packed with circuits that an electron microscope is needed to photograph their details. The scale of these photographs is of order 1 μm (10^{-6} m). Visible light has wavelengths in the range 0.4 to 0.7 μm. Hence, substantial reductions in the size of integrated circuits would reduce their part dimensions to submicrometer size and make the parts small compared with a wave of visible light. Already, ultraviolet light is routinely used in producing integrated circuits and the use of even shorter wavelength radiation is being explored.

Consequently, further increases in computer speed and reductions in computer size will require new physics, new chemistry, and new approaches to computer architecture and systems programming. Further cost reductions will probably come from new approaches in circuit and logic design. Already the cost of designing an integrated circuit can equal or exceed the cost of producing it, even when the design cost is shared by thousands of mass-produced copies. Thus, the computer engineers of the future have a real challenge if progress in speed, size, and cost is to continue.

In the future, big computers will get more powerful and small computers will get more numerous. As an example of why even

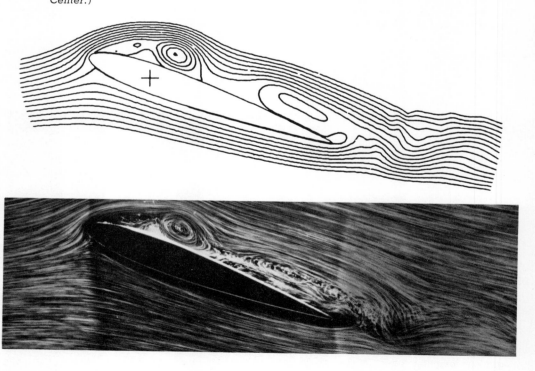

Computer calculation of airflow over a two-dimensional model of an airfoil compared with streamlines observed with corresponding model (at the same Reynold's number) in a water tunnel. (*NASA—Ames Research Center.*)

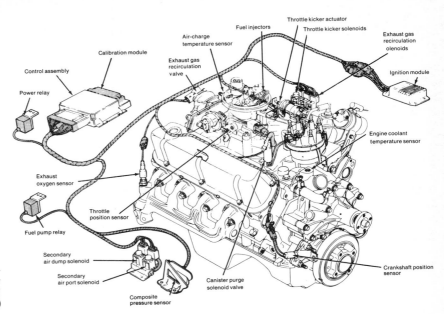

Computer control of an automobile engine to optimize efficiency and minimize pollution. (*Ford Motor Company.*)

today's super computers are not powerful enough, consider the computer calculation of air flow over an airfoil, shown on the facing page. The calculation (done with a CYBER 76 supercomputer) aimed to locate sources of energy-wasting drag. It used the entire high-speed memory of the computer and required one billion calculations. Computer time was 65 s. Present-day calculations come close to reproducing the experimental measurements made on simple (two-dimensional) shapes. Future progress, however, requires that calculations be done in three dimensions (chord, thickness, and span) and account for small obstacles to smooth air flow (possibly on the scale of individual rivets or heat-shield tiles). It is estimated that such calculations would require a computer with a memory capacity 3000 times larger than the largest computers available today. Yet the potential of time, energy and money savings are so great that NASA has let it be known that they will buy a computer as powerful as required with no price limits. So far no such computer has been designed.

At the opposite end of the computer-size scale it is likely that every household will in the near future contain dozens of small computers for controlling such things as washing machines, thermostats, and even home entertainment equipment. In our energy-hungry air-polluted society, however, one of the most exciting developments is the replacement of the "Rube Goldberg" carburetor and ignition controls in the car with computer-con-

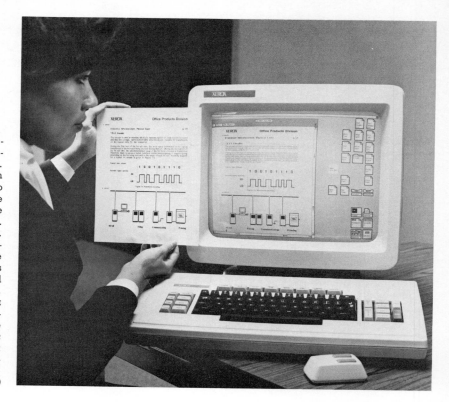

An input and output device designed to be one of many connected to present-day computers. Output is displayed on a television screen, but it may be disposed to a printer for permanent copy. The printer uses a laser to produce clear copy at electronic speed. Input is particularly versatile. Words, numbers, and mathematical symbols are input from the typewriter-style keyboard. Actions are controlled by the so-called mouse on the desk. If the mouse, for example, is used to move a pointer from the image of a document on the screen to the image of a file cabinet, the electronic equivalent of filing the document occurs instantly and automatically. (Xerox, Palo Alto Research Center.)

trolled fuel-injection and ignition systems. Such systems can allow the car to be started easily and run economically while emitting minimal pollutants.

Thus, computer engineers of the near future will be working toward progress in two complementary directions.

- For *computation* in engineering, science, and management, supercomputers with even faster arithmetic and still larger memories will be produced and used.
- For *control* in the home, the car, and industry, microcomputers able to survive the heat, moisture, and hard knocks of everyday life are already here. They will increase, with most users remaining unaware that a computer is helping them.

PROBLEMS

3.2 The purpose of this and the following problem is to help you place computing developments in historical perspective. On a large sheet of paper, lay out a time line covering the four centuries from 1600 to 2000. For this problem, use a scale of about one century = 20 cm. Place the principal events in computing on the time line and for comparison, mark such events in technology as the invention of the microscope, telescope, marine chronomometer, cotton gin, steam engine, steam boat, steam railroad, telephone, telegraph, wireless telegraph, airplane, and jet engine. If you wish, you can also include the lifespans of famous historical figures such as presidents, composers, artists, and military heros. If you are expert in any field of history (such as the history of your state), mark events that you are familiar with on the time line also. The key dates in computing history are in this book. General historical references, such as the *World Almanac*, will give dates for the other historical events.

3.3 To place the recent history of computers in perspective, lay out a time line spanning 1945 to 1985 at a scale of one decade = 10 cm. Indicate the approximate time of: completion of ENIAC; invention of the stored-program computer and magnetic memory; mass production of computers; delivery of IBM 704 and CDC 6600; availability of first computer under $10,000, hand-held scientific calculator, and scientific calculator under $25; first retail sales of computers. To put these events in historical perspective, indicate the dates of other events such as commercialization of jet aviation, nuclear power, transistor, laser, and video disk. You might also indicate such events in space as the first artificial satellite and manned satellite; manned exploration of moon; launching of weather satellite; and probes of Moon, Mars, and Saturn. Again, the *World Almanac* will be helpful.

3·3 COMPUTER BASICS

In the next few sections we will explore how a computer actually works. Even in this introductory description, you will find out that the complexity of computers tends to be overwhelming. Indeed, if every computer user were required to understand the computer and electronics theory completely, there would probably be far fewer computer users. The external simplicity of modern computers and calculators conceals the intricate details. When you press 2 + 2 = on your calculator and the display shows you a 4, it is easy to dismiss the complex process taking place inside.

Only a handful of readers of this section will ever work with computers at the level of the ones and zeros (1s and 0s) to be discussed here. But engineering students usually want to know how things work, so learning how computers add and subtract and how sequences of these simple operations can be programmed to perform useful calculations will no doubt be of interest. For those who wish to dig deeper on their own, we can recommend handbooks written by and for computer manufacturers [2, 3, 4, 5] and computer science texts [6, 7].

ORGANIZATION AND DESIGN

The typical modern computer system uses the organization, called "the von Neumann architecture," proposed by John von Neumann in the 1940s. The five major subsystems are arranged as in Fig. 3.1. The machine receives instructions and data *from* the outside world via the input unit, which may often resemble a typewriter keyboard. It transmits data or instructions *to* persons and devices in the outside world via the output unit which may often resemble a television screen or a typewriter. Modern computers commonly have multiple input and output devices, some of which may take in information from instruments or control the movements of machines.

The memory in the von Neumann architecture contains both programmed instructions and data. Some applications such as census tabulations or tax calculations may require storage for a few thousand instructions and hundreds of millions of data items. In contrast, some scientific programs may require a relatively small amount of input data, use perhaps hundreds of thousands of instructions, and produce millions of results, of which some are stored and others are sent through the output unit as tables, graphs, or commands. A master computer program, called "the operating system," contains instructions which the control unit uses to coordinate the other four subsystems.

It should be apparent that a computer system is too complex to be designed by a single person. A team of dozens, sometimes

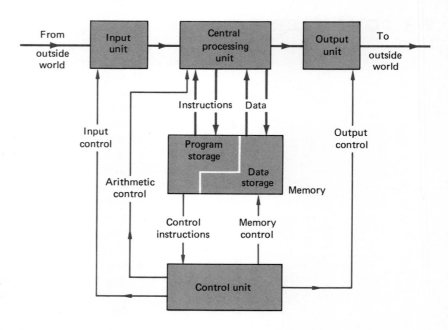

FIGURE 3·1
A digital computer using von Neumann architecture.

hundreds, of people, mostly electrical and computer science engineers but also other engineers and scientists with a variety of backgrounds, is used. Often the team members do not even work in the same city but may be located in different states or different countries. Included in the team are system architects who decide on the basic configuration of the machine, the operation of the central processing unit (CPU), the methods for arithmetic operations, etc. Circuit designers develop the electronic circuits for the CPU and the memory as well as those for the input and output devices. The logic designers apply mathematical concepts and ingenuity to use the electronic circuits efficiently and conveniently. Systems programmers design the operating system program, which acts as master control for the computer system, and compiler programs, which translate the languages of the computer user (FORTRAN, BASIC, COBOL, etc.) into the 1s and 0s which must actually be used in computer operation.

NUMBER SYSTEMS

The first decision which a system architect must make in computer design is to select the number system to be used. Our everyday number system, the familiar decimal system, is based on powers of 10. The decimal system was also used for many years in computers, but then, motivated by von Neumann's work, the binary system became dominant. For nearly two decades, all computers were purely binary in their innards, although input and output remained decimal. More recently (about 1979), some microcomputer system architects have included both binary and special binary-coded-decimal instructions as basic CPU functions, to ease the work of systems programmers in dealing with decimal input and output. Other number systems which are important to programmers are the *octal* (base 8) and *hexadecimal* (base 16) systems used as shorthand notations for binary numbers. Number systems, which we now examine, are thus a fundamental part of computer design and also computer programming.

All the number systems we have just mentioned are *positional* notation systems. In each system, there is a zero symbol and a different symbol for each of the integers up to, but not including, the base. In the decimal system (base 10) the digits are 0 1 2 3 4 5 6 7 8 9; in the binary system (base 2) the digits are only 0 and 1; in the octal system (base 8) the digits are 0 1 2 3 4 5 6 7; and in the hexadecimal (base 16) system the digits are 0 1 2 3 4 5 6 7 8 9 A B C D E F. In positional number systems each digit has two essential attributes:

Its value
Its position, which indicates the *power* of the base it multiplies

For example, in decimal we understand that

$$2001 = 2 \times 10^3 + 0 \times 10^2 + 0 \times 10^1 + 1 \times 10^0$$

Similarly, in binary, powers of 2 are used; in octal, powers of 8; and in hexadecimal, powers of 16.

In a computer or calculator, the digits are represented by *devices*, and position on the paper is replaced by position in the machine. The *value* of the digit must correspond to a distinct *state* of a digit device. Hence, in a decimal machine, the digit devices must have 10 distinct states (corresponding to 0 and 1 through 9); but the digit devices in a binary machine need have only two states (corresponding to 0 and 1). In early mechanical and electromechanical calculators, as in automobile odometers, the 10 states were represented by 10 distinct angles of a wheel. That is, 0 was represented by 0°, 1 by 36°, 2 by 72°, and 9 by 324°. An additional 36° rotation brought the wheel back to zero position and generated a "carry" to the next position. The advantage of this use of the decimal system was that the machine and its human users worked in the same system. The digits 0 through 9 were frequently marked on the digit wheel, allowing it to function simultaneously as a computing device and as an output device.

In early electronic computers, the machine organization continued to be decimal in its architecture and at the programming level, although the electronic circuits always used two-state (on-off) devices. Thus the ENIAC computer worked with numbers of 10 decimal digits, and each digit, in turn, consisted of 10 two-state devices called flip-flops.[1] As in mechanical calculators, use of the decimal system allowed the machine to operate in a number system that its human users were familiar with. In his 1946 reports, von Neumann had pointed out that there were cost advantages to the use of the binary system: 33 flip-flops in binary could achieve about the same precision as 40 flip-flops in decimal. But even though von Neumann's first machine was binary, other manufacturers continued to build decimal machines, the last of which, the IBM 7070, was announced commercially in 1959. Eventually the efficiency and logical simplicity of the binary system were widely recognized, and now, every computer uses the binary system in its CPU.

THE BINARY SYSTEM

In the binary system, the digits are 0 and 1, and the digit positions correspond to powers of 2. The binary digit is called a bit (a con-

---

[1] Actually, only four flip-flops are really necessary.

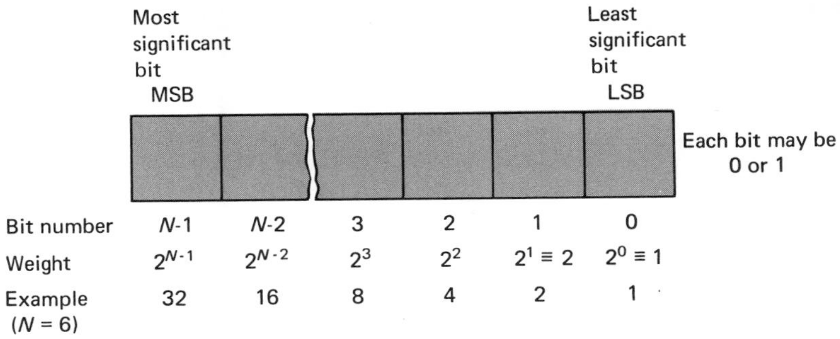

FIGURE 3·2
A binary number of N bits.

Bit number	N-1	N-2	3	2	1	0
Weight	2^{N-1}	2^{N-2}	2^3	2^2	$2^1 \equiv 2$	$2^0 \equiv 1$
Example ($N = 6$)	32	16	8	4	2	1

Most significant bit MSB. Least significant bit LSB. Each bit may be 0 or 1.

traction of binary digit). As shown in Fig. 3.2 the bits are numbered from right to left. The rightmost bit is called the least significant bit (LSB)—it has the smallest effect on the value of the binary number. The leftmost bit is called the most significant bit (MSB)—it has the largest effect on the value of the binary number. Each bit has a weight which is the power of 2 corresponding to its bit position; for example, bit 2 has a weight of 2^2, bit 3 has a weight of 2^3, and so on. The LSB has position zero and a weight of $2^0 \equiv 1$. An N-bit binary number may take on 2^N different values, from zero to $2^N - 1$. For example, the largest 6-bit binary number is 111 111, which equals $(1 \times 2^5) + (1 \times 2^4) + (1 \times 2^3) + (1 \times 2^2) + (1 \times 2^1) + (1 \times 2^0) = 63$. The smallest 6-bit binary number is 000 000, which equals 0. Thus a 6-bit binary number may take on $2^6 = 64$ different values.[2]

Binary-to-Decimal and Decimal-to-Binary Conversions

Decimal-to-binary conversions are often needed as information is input to the computer and binary-to-decimal conversions are often needed on output. Binary-to-decimal conversions work with the simple rule: multiply each bit by its weight and sum the products. For example, consider the conversion of 001 011 to decimal.

MSB					LSB	
0	0	1	0	1	1	Binary number
32	16	8	4	2	1	Weight
0	0	8	0	2	1	Product of bit × weight

The decimal value is $8 + 0 + 2 + 1 = 11_{10}$ where the subscript 10 indicates base 10. We will use subscript 2 to indicate base 2 (bi-

---

[2] In these examples, all numbers are positive. We will consider negative numbers in Sec. 3.4.

nary), subscript 8 to indicate base 8 (octal), and subscript 16 to indicate base 16 (hexadecimal).

Decimal-to-binary conversion is done by the following rule, based on successive division. With the number in decimal form, divide by 2. There will be either a zero remainder or a remainder of 1. This remainder becomes the least significant bit (LSB). The quotient is divided by 2, and the new remainder becomes the next bit. The process continues until a zero quotient is obtained. Consider conversion of 57_{10} to binary:

57 divided by 2 is 28, remainder 1: set bit 0 = 1
28 divided by 2 is 14, remainder 0: set bit 1 = 0
14 divided by 2 is 7, remainder 0: set bit 2 = 0
 7 divided by 2 is 3, remainder 1: set bit 3 = 1
 3 divided by 2 is 1, remainder 1: set bit 4 = 1
 1 divided by 2 is 0, remainder 1: set bit 5 = 1

Because this quotient is zero, all higher-order bits will be zero. So

$$57_{10} = \boxed{\underset{\text{MSB}}{1} \mid 1 \mid 1 \mid 0 \mid 0 \mid \underset{\text{LSB}}{1}}$$

The result is easily checked by a binary-to-decimal conversion:

$(1 \times 32) + (1 \times 16) + (1 \times 8) + (0 \times 4) + (0 \times 2) + (1 \times 1) =$
$32 + 16 + 8 + 0 + 0 + 1 = 57_{10}$

Some additional practice problems are given at the end of this section.

Conversions to and from the Octal and Hexadecimal Systems

The octal and hexadecimal systems are used in the computer field as shorthand notation for binary numbers. They are like shorthand in that octal and hexadecimal (hex) numbers have fewer digits than do the corresponding binary numbers and are more easily written and read by most people than are binary numbers. For example, 1980_{10} is 011 110 111 100 in binary—12 digits long. It is 3674 in octal—only 4 digits long. In hex, it is 7BC—only 3 digits. Conversion of binary to and from octal or hex is very simple. In practice, machines actually compute in binary, but people who deal with machines at the hardware and systems level often think of machine operations as octal or hex. Conversion is easily done using the binary-octal and binary-hexadecimal conversion tables.

TABLE 3·1
BINARY-TO-OCTAL CONVERSION TABLE

Binary	000	001	010	011	100	101	110	111
Octal Digit	0	1	2	3	4	5	6	7

If the number of binary digits is a multiple of 3, the binary number can be divided into 3-bit groups and Table 3.1 used to convert it to octal. Similarly, if the number of binary digits is a multiple of 4, the binary number can be divided into 4-bit groups and Table 3.2 used to convert it to hex. Hence, a 12-bit number can be converted either to octal or hex, depending on how it is divided. For example, the binary number 011 110 111 100 using four 3-bit groups and Table 3.1 is 3674:

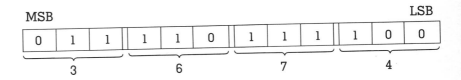

Similarly, that binary number is also $7BC_{16}$ in the hexadecimal conversion using three 4-bit groups and Table 3.2:

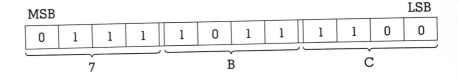

The octal number 3674 and the hexadecimal number 7BC are much easier to remember and discuss than is the binary number 011 110 111 100.

Binary-Coded Decimal (BCD)
Note in Table 3.2 that the first 10 hexadecimal digits are identical

TABLE 3·2
BINARY-TO-HEXADECIMAL CONVERSION TABLE

Binary	0000	0001	0010	0011	0100	0101	0110	0111	1000	1001	1010	1011	1100	1101	1110	1111
Hexadecimal Digit	0	1	2	3	4	5	6	7	8	9	A	B	C	D	E	F

to the 10 decimal digits 0 through 9. The corresponding 4-bit binary groups are a natural way to represent decimal digits in a computer. Binary-coded decimals (BCD) are thus individual decimal digits stored in a binary form. Natural-binary-coded decimals are decimal digits stored by the 4-bit binary form. This is an especially convenient form—and one reason that the latest computers are designed to use binary numbers which are a multiple of 4 bits in length: in microcomputers 8-bit and 16-bit numbers are common and can store 2 and 4 decimal digits, respectively. Larger binary numbers of 32-, 48-, and 64-bit lengths are common in large computers. These lengths are also compatible with the 8-bit codes used in transmitting data over telephone lines.

PROBLEMS

3.4 In decimal (base 10) we understand the number 2001 to mean

$$2 \times 10^3 + 0 \times 10^2 + 0 \times 10^1 + 1 \times 10^0$$

By analogy, what do we understand 2001 to mean:
(a) In octal (base 8)?
(b) In hexadecimal (base 16)?

3.5 How is the value of a 6-bit binary number affected
(a) If the MSB is changed from 0 to 1
(b) If the LSB is changed from 0 to 1
Express your answers in the decimal system.

3.6 How many distinct positive (or zero) numbers can be represented by an 8-bit binary number? What is the difference between successive numbers?

3.7 The relative precision of a binary number is the following ratio: (difference between successive numbers)/(largest possible number).
(a) What is the relative precision of an 8-bit binary number?
(b) What is the smallest number of binary bits which gives a relative precision smaller than 0.01 (1 percent)?
(c) Approximately how many bits are required to obtain a relative precision of 0.001 (0.1 percent)?

3.8 The origin of the term "weight" used in binary-to-decimal conversions stems from a precomputer use of the binary system. As shown in Fig. P3.8, a balance was supplied with weights whose mass increased in binary progression. For such a balance, devise a systematic method of binary-to-decimal conversion. Assume that the unknown has an integer mass between 1 and 63 g, inclusive. Also,
(a) What is the smallest mass that can be measured?
(b) What is the largest mass that can be measured?
(c) What weights are needed to balance a 45-g mass?

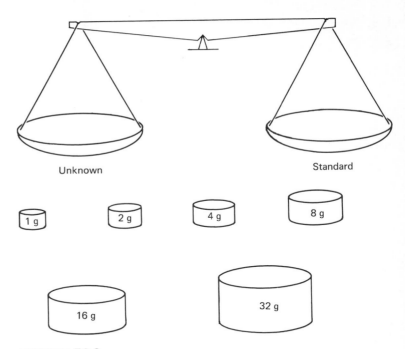

FIGURE P3·8

3.9 An advantage of the balance shown in Fig. P3.8 is that all the weights can be checked using only a single 1-g standard. Devise such a check procedure. You might start by balancing the 1-g standard against the balance's own 1-g weight.

3.10 Convert to decimal:
(a) $0\ 011_2$
(b) $0\ 110_2$
(c) $1\ 011_2$
(d) $1\ 110_2$

3.11 Consider the decimal numbers 15_{10}, 45_{10}, 68_{10}, and 77_{10}.
(a) How many bits must a binary number have if it is to be capable of expressing all these numbers?
Using the number of bits determined in a, convert the following to binary. Check your work by binary-to-decimal conversion.
(b) 15_{10} (c) 45_{10} (d) 68_{10} (e) 77_{10}

3.12 Convert the binary numbers below to octal; check by octal-to-decimal conversion.
(a) $001\ 111_2$ (b) $101\ 101_2$ (c) $111\ 111_2$ (d) $110\ 010_2$

3.13 Convert the binary numbers below to hexadecimal; check by hexadecimal-to-decimal conversion.
(a) $0100\ 0100_2$ (b) $0100\ 1101_2$ (c) $1111\ 1111_2$
(d) $1010\ 1111_2$

3·4 COMPUTER ARITHMETIC: ADDITION AND SUBTRACTION

Because addition and subtraction illustrate sufficiently the power and the complexity of the modern computer, we omit multiplication and division in our description of binary arithmetic. Moreover, multiplication and division in a computer are often carried out by repeated additions or subtractions.

ADDITION

Addition in binary is based on the following set of "number facts":

$$0 + 0 = 0 \quad \text{with no carry}$$
$$0 + 1 = 1 \quad \text{with no carry}$$
$$1 + 0 = 1 \quad \text{with no carry}$$
$$1 + 1 = 0 \quad \text{with a carry of 1}$$

To apply these rules, consider adding 001 010 to 000 101.

$$\begin{array}{r} 001\ 010 = (10)_{10} \\ +\ 000\ 101 = \underline{(5)_{10}} \\ \hline 001\ 111 = (15)_{10} \end{array}$$

There were no carries in this simple problem. An addition which does involve carries is illustrated in Example 3.1.

SUBTRACTION

The method used for binary subtraction in computers is quite different from that used in regular subtraction, which has one major problem: what happens when a big number is to be subtracted from a little number. Sometimes this results in borrowing, which can get complicated. Consider this base-10 example:

$$\begin{array}{r} 2000_{10} \\ -\ 999_{10} \\ \hline ? \end{array}$$

Working as a computer step by step, you must borrow when you take 9 from 0 and then make additional borrows on to the left. Also, in the problem

$$\begin{array}{r} 999_{10} \\ -\ 2000_{10} \\ \hline ? \end{array}$$

Example 3·1
Use the number facts to perform the binary addition $z = x + y$, with $x = 001\ 101$ and $y = 001\ 011$.

Solution
Add from right to left, just as in decimal. Start with the least significant bit (LSB) and work your way, bit by bit, to the most significant bit (MSB).

Bit Number	5	4	3	2	1	0
Carries		1	1	1	1	
x	0	0	1	1	0	1
y	0	0	1	0	1	1
z	0	1	1	0	0	0

Bit 0: $1 + 1$ gives a zero in the sum and gives a carry of 1.

Bit 1: Because of the carry, there are three binary digits to add in this and all higher-number digits; here $1 + 0$ gives 1 with no carry (yet), but $1 + 1$ gives 0 in the sum with a carry of 1.

Bit 2: $1 + 1$ gives 0 with a carry of 1; adding 0 does not change anything.

Bit 3: Here, because of the carry, we have three 1s to add. But $1 + 1$ gives 0 with a carry of 1; adding 1 to 0 gives 1 in the sum but does not affect the carry.

Bit 4: $1 + 0 + 0$ gives 1 in the sum and no carry.

Bit 5: $0 + 0$ gives 0 in the sum.

As a check, we can convert everything to decimal:

$$(001\ 011)_2 = (8 + 2 + 1)_{10} = 11_{10}$$
$$(001\ 101)_2 = (8 + 4 + 1)_{10} = 13_{10}$$
$$(011\ 000)_2 = (16 + 8)_{10} = 24_{10}$$

you must recognize the result as negative and rearrange the problem using $(999 - 2000) = -(2000 - 999) = -1001$. In such an example, we recognize a special case—a skill that computers do not yet possess. A computer *can* tell that 2000 is larger than 999, but only by subtracting 2000 from 999 and testing the sign of the result. We know that 2000 is bigger than 999 *without* subtracting. Nobody is sure how we do this. Because of the many special cases which can arise, the subtraction methods which most students learn by practice and drill are not at all suitable for computers. Instead, computers do subtraction by adding complements. The use of complements is an old idea—dating back to mechanical calculators—but it may seem devious at first, so we will explain it first in decimal, then in binary.

COMPLEMENTS IN DECIMAL

To see where the idea of complements comes from, consider a three-digit reversible mechanical counter (such as the trip odometer in a car) capable of being rotated forward or backward. If you rotate it backward and forward around zero, its readings might look like:

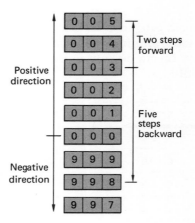

We can use the successive readings as a scale—a kind of mechanical number line. The readings

$$\boxed{0\ 0\ 0}, \boxed{0\ 0\ 1}, \boxed{0\ 0\ 2}, \boxed{0\ 0\ 3},$$

and so on correspond directly to zero and the *positive numbers* 1, 2, 3, etc. Since they represent counting backward from zero, it is natural to consider the three-digit numbers

$$\boxed{9\ 9\ 9}, \boxed{9\ 9\ 8}, \boxed{9\ 9\ 7},$$

and so forth as representing the *negative* numbers -1, -2, -3, ...

On our mechanical number line, addition corresponds to stepping forward and subtraction corresponds to stepping backward. For example, taking two steps forward from

$$\boxed{0\ 0\ 3} \text{ gives } \boxed{0\ 0\ 5},$$

corresponding to 3 + 2 = 5. Similarly, five steps backward from

| 0 | 0 | 3 | gives | 9 | 9 | 8 |,

corresponding to 3 − 5 = −2. The number 998 is called the three-digit 10's complement of 2. In general, the three-digit 10's complement of a number is found by subtracting it from 1000_{10}. On our mechanical number line, then, negative numbers are represented by their three-digit 10's complements.

Because the complement system has no explicit "minus sign," we must decide what is and is not a negative number. For example, is 505 a positive number or the 10's complement of 495? In a three-digit decimal system, with 1000 different numbers (000 through 999), the usual convention uses half the possible combinations for positive and half for negative numbers. We regard 000 through 499 as the corresponding positive numbers and 999 through 500 as the negative numbers −1 through −500. A programmer using complements must be aware of this convention. A similar situation occurs when using complements in binary.

COMPLEMENTS AND SUBTRACTION IN BINARY

Complements

In binary, two complement systems are common. They are called 2's complements and 1's complements.[3] The 1's complement representation saves a few nanoseconds per subtraction, so it is used in machines designed for the highest possible speed. Since 2's complements are frequently used in smaller computers, we will use them in our examples here.

We will consider a 6-bit number. There are $2^6 = 64$ possible numbers. Of these, the 32 numbers 000 000 through 011 111 are considered positive, and the 32 numbers 111 111 through 100 000 are considered negative. In this representation, bit 5, the most significant bit (MSB) is a "sign bit." If it is 0, the number is positive; if 1, the number is negative. Logically, 000 000 is 0. Since its sign bit is zero, 000 000 is also considered positive.

To form the 2's complement of a binary number

1. Change all 1s to 0s and 0s to 1s—forming the 1's complement
2. Add 1 to the LSB of the 1's complement

For example, in forming the 2's complement of 010 010, we have

---

[3] Each method has a quirk. In 2's complement, the most negative number differs slightly in magnitude from the most positive number. In 1's complement, there are two 0s, called +0 and −0.

```
    010 010     binary number
    101 101     1's complement
   +000 001     add 1 to LSB
    101 110     2's complement
```

Subtraction

Subtraction in binary is done by adding complements. We will give the rules and illustrate them by an example, but first we must define *carry out*. A *carry out* occurs whenever a carry to the left of the MSB occurs, as for example, in adding $x = 001\ 000$ to $y = 111\ 000$.

	MSB					LSB
Carries	1	1	1			
y		1	1	1	0	0	0
x		0	0	1	0	0	0
Sum		0	0	0	0	0	0

The 1 carried to the left of the MSB is a carry out.

The rule for binary subtraction is as follows: To compute $d = m - s$ add the 2's complement of s to m and then

1. If there is a carry out, the difference d is positive or zero. Use the carry out as a signal that the result is nonnegative. Otherwise, ignore it. The rest of the result is d.
2. If there is no carry out, the difference d is negative. The result is then the 2's complement of d.

The mathematical proof of these rules can be found in Ref. 7, pages 281 to 289. Note that if the difference is negative, the result is in 2's complement form. This is simply the form needed for further calculations.

The subtraction process is illustrated by Example 3.2, where we see how a computer would carry out some simple subtractions using binary numbers and 2's complements. Note in the example how borrowing is unnecessary and how negative, zero, and positive results are signaled by the presence or absence of a carry out.

3·5 COMPUTER ARITHMETIC: HARDWARE FUNDAMENTALS

In this section we will show how binary arithmetic is carried out electronically.

Example 3·2

Perform the following subtractions in binary using 6-bit 2's complements:

$26_{10} - 25_{10} = ?$ $25_{10} - 26_{10} = ?$ and $26_{10} - 26_{10} = ?$.

Solutions

$$26_{10} - 25_{10} = ?$$
$$26_{10} = 011\ 010_2$$
$$25_{10} = 011\ 001_2$$
$$\text{2's complement } 25_{10} = 100\ 111_2$$

$$\begin{array}{r} 011\ 010_2 \\ +\ 100\ 111_2 \\ \hline (000\ 001)_2 \end{array}$$
Carry out 1

Carry out is discarded but implies that the result is positive or zero; $26_{10} - 25_{10} = 000\ 001_2$.

$$25_{10} - 26_{10} = ?$$
$$25_{10} = 011\ 001_2$$
$$\text{2's complement } 26_{10} = \underline{100\ 110_2}$$
$$111\ 111_2$$

No carry out implies that the result is negative; 111 111 is the 2's complement representation of -1.

$$26_{10} - 26_{10} = ?$$
$$26_{10} = 011\ 010_2$$
$$\text{2's complement } 26_{10} = \underline{100\ 110_2}$$
Carry out 1 $(000\ 000)_2$

The carry out is discarded; the result is zero.

Everything in Example 3.2 works out as expected, including the last subtraction, which leads to zero.

PHYSICAL REPRESENTATION OF BINARY DIGITS

The success of electronic digital computation arises from the use of two-state, or on-off, devices. That is, the physical representation of a binary digit (bit) is dependent on the presence or absence of some physical quantity and is independent, or nearly independent, of its magnitude. Examples of physical quantities which are used to represent bits are:

Punched cards and paper tape	Presence of a hole represents a 1; absence of a hole represents a 0
Computer and calculator keyboard	Closed switch represents a 1, open switch represents a 0 (or vice versa)
Transistor logic circuit	Conducting transistor represents a 1; nonconducting transistor represents a 0 (or vice versa)
Magnetic core memory	Clockwise magnetization represents a 1; counterclockwise magnetization represents a 0 (or vice versa)

In all these cases, every effort is made to avoid in-between states which are neither 1 nor 0. Vice versa indicates that some representations are internal to the computer: members of the design team can decide among themselves what is to be a 1 and what a 0. When devices designed by separate design teams are to be interconnected, however, there must be a clear standard about 1 and 0. The standards are typically technical specifications (as discussed in Sec. 1.4) that members of the computer industry agree on among themselves. An obvious standard would be $+1.0$ V is a 1, 0.0 V is a 0, but for technical reasons, this obvious standard is impractical and is never used. As an example of a practical standard, let us consider the standards for transistor-transistor logic (TTL) voltage levels. These standards are often followed in interconnecting computer circuits when digital data must be transmitted from one part of a computer to another.

TTL Voltage Levels
In TTL logic, a potential of approximately $+5$ V represents a 1; a potential of approximately 0.0 V represents a 0. The meaning of "approximately" is spelled out in detail, however, to make the operation of TTL logic nearly independent of the *magntiude* of the potentials actually used to transmit 1s and 0s. The tolerances are shown in Fig. 3.3. If a device is in the 1 state, its output is permitted to be any voltage between $+2.4$ and $+5.0$ V. However, the device receiving this signal recognizes any voltage greater than 2.0 V as a 1. Thus, in the worst case there is an 0.4-V tolerance, called the noise immunity, between the lowest allowed output voltages and the input voltages recognized as 1. A similar 0.4-V noise immunity is specified for the 0 state.

There must be a 2.0-V change in output voltage between the 0 and the 1 state, but an input circuit in the worst case can recognize a swing of only 1.2 V as a change from 0 to 1. You can see how the engineers who drew up the TTL standard have made system operation substantially independent of variations in individual devices even under worst-case conditions. In our further discussion, we use a 1 and 0 as shorthand for any voltage in the respective ranges.

LOGIC AND LOGIC CIRCUITS
Boolean logic is based on propositions which, like binary digits, have no in-between states: they are either true or false.

One of the basic operating principles of digital computers is that arithmetic is done by use of logic circuits in which the binary digits 1 and 0 replace the true and false states of Boolean logic.

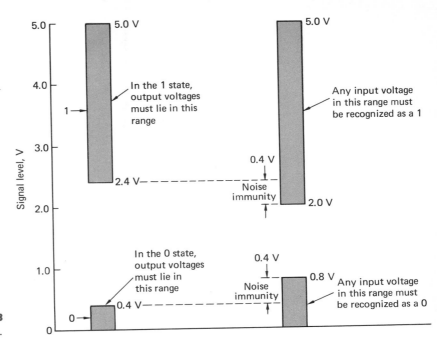

FIGURE 3·3
TTL voltage levels.

Let us first introduce the elements of Boolean logic and then see how logic circuits are applied to arithmetic operations, such as binary addition.

Consider the following propositions.

A. Tomorrow will be a cold day.
B. Tomorrow will be a rainy day.

Then consider these compound propositions:

A and B.
A or B.
Either A or B.

The proposition "A and B" in words is "Tomorrow will be a cold day *and* tomorrow will be a rainy day." It is true if tomorrow is *both* cold and rainy.

The proposition "A or B" is true if tomorrow is cold or if it is rainy *or* if it is both cold and rainy. The proposition "A or B" is false only if tomorrow is not cold *and* not rainy.

The proposition "either A or B" is true if tomorrow is cold or if it is rainy, but the proposition is false if tomorrow is *both* cold and rainy. Because it excludes one possibility, the proposition "either A or B" is an example of the EXCLUSIVE-OR operation in logic.

Gates and Logic Circuits

In electronic technology today, logical operations such as AND, OR, and EXCLUSIVE-OR are embodied in readily available integrated-circuit devices. These are referred to as "logic gates," which cost only a few cents per gate, even in retail quantities. In logic circuits, 1 and 0 replace true and false, but the specification of the circuit action is still referred to as a "truth table." The truth table defines the output of a logic circuit in terms of its inputs.

The symbols for logic gates have been standardized by ANSI Standard Y32.14 [8]. For example, consider an OR gate which has the following symbol.

Its truth table is as follows:

A	B	Output
1	1	1
0	1	1
1	0	1
0	0	0

If either A or B or both are equal to 1, the output is 1; the output is 0 only if *both* A and B are zero. The operation of the logic gate just described, as well as gates for the other logic operations we have mentioned, are summarized in Fig. 3.4. In the next section, we will see how the logic gates can be combined to perform arithmetic operations.

Operation	Symbol	Truth table
AND	(A, B → Output)	A B Output 1 1 1 0 1 0 1 0 0 0 0 0
OR	(A, B → Output)	A B Output 1 1 1 0 1 1 1 0 1 0 0 0
EXCLUSIVE-OR	(A, B → Output)	A B Output 1 1 0 0 1 1 1 0 1 0 0 0

FIGURE 3.4 Some typical logic gates and their truth tables.

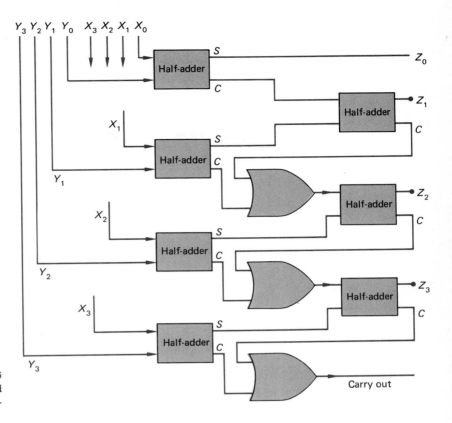

FIGURE 3·6
A 4-bit half-adder constructed from half-adders and OR gates.

ual bits of the binary input X. The inputs Y_3, Y_2, Y_1, Y_0 represent the binary input Y. The adder computes $Z = X + Y$, where Z_3, Z_2, Z_1, Z_0 are the bits for Z.

Figure 3.7 shows the 4-bit full adder adding the binary representations of $X = 3_{10}$ and $Y = 5_{10}$ and obtaining a binary representation of $Z = 8_{10}$. The addition proceeds as follows:

Work down the left column of half-adders and fill in the sum and carry outputs.
Starting with the top half-adder in the second column, fill in the sum and carry outputs. Now both inputs of the top OR gate are known. Fill in its output.
Since both inputs to the second half-adder are known, fill in its sum and carry outputs. Now both inputs to the second OR gate are known. Fill in its output.
Finally, fill in the outputs of the last half-adder in the second column and fill in the output of the last OR gate.

The solution process just outlined corresponds to the actual physical behavior of this type of adder. The paper solution pro-

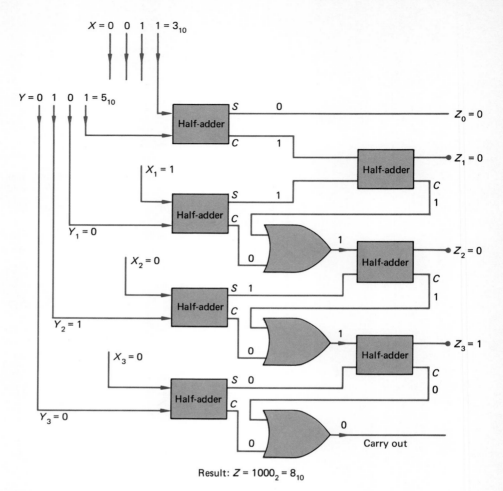

FIGURE 3·7
Full-adder operation. $X = 3_{10}$, $Y = 5_{10}$, $Z = X + Y$.

ceeds from the LSB to the MSB. In the laboratory, it is found that the LSB reaches its correct value first, and the result ripples bit by bit to the MSB. The typical delay in this circuit might be 0.1 μs.

Subtraction is performed with the full adder by carrying out the complement operation discussed in Sec. 3.4 before carrying out the addition.

LEVELS OF INTEGRATION

In today's technology, many gates can be combined and fabricated from a single chip of silicon. The levels of integration are often classified as follows:

Small-scale integration (SSI): fewer than 12 gates per chip
Medium-scale integration (MSI): 12 to 100 gates per chip

guage occasionally. Third, to satisfy the how-it-works curiosity of some readers and possibly to interest others in programming at this level. There are hobbyist magazines which cover this subject, of which BYTE has the largest circulation. For an easy introduction which ties together set theory, Boolean algebra, and digital electronics, see Ref. 10.

PROBLEMS

3.25 In the cell-growth problem presented in Sec. 3.7, assume that growth is allowed to continue for 32 generations. How would this affect: (a) the flowchart, (b) the FORTRAN program, (c) the BASIC program?

3.26 For those with access to a computer having a BASIC interpreter or a FORTRAN compiler, modify the BASIC or FORTRAN program to print results after each cell doubling and run the modified program on your computer.

3.27 For those with access to a programmable calculator, follow the flowchart (Fig. 3.8) and program your calculator to solve the cell-growth problem. The printing of results may be replaced by display of generation count, followed by a pause, followed by display of cell count, followed by a stop.

3.28 Determine which of the machine-language instructions in Fig. 3.13 would be modified if cell growth were allowed to continue for 32 generations. Write the modified machine-language instruction as a 10-digit octal number.

REFERENCES

1. Ralson, Anthony, *Introduction to Programming and Computer Science*, New York, McGraw-Hill, 1971.
2. Digital Equipment Corporation, *Small Computer Handbook Series*, Maynard, Mass., Digital Equipment Corporation, 1968.
3. Osborne, Adam, *An Introduction to Microcomputers: The Beginner's Book*, New York, Osborne/McGraw, 1979.
4. Leventhal, Lance, *8080A/8085 Assembly Language Programming*, Berkeley, Calif., Osborne and Associates, 1978.
5. Spracklen, Kathe, *Z-80 and 8080 Assembly Language Programming*, Rochelle Park, N.J., Hayden, 1979.
6. Abd-alla, Abd-Elfattah M., and Arnold C. Meltzer, *Principles of Digital Computer Design*, vol. 1. Englewood Cliffs, N.J., Prentice-Hall, 1976.
7. Hellerman, Herbert, *Digital Computer Systems Principles*, New York, McGraw-Hill, 1967.
8. American National Standards Institute, *IEEE Standard Graphic Symbols for Logic Diagrams*, IEEE Std 91-1973, ANSI Y32.14-1973.

9. Dahl, O. J., E. W. Dijkstra, and C. A. R. Hoare, *Structured Programming*, Academic, New York, 1972.
10. Bunce, Dan, and Art Schwartz, "Some Musings on Boolean Algebra," *BYTE*, vol. 3, no. 2, February 1978, pp. 25–29.

SUGGESTED READINGS

Bowden, B. V., *Faster than Thought*, New York, Pitman, 1953.

Dertouzos, M. L., and J. Moses, eds, *The Computer Age: A Twenty-Year View*, Cambridge, Mass., The MIT Press, 1980.

Moseley, M., *Irascible Genius: A Life of Charles Babbage, Inventor*, London, Hutchinson, 1964.

Mauchly, J. W., "Mauchly on the Trials of Building ENIAC," *IEEE Spectrum*, vol. 12, no. 4, April 1975, pp. 70–76.

Siewiorck, Daniel P., Gordon C. Bell, and Allen Newell, *Computer Structures: Principles and Examples*, New York, McGraw-Hill, 1982.

Taub, A. H. (ed.), *John von Neumann: Collected Works*, Oxford, Pergammon, 1963.

PART THREE

BASIC TOPICS IN ENGINEERING THEORY

The following events may at first seem deceptively simple, but they are actually quite complicated and difficult to describe accurately: (1) A wooden stick thrown into a pond hits the water with a splash, bobs up, makes ripples that spread and gradually dissipate, and then floats quietly on the surface. (2) An electric fan turned on in a room on a warm day. Its switch applies voltage to the motor to rotate the blades. The motor speed increases until a constant speed is reached. (3) A small heater in a tent is lighted on a chilly evening. The temperature inside increases to a steady level that is maintained for the rest of the evening.

Engineers can ask many questions about these events: how deep does the stick go when it enters the water; how high does the splash reach; what portion of the stick is under water at equilibrium? What is the maximum current required by the fan motor at startup; how much power is needed to operate the fan; how does blade length affect the fan's performance? Finally, what size heater is required to keep the tent comfortable; how much does heater size depend on outside temperature; what influence does wind speed have on the tent temperature?

Engineering principles along with practice can help you answer these questions. Thus a thorough engineering education combines discussion of theory with instructions about how to apply theory to solve problems in a particular field of engineering. Part Three covers the basic principles used in various engineering disciplines.

Inasmuch as each event mentioned above involves changes that take place during a transient period until a steady state, or equilibrium, is reached, we will focus on equilibrium in our discussion of statics (Chap. 4), electric circuits (Chap. 5), and fluid-thermal problems (Chap. 6). Transient, or dynamic, conditions, will be handled separately in Chap. 7.

4 STATICS

4·1 FORCES

The first topic in our discussion of engineering theory, very appropriately called *statics*, is the area of engineering mechanics that deals with mechanical equilibrium. It is the study of bodies which are held in equilibrium through the interaction of applied forces and moments.

We define a body in equilibrium as one at rest or moving at a constant velocity. There are two requirements which must be met for a body to be in equilibrium. First, the sum of all the forces acting on the body must be zero; and second, the sum of all the moments acting on it must be zero. In this section we will describe the analysis of forces acting on bodies. We will discuss moments in Sec. 4.2.

Let's begin our discussion of forces by considering the simplified motor vehicle model in Fig. 4.1. We consider the vehicle to have rear-wheel drive and to be moving along a straight, level road at constant velocity. It is in equilibrium. Figure 4.1 is called a

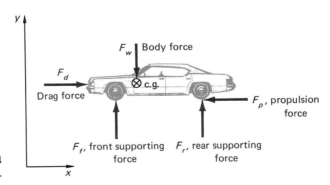

FIGURE 4·1
Free-body diagram of a vehicle.

free-body diagram since it shows the body of interest free of contact with any other objects. All the forces which act on the body and support it are instead shown by arrows which indicate the direction and point of application for each force. This free-body diagram is a model of the real physical situation and can be easily described in mathematical terms. Included in Fig. 4.1 is a coordinate system which defines x as the horizontal direction (with +x to the right) and y as the vertical direction (with +y upward). This type of coordinate system is important to help define the directions of the applied forces. For example, our equilibrium requirement that the sum of applied forces must be zero considers the x and y axes separately, treating forces acting in the plus coordinate direction as positive and in the minus coordinate direction as negative. The body force F_w shown in Fig. 4.1 is the downward force due to gravity acting on the vehicle. It is, of course, normally referred to as the vehicle weight.

We usually treat the mass or weight of a body by assuming it to be concentrated at the center of gravity. Correspondingly, force F_w in Fig. 4.1 is acting at the vehicle's center of gravity. You will find in your later studies that the center of gravity of a body may be located through the application of integral calculus [1].

When the summation of forces is carried out for a two-dimensional problem of the type shown in Fig. 4.1, we have equilibrium only when the forces in both the x and y directions sum to zero. Using summation notation, the equations for each of the two directions are written as

$$\sum_{i=1}^{n} F_{xi} = 0 \tag{4.1}$$

and

$$\sum_{j=1}^{m} F_{yj} = 0 \tag{4.2}$$

where i takes on values from 1 to n so that F_{xi} can represent each of the n forces acting in the x direction. Similarly, j takes on values from 1 to m to include each force in the y direction. This notation may seem difficult but simply means to add all the forces acting in each direction. If $n = 2$ and $m = 3$, these equations become

$$F_{x1} + F_{x2} = 0$$

and

$$F_{y1} + F_{y2} + F_{y3} = 0$$

where proper signs must be included with each force in an actual problem. Returning to Fig. 4.1, we find that the two equations for the vehicle are

$$F_d - F_p = 0 \tag{4.3}$$

and

$$F_f + F_r - F_w = 0 \tag{4.4}$$

for the x and y directions. A minus sign occurs for F_p in the first equation because it acts in the minus x direction. Also, a minus sign appears for F_w since it acts in the minus y direction.

If we were now told that the mass of this vehicle was 1000 kg and that a propulsion force F_p of 500 N is required at a constant speed of 100 km/h, Eqs. (4.3) and (4.4) immediately show that

$$F_d = F_p = 500 \text{ N} \tag{4.5}$$

and

$$F_f + F_r = F_w = (1000)(9.18)$$

or

$$F_f + F_r = 9180 \text{ N} \tag{4.6}$$

Equation (4.5) tells us directly that at equilibrium the drag force and propulsion force must be equal. Unfortunately, in the y direction, Eq. (4.6) cannot tell us specifically how much force the front and rear wheels each support; it tells us only that the sum of F_f and F_r must support the vehicle body force or weight. We need another equation to find each vertical force explicitly. We will see in Sec. 4.2 that this additional equation is one which sums moments in the way that we have summed forces here.

In Eq. (4.5) we learned for the simple vehicle model that the drag and propulsion forces are equal at equilibrium. The propulsion force of a vehicle is controlled by its throttle (accelerator pedal) position. At highway speeds for a constant throttle position, the vehicle propulsion force is essentially constant. The drag force depends on vehicle speed. An increase in speed increases the drag force; a decrease in speed decreases the drag force. A representative graph of the drag force versus speed is shown in Fig. 4.2. If a constant propulsion force is provided, the vehicle speed will increase until F_d just matches F_p. If a slight decrease in F_p later occurs, the larger force F_d will then tend to slow the ve-

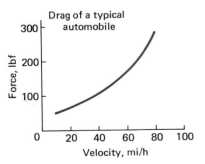

FIGURE 4·2 Typical curve showing drag force vs. speed.

hicle until F_d decreases to equal or balance the lower value for F_p. This illustrates the way a natural balance of physical forces occurs in actual equilibrium conditions.

CABLES

Now we will move on to the statics problem of Fig. 4.3a, where the cables fastened to the fixed supports at points A and C are supporting the weight of 1000 lbf suspended from the ring B. The problem here is to calculate the forces F_{AB} and F_{BC} in the two cables, with cable weight neglected. A cable, rope, or chain can only support loading in tension. It will always be pulled tight when loaded; it will always collapse if pushed on or bent by a load. Accordingly, the force applied to an object by a cable will always be along the direction of the cable and will always pull on the object. The free-body diagram of Fig. 4.3b has been drawn to serve as a model in writing equations for F_{AB} and F_{BC}. It shows F_{AB} and F_{BC} pulling on the ring at the appropriate angles and the 1000 lbf acting downward. In Fig. 4.3c and d are free-body diagrams of the cables, showing both loaded in tension. Note that the forces pulling on ring B also pull on the cables but in opposite directions and that the fixed supports must pull with equal and opposite force at the other end of the cables in order for the cables to be in equilibrium. This is true for cables of negligible weight but is, of course, somewhat different in cases where cable weight is significant.

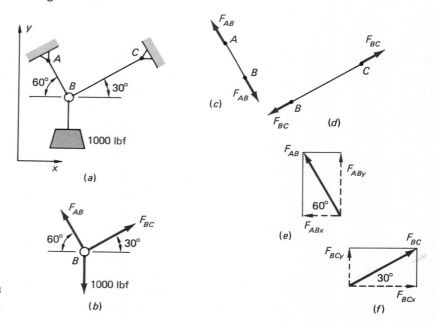

FIGURE 4·3
A cable problem.

To solve the problem at hand and obtain F_{AB} and F_{BC}, we look again at Fig. 4.3b and sum forces in the x and y directions. However, neither F_{AB} nor F_{BC} is in the x or y direction, and so we must "resolve" each force into its components in the x and y direction. That is, we must find the forces which will have the same effect as the original force when simultaneously applied in the x and y direction. Since x and y are perpendicular, the resolution of each force can be performed by forming the rectangles shown in Fig. 4.3e and f. The ratios of the component forces to the original forces are the same as the ratios of the lengths of the corresponding lines in the rectangles. Thus

$$\frac{F_{AB_y}}{F_{AB}} = \sin 60 = \cos 30 \tag{4.7}$$

$$\frac{F_{AB_x}}{F_{AB}} = \cos 60 = \sin 30 \tag{4.8}$$

$$\frac{F_{BC_x}}{F_{BC}} = \cos 30 = \sin 60 \tag{4.9}$$

$$\frac{F_{BC_y}}{F_{BC}} = \sin 30 = \cos 60 \tag{4.10}$$

We can now apply the components of forces F_{AB} and F_{BC} in summing the forces acting on ring B in the x and y directions of Fig. 4.3b. In the x direction,

$$-F_{AB_x} + F_{BC_x} = 0$$

or

$$-F_{AB} \cos 60 + F_{BC} \cos 30 = 0 \tag{4.11}$$

and in the y direction,

$$F_{AB_y} + F_{BC_y} - 1000 = 0$$
$$F_{AB} \sin 60 + F_{BC} \sin 30 - 1000 = 0 \tag{4.12}$$

Equations (4.11) and (4.12) are now two simultaneous equations which must both be satisfied for equilibrium. When the equations are solved, the forces in the cables are determined. By substituting the sine and cosine values in these equations, we rearrange them to obtain

$$-0.500 \, F_{AB} + 0.866 \, F_{BC} = 0$$
$$0.866 \, F_{AB} + 0.500 \, F_{BC} = 1000$$

which are easily solved to find

$$F_{AB} = 866 \text{ lbf}$$
$$F_{BC} = 500 \text{ lbf}$$

as the forces in the cables. Other problems with loaded cables are presented in Examples 4.1 and 4.2.

Example 4·1

Cable *EG* and *GH* are supporting the weight of 1000 lbf, as shown in Fig. 4.4a. Find the forces F_{EG} and F_{GH} in the cables.

Solution

We begin by drawing the free-body diagram of Fig. 4.4b. We then sum forces in the x and y directions using the components of F_{EG} and F_{GH}.

$$F_{GH_x} - F_{EG_x} = 0$$
$$F_{GH_y} + F_{EG_y} - 1000 = 0$$

Realizing that

$$F_{GH_x} = F_{GH} \cos 50 \quad \text{and} \quad F_{EG_x} = F_{EG} \cos 40$$
$$F_{GH_y} = F_{GH} \sin 50 \quad \text{and} \quad F_{EG_y} = F_{EG} \sin 40$$

we find

$$0.643 F_{GH} - 0.766 F_{EG} = 0$$
$$0.766 F_{GH} + 0.643 F_{EG} = 1000$$

The simultaneous solution to these equations is $F_{EG} = 643$ lbf and $F_{GH} = 766$ lbf.

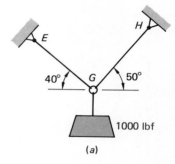

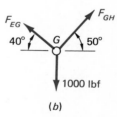

FIGURE 4·4

Example 4·2

Cable AB, shown in Fig. 4.5a, connected to the fixed point A and ring B, supports both the 100-kg and 75-kg masses. Determine the angular position θ of the cable where it will support the two masses in the equilibrium position shown and also determine the tensile load in the cable.

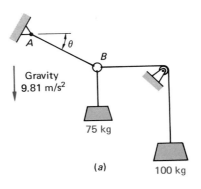

Solution

Since the pulley is frictionless, the 981-N force supporting the 100-kg mass pulls the ring B to the right as in its free-body diagram of Fig. 4.5b. The force F_{AB} is the cable force with components as shown in Fig. 4.5c. Summing the forces for the x direction in the free-body diagram,

$$981 - F_{AB_x} = 0$$

and for y,

$$F_{AB_y} - 736 = 0$$

Thus,

$$F_{AB_x} = 981 \text{ N} \quad \text{and} \quad F_{AB_y} = 736 \text{ N}$$

From the component diagram

$$\tan \theta = \frac{F_{AB_y}}{F_{AB_x}}$$

and so $\theta = 36.9$ deg. Also, for the cable force,

$$F_{AB} = \frac{F_{AB_x}}{\cos \theta}$$

$$F_{AB} = 1230 \text{ N}$$

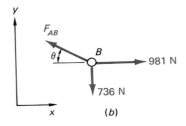

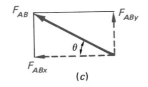

FIGURE 4·5

PINNED JOINTS

Let's move now to another class of problems for which the equilibrium of forces frequently provides enough information to obtain solutions. These are problems involving pinned joints, such as the mechanism of Fig. 4.6a. Here the pinned joints are formed by passing pins or shafts through holes in the members; ideally, the joint rotates freely. Only a simple force (i.e., pushing or pulling) may be applied to a member at a pinned joint, not a bending or twisting action. In a member having only two pinned joints, force can normally be transmitted only on a line between pins and along the member itself. This may be a tensile force, as in a cable, but in general the member is solid and can also support compressive loading. That is, the member may be squeezed rather than stretched as it is in tension. Often the pin or shaft is supported by a well-designed bearing within the member and not simply a hole. Either a ball bearing or an oil-lubricated bearing may be used. An example is the connecting-rod bearing in an automobile engine which supports the "pin" joining the crankshaft and connecting rod.

The problem in Fig. 4.6a is to determine the vertical force F which must be applied to the mechanism to counteract the 1000 lbf shown. If force F is too small, the 1000 lbf will move the sliding member where pin A is located along the base and toward C, pushing pin B upward and folding the mechanism. If force F is larger than required, the sliding member will move outward along the base, forcing pin B downward, and the mechanism will extend. The free-body diagram of Fig. 4.6b is for the sliding

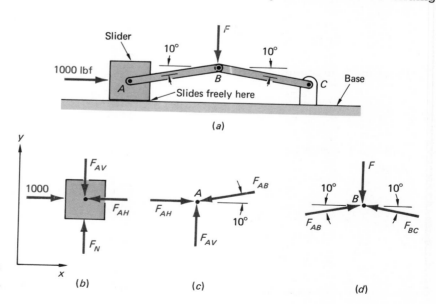

FIGURE 4·6
Forces in a mechanism.

180 STATICS

member. Since the block slides freely on the base without friction, only the normal force F_N acts along the bottom and, of course, the 1000 lbf acts on the side. The force applied by pin A to the block has components F_{AH} and F_{AV}. By summing x direction forces, we see easily that for equilibrium, $F_{AH} = 1000$ lbf; and by a y-direction summation, $F_{AV} = F_N$. Figure 4.6c shows a free-body diagram of pin A with the components F_{AH} and F_{AV} now applied by the slider to the pin. Figure 4.6b and c illustrates a basic principle of analyzing interconnected parts. When the pin pushes on the slider with force F_{AH}, the slider pushes back on the pin with the same force, but the arrows in the two free-body diagrams have opposite direction. The force in link AB must be along the line from A to B; the force F_{AB} thus appears in the free-body diagram of Fig. 4.6c inclined at 10° from the horizontal.

We assume that member AB is in compression, so F_{AB} is shown pushing on the pin. Summing forces in the x direction here yields

$$F_{AH} - F_{AB} \cos 10° = 0$$

Thus,

$$F_{AB} = \frac{F_{AH}}{\cos 10°} = 1015 \text{ lbf}$$

Summing in the y direction, we find

$$F_{AV} - F_{AB} \sin 10° = 0$$

So that,

$$F_N = F_{AV} = F_{AB} \sin 10° = 176 \text{ lbf}$$

is the force between the slider and the base. Now, focusing on the free-body diagram of pin B in Fig. 4.6d, we sum forces in the x and y directions to obtain

$$F_{AB} \cos 10° - F_{BC} \cos 10° = 0 \tag{4.13}$$

and

$$F_{AB} \sin 10° + F_{BC} \sin 10° - F = 0 \tag{4.14}$$

From Eq. (4.13) we see $F_{BC} = F_{AB} = 1015$ lbf and then from Eq. (4.14),

$$F = 2F_{AB} \sin 10° = 353 \text{ lbf}$$

is the required force to counteract the 1000 lbf at the slider. This indicates a substantial mechanical advantage, since the smaller force of 353 lbf can counteract the larger force of 1000 lbf. As the pinned members AB and BC are more closely aligned, the mechanical advantage increases. For example, if our 10° angle is reduced to 5°, only 175 lbf is required to counteract the 1000 lbf. A well-known brand of self-locking pliers makes use of a mechanism based on this concept. It is described further in Example 4.4.

TRUSSES

A second group of problems which can be analyzed by pinned-joint techniques has to do with truss structures, one of the most common types of structures in use today. They are based on a set of three members joined by pins to form a structural element of triangular shape. This three-member element is rigid and will not deform substantially unless the materials are overloaded. Truss structures consist of a number of interconnected triangular elements. Unlike pinned-joint mechanisms where motion between members is required and the pinned joints are important to allow such motion, trusses do not normally contain actual pinned joints, since motion is neither necessary nor desired. However, the calculation of forces in truss structures based on the assumption of pinned joints produces acceptable results and is commonly used.

A simple truss consisting of only one triangular element is shown in Fig. 4.7a. It is loaded at the top with a vertical force of 100 N. In this figure, note that pin A is attached firmly to a solid base but there is a pair of rollers at pin C. The rollers prevent any horizontal force from being developed at pin C but allow for a vertical force. This type of end condition may be needed in a real structure so that member AC can be free to expand or contract with temperature changes. However, it is also necessary in analysis to allow a solution by the use of statics and to prevent a statically indeterminate problem, as is explained in Sec. 4.3.

We will now analyze our simple structure utilizing the free-body diagrams of Fig. 4.7b. Here we see the vertical and horizontal forces, F_{AV} and F_{AH}, respectively, at pin A. They are provided by the base and are called reaction forces. We assume that member AB is in compression and members AC and BC are in tension. Thus force F_{AB} appears pushing on pins A and B, while forces F_{BC} and F_{AC} pull on the pins where they act. Because of the rollers, at pin C, only the vertical reaction force F_{CV} is possible. Summing forces at pin B yields

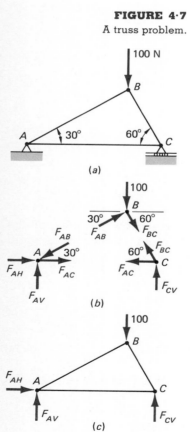

FIGURE 4·7
A truss problem.

In x: $$F_{AB} \cos 30 + F_{BC} \cos 60 = 0 \qquad (4.15)$$
In y: $$F_{AB} \sin 30 - F_{BC} \sin 60 - 100 = 0 \qquad (4.16)$$

From Eq. (4.15), we can easily find that $F_{BC} = -1.73 F_{AB}$; and substituting into Eq. (4.16), we obtain

$$F_{AB} \sin 30 + 1.73 F_{AB} \sin 60 = 100$$

which gives $F_{AB} = 50.0$ N and thus $F_{BC} = -86.5$ N. This negative result for F_{BC} simply means that our assumption of tension for member BC was incorrect and the member is actually under compression. We continue now, summing forces at pin C.

In x: $\qquad -F_{AC} - F_{BC} \cos 60 = 0 \qquad$ (4.17)
In y: $\qquad F_{BC} \sin 60 + F_{CV} = 0 \qquad$ (4.18)

Substituting our result for F_{BC} in Eq. (4.17), we obtain

$$F_{AC} = -(-86.5) \cos 60 = 43.3 \text{ N}$$

Similarly, from Eq. (4.18), the vertical reaction force $F_{CV} = 74.9$ N. Returning to pin A and summing forces,

In x: $\qquad F_{AH} + F_{AC} - F_{AB} \cos 30 = 0 \qquad$ (4.19)
In y: $\qquad F_{AV} - F_{AB} \sin 30 = 0 \qquad$ (4.20)

From Eq. (4.19) and our results for F_{AB} and F_{AC}, we find

$$F_{AH} = 50.0 \cos 30 - 43.3 = 0.0 \text{ N}$$

and from Eq. (4.20), $F_{AV} = 25.0$ N.

Our final results then show the forces in the members of the simple truss to be

$$F_{AB} = 50.0 \text{ N (compression)}$$
$$F_{BC} = 86.5 \text{ N (compression)}$$
$$F_{AC} = 43.3 \text{ N (tension)}$$

A convenient check on your result is provided by realizing that the forces acting on the overall structure must also satisfy equilibrium requirements. The free-body diagram of the structure is shown in Fig. 4.7c, where we sum forces in x to find

$$F_{AH} = 0$$

and since we determined F_{AH} to be zero, this checks our result.

We also sum forces in y to obtain

$$F_{AV} + F_{CV} - 100 = 0$$
$$25.0 + 74.9 - 100 = 0$$
$$-0.1 = 0$$

This checks our result within the accuracy to which we carried out the calculations.

Recall now the slight problem we had with assuming member BC to be in tension when it is actually in compression. Of course,

Example 4·3

The truss shown in Fig. 4.8a is to be analyzed, with the load considered to be the force F as shown. In terms of F, determine the forces for each of the members and the reaction forces at pins A and D.

Solution

We assume that all members are in tension and draw free-body diagrams for each pin, as shown in Fig. 4.8b. We then go from pin to pin summing forces and solving for unknown forces as it becomes possible. Beginning at pin C,

In x: $\quad F_{CD} - F_{AC} = 0$
In y: $\quad F_{BC} - F = 0$
$\quad\quad\quad F_{BC} = F$

At B,

In x: $\;-F_{AB} \sin 45 + F_{BD} \sin 60 = 0$
$$F_{BD} = \frac{\sin 45}{\sin 60} F_{AB} = 0.816 F_{AB}$$

In y:
$$-F_{BC} - F_{AB} \cos 45 - F_{BD} \cos 60 = 0$$
$$-0.816 F_{AB} \cos 60 - F_{AB} \cos 45 = F$$
$$-1.12 F_{AB} = F$$
$$F_{AB} = -0.897F$$
$$F_{BD} = -0.732F$$

At D,

In x: $\quad -F_{BD} \cos 30 - F_{CD} = 0$
$$F_{CD} = -(-0.732F) \cos 30 = 0.634F$$

Also, since $F_{AC} = F_{CD}$:
$$F_{AC} = 0.634\ F$$

In y: $\quad F_{BD} \sin 30 + F_{DV} = 0$
$$F_{DV} = -(-0.732F) \sin 30 = 0.366F$$

At A,

In x:
$$F_{AB} \cos 45 + F_{AC} + F_{AH} = 0$$
$$F_{AH} = -0.634F - (-0.897F) \cos 45$$
$$F_{AH} = 0F$$

our negative result told us about the incorrect assumption. This happens often with assumed directions for forces, and if we had chosen F_{AV} or F_{CV} to be in the wrong direction, we would also have obtained a negative result. In more complex structures, it is more convenient to simply assume that all members are in tension rather than try to keep track of various assumptions. Then a positive result for a member means that it actually is in tension and a negative result means that it is in compression. This procedure is followed in Example 4.3.

Example 4·3 (continued)

In y:
$$F_{AB} \sin 45 + F_{AV} = 0$$
$$F_{AV} = -(-0.897F) \sin 45$$
$$= 0.634F$$

The results of these calculations are summarized by Fig. 4.8c, where the negative numbers indicate members in compression. Note that since only member BC supports pin C vertically, it carries the full applied force F as a tensile load. Member BC then pulls down on pin B, putting members AB and BD into compression. These in turn push outward on pins A and D, pulling members AC and CD into tension. The forces within AC and CD must be equal since these are the only horizontal forces occurring at pin C and must balance each other to be in equilibrium.

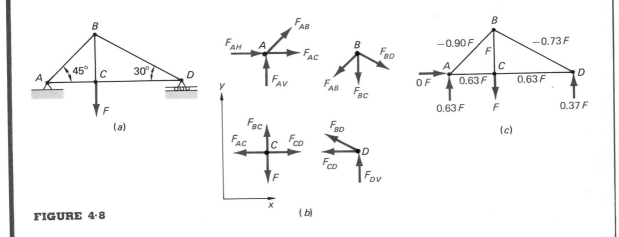

FIGURE 4·8

PROBLEMS

4.1 In Fig. P4.1, write the equations summing forces in the x and y directions and then solve the resulting equations to determine the unknown forces. In each case, determine both F_1 and F_2.

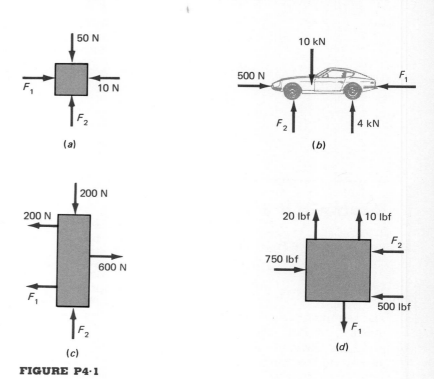

FIGURE P4·1

4.2 In each case in Fig. P4.2, resolve the given force into its horizontal and vertical components by determining F_x and F_y.

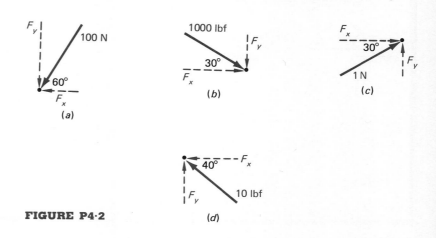

FIGURE P4·2

4.3 Determine the tension force in cables AB and BC in Fig. P4.3.

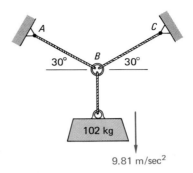

FIGURE P4·3

4.4 Find the tension force in cables JK and KL in Fig. P4.4.

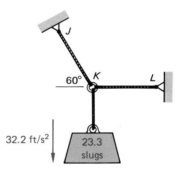

FIGURE P4·4

4.5 Determine the forces in each member of the trusses in Fig. P4.5. Indicate in each case whether the force is tension or compression.

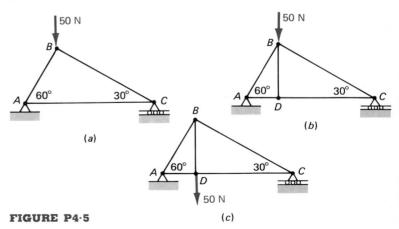

FIGURE P4·5

4.6 For the trusses in Fig. P4.6, determine the reaction forces at each support and the forces in each truss member. Indicate which members are in tension and which are in compression.

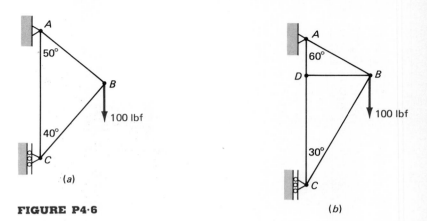

FIGURE P4·6

4.7 Determine the forces in each member of the trusses in Fig. P4.7. You may find it easiest to begin the solutions at the supporting points A and B where fewer unknowns are found. In both cases below, the load is centered so that each of the supporting points furnishes one-half the required vertical force.

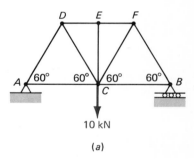

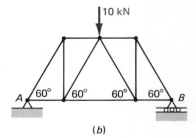

FIGURE P4·7

4·2 MOMENTS

The tendency of a force to rotate a body about a given point is indicated by the moment of the force about that point. The moment is equal to the product of the force's magnitude times the distance between the point and the force as measured along a line through the point and perpendicular to the line of action of the force. Thus moments have the dimension of force times distance (FL), with SI units of newton-meters (N·m) and conventional American units of foot-pounds (ft·lbf). In the American system, however, the order is often reversed (lbf·ft) to avoid confusion with energy units. For a planar body to be in equilibrium and not have a tendency to rotate, the sum of the moments of all forces acting in the plane of the body must be zero at all points within the plane.

As an example, consider Fig. 4.9a, where body AB is in the plane of the paper and is shown pinned and free to rotate at C. The 10-N force acting 0.50 m from pin C produces a moment of 5.0 N·m counterclockwise about pin C. This is because the distance of 0.50 m is through C along a perpendicular to the line of action of the force. At B the force of 7.1 N also produces a moment about C. To compute this moment however, triangle BDC may be constructed so that line CD is the perpendicular through C to the line of action of the force. The length of CD is 1.0 sin 45° = 0.71 m, so the moment of this force about pin C is (7.1 N)(0.71 m) = 5.0 N·m in the clockwise direction.

If there are p moments acting about the point of interest, our requirement for equilibrium is stated mathematically as

$$\sum_{k=1}^{p} M_k = 0 \qquad (4.21)$$

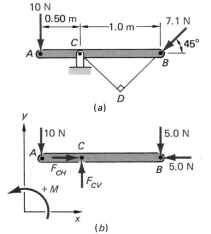

FIGURE 4·9

Just as in our summation of forces, it is necessary to establish a sign convention for this summation. Here we will consider counterclockwise moments to be positive. The summation of moments about C in Fig. 4.9a is thus,

$$\sum_{k=1}^{p} M_k = M_1 + M_2$$
$$= 10(0.50) - 7.1(0.71)$$
$$= 5.0 - 5.0$$
$$= 0 \qquad (4.22)$$

This satisfies one condition for equilibrium; body AB will not have a tendency to rotate. In addition, however, the summation of forces on body AB must also be zero and the necessary reaction force to ensure this is provided by pin C. In Fig. 4.9b we have a free-body diagram of AB with the force at B resolved into its vertical and horizontal components. From this figure we can easily sum forces in the x and y directions to find that $F_{CH} = 5.0$ N and $F_{CY} = 15$ N are the components of the reaction force at pin C which hold the body in equilibrium.

With Fig. 4.9b and the results for F_{CH} and F_{CV}, let us consider again the summation of moments. Our choice of summing moments about pin C in Eq. (4.22) was very convenient. The force at pin C does not produce a moment about C, since it acts at zero distance. Thus we were able to sum moments before calculating the forces at C. However, when a planar body is in equilibrium, the sum of the moments at any point in the plane is zero. We will check this by summing moments again, this time about point A. As an exercise, you can check the sum about B yourself. In using the free-body diagram to sum moments at point A, we find that the resolution of forces at B and C into components is an aid to calculating their moments. In fact, it is not actually necessary to construct any triangles similar to BDC in Fig. 4.9a. The vertical component F_{CV} at point C produces a moment about A equal to (15 N)(0.50 m). Here the distance is measured along AC and is perpendicular to the line of action of the component F_{CV}. The component F_{CH} has its line of action passing directly through A, so that the distance is zero in the moment calculation and this component does not produce a moment about A. Similarly, the vertical component of the force at B produces a moment about A of (5 N)(1.5 m), whereas the horizontal component has zero effective distance and does not produce a moment about A. The summation of moments about point A, with counterclockwise positive, is now

$$M = (15)(0.50) - (5.0)(1.5)$$
$$= 7.5 - 7.5 = 0$$

This confirms our previous equilibrium calculation for moments made about point C. The same zero summation will result for any point in the xy plane of Fig. 4.9b.

For another application of the summation of moments in equilibrium calculations, let us return to the simple motor vehicle model discussed in Sec. 4.1 and Fig. 4.1. When we left our results in Eqs. (4.5) and (4.6), we were not able to identify the specific supporting forces provided at the front and rear of the vehicle. Figure 4.10 is a repeat of Fig. 4.1, with specific force magnitudes and dimensions for the wheel base and center-of-gravity location. We have assumed for simplicity that our drag force acts at the same height as the center of gravity. With these details we can now make a summation of moments about any of various points in the

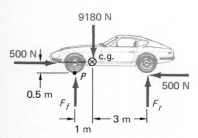

FIGURE 4·10
Free-body diagram of a vehicle.

Example 4·4

Figure 4.11a shows a special set of pliers holding a small nut. If the force F is applied at the handle, determine the force F_{SQ} squeezing the nut. Assume that F_{SQ} acts 1.50 in from pin A.

Solution

A free-body diagram of member BCE is shown in Fig. 4.11b. The member CD is a pinned member loaded only at its pins: the force F_{CD} is thus known to act along it, as is shown. We determine F_{CD} by summing moments about pin B. Note that only the vertical component of F_{CD} produces a moment about pin B, so

$$F_{CD} \sin 5°(1) - 6F = 0$$

and

$$F_{CD} = 68.8F$$

Now we sum forces acting on BCE

In x: $\quad F_{BH} - F_{CD} \cos 5° = 0$
$$F_{BH} = 68.5F$$
In y: $\quad F_{BV} + F_{CD} \sin 5° - F = 0$
$$F_{BV} = -5F$$

We now move to Fig. 4.11c, the free-body diagram of the jaw of the pliers. Taking moments about A, we find that only the horizontal component at B contributes, so we need not determine the forces at A; we thus have

$$F_{BH}(1) - 1.50 F_{SQ} = 0$$
$$F_{SQ} = 45.7F$$

which is a very large force multiplication.

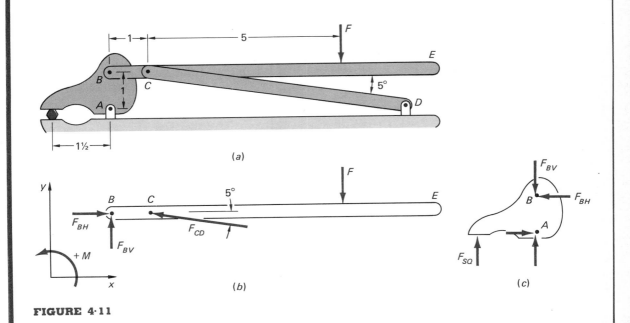

FIGURE 4·11

figure. By choosing the summation about the point of application of the front-wheel support force, labeled P, it is possible to obtain F_r conveniently. The summation, with counterclockwise moments considered positive, yields

$$(-500)(0.50) - (9180)(1.0) + 4.0F_r = 0$$

and

$$F_r = 2360 \text{ N}$$

Recalling the summation of vertical forces in Eq. (4.6), we find that

$$F_f + F_r = 9180 \text{ N}$$

and thus

$$F_f = 9180 - 2360 = 6820 \text{ N}$$

This now defines the front and rear supporting forces of our vehicle for the given drag and propulsion conditions. To finally obtain this result, we had to use the equations of equilibrium for both forces and moments. In Example 4.4, we applied both principles again, this time to calculate the gripping action possible in pliers utilizing a special mechanism for their operation.

PROBLEMS

4.8 In Fig. P4.8a to d, demonstrate that each satisfies the equilibrium requirement that the sum of the moments equals zero.

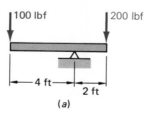

(a)

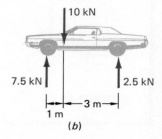

(b)

FIGURE P4·8

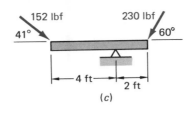
(c)

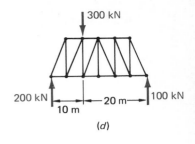
(d)

4.9 Based on the summations of moments and forces, solve for the unknown forces shown in each case in Fig. P4.9.

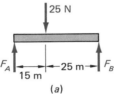
(a)

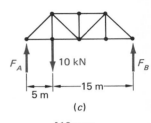
(c)

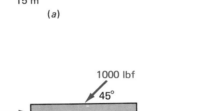
(b)

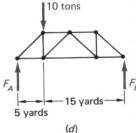
(d)

FIGURE P4·9

4.10 Figure P4.10 is a simple model representing a pair of standard pliers. For an applied force F, determine the value of F_{SQ} squeezing the nut. Compare your result with that of Example 4.4. What is the advantage of that special design?

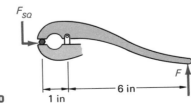

FIGURE P4·10

4.2 MOMENTS **193**

4·3 GENERAL COMMENTS

In this brief introduction to statics, we have presented only the most basic concepts in order to demonstrate the techniques and a few applications of calculations based on mechanical equilibrium. Mechanical and civil engineers very often utilize the principles of statics we have presented here. The study of statics, however, also forms an important part of the general knowledge needed by electrical, chemical, and other types of engineers as you will notice later in this book.

In this discussion, we have focused primarily on problems in two dimensions, where all forces are considered to act in the same plane. Such forces are called *coplanar*. In your further study of statics you will study equilibrium in three dimensions. You will also find a mathematical method for dealing with forces by treating them as *vectors*, a name given to physical quantities for which direction as well as magnitude is important. The techniques of vector algebra considerably simplify the treatment of forces, positions, and moments in both two- and three-dimensional problems. You will also learn to deal with bodies of irregular shape and distributed loads when forces are not simply applied at a single point. This requires a knowledge of integral calculus. References 1 to 3 are examples of books containing much more detailed information on statics. These or similar references should be available at your university library.

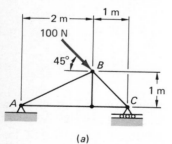
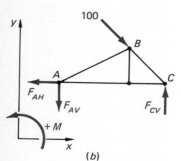

FIGURE 4·12
A simple truss problem.

STATICALLY INDETERMINATE PROBLEMS

Statics, unfortunately, does not provide the solution to all our problems of mechanical equilibrium. Many real problems are statically indeterminate, that is, the equations of static equilibrium do not alone provide enough information to solve for all the unknown forces in the problem. You should recall that we mentioned this type of difficulty back in Sec. 4.1 when we introduced the roller-mounted support shown in Fig. 4.7a.

To discuss the concept of static indeterminacy in more detail, let's return to the mounting problem. Figure 4.12a shows a simple structure with one fixed end and one end supported on rollers. The free-body diagram of Fig. 4.12b shows the forces acting on the structure, so we can readily write equations for the overall structure. Summing forces on the structure, we find

In y: $\qquad -100 \sin 45 - F_{AV} + F_{CV} = 0 \qquad$ (4.23)

In x: $\qquad 100 \cos 45 - F_{AH} = 0 \qquad$ (4.24)

and summing moments about pin C, we have

$$(1)(100 \sin 45) - (1)(100 \cos 45) + 3F_{AV} = 0 \qquad (4.25)$$

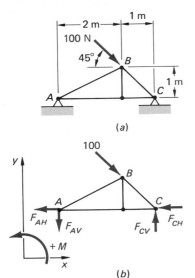

FIGURE 4·13
Statically indeterminate mounting.

Equations (4.23) to (4.25) involve three unknowns—F_{AV}, F_{CV}, and F_{AH}. We can easily find from these equations that $F_{AV} = 0$, $F_{AH} = 70.7$ N, and $F_{CV} = 70.7$ N. This is characteristic of statically determinate problems. The number of unknown forces matches the number of static equilibrium equations.

Now consider Fig. 4.13a, which is identical to our previous problem except that the right-hand support is not on rollers. The additional horizontal component force F_{CH} now appears on the structure's free-body diagram in Fig. 4.13b. Summing forces now, we find

In y: $$-100 \sin 45 - F_{AV} + F_{CV} = 0 \qquad (4.26)$$
In x: $$100 \cos 45 - F_{AH} - F_{CH} = 0 \qquad (4.27)$$

and taking moments about pin C, we get

$$(1)(100 \sin 45) - (1)(100 \cos 45) + 3F_{AV} = 0 \qquad (4.28)$$

Here we have three equations again, but now there are four unknowns: F_{AV}, F_{CV}, F_{AH}, and F_{CH}. We can, of course, solve Eq. (4.28) to obtain $F_{AV} = 0$, and we can substitute this into Eq. (4.26) to find $F_{CV} = 70.7$. However, rearranging Eq. (4.27) yields only

$$F_{AH} + F_{CH} = 70.7 \qquad (4.29)$$

and we do not have an explicit solution for F_{AH} and F_{CH}. You may think that adding another equation by summing moments about pin B is useful. The resulting equation is

$$-1F_{AH} - 1F_{CH} + 2F_{AV} + 1F_{CV} = 0 \qquad (4.30)$$

Then, substituting the results we have for F_{AV} and F_{CV} into this equation yields

$$F_{AH} + F_{CH} = 70.7 \qquad (4.31)$$

which is the same as Eq. (4.29). We have not gained a new, useful equation; similar discouraging results will occur for a summation of moments about any additional point. The difficulty is that we have too many unknowns for the number of static equilibrium equations available. This is a characteristic of statically indeterminate problems. To actually determine F_{AH} and F_{CH}, we must learn about the rigidity of the mounts at A and C. The more rigid mount will carry the larger horizontal load.

Statically indeterminate problems are solved by considering deformation of the members or supports. Methods of strength of materials, an area of mechanics dealing with stress and deforma-

4.3 GENERAL COMMENTS

tion, are used. The techniques of strength of materials are also important in determining if a material is strong enough and suitable for a particular application. Representative books dealing with this subject are References 4 to 6.

REFERENCES

1. Shames, I. H., *Engineering Mechanics: Statics,* Prentice-Hall, Englewood Cliffs, NJ, 1980.
2. Beer, F. P., and E. R. Johnston, Jr., *Vector Mechanics for Engineers: Statics and Dynamics,* McGraw-Hill, New York, 1976.
3. Tuma, J. J., *Statics,* Quantum, New York, 1974.
4. Shames, I. H., *Introduction to Solid Mechanics,* Prentice-Hall, Englewood Cliffs, NJ, 1975.
5. Popov, E. P., *Mechanics of Materials,* Prentice-Hall, Englewood Cliffs, NJ, 1976.
6. Crandall, S. H., N. C. Dahl, and T. J. Lardner, *An Introduction to the Mechanics of Solids,* McGraw-Hill, New York, 1972.

5 ELECTRIC CIRCUITS

Our discussion of engineering concepts now continues with consideration of some basic electric circuits and electromagnetic components. These topics have taken on increased importance in recent years for all types of scientists and engineers, because the use of electric instrumentation and electric controls has become more and more commonplace.

Electric circuits may contain both active and passive components. Active components are batteries, transistors, vacuum tubes, amplifiers, and generators, for example—components that supply or control electric energy within a circuit. Components such as resistors, capacitors, and inductors are passive components, ones that do not introduce electric energy into a circuit but either store or dissipate existing electric energy. Active components known as transistors and vacuum tubes are called electronic, and an electric circuit containing at least one transistor or tube may be called an electronic circuit. Devices which contain electronic circuits, such as radios and televisions, are, of course, called electronic. Electric motors are never electronic, because they do not themselves contain transistors or tubes. But, motor control circuits are often electronic, since they usually contain transistors.

In this chapter we introduce you to the basic ideas of electric circuitry from an engineering viewpoint. We will not discuss electronic components but will focus on electric-resistance circuits. We will also discuss electromagnetic concepts, so that you can begin to understand the operating principles of electric motors and generators which play a large role in modern society.

5·1 CURRENT AND RESISTANCE

Electric current is a measure of the flow rate of electric charge in a conductor. The unit of electric charge is the coulomb, where one coulomb represents the combined charge of 6.25×10^{18} protons. Also, a negative charge corresponding to minus one coulomb is the charge of 6.25×10^{18} electrons. The unit of electric current is the ampere; one ampere is defined as a flow of charge of one coulomb per second.

Figure 5.1 represents a typical metallic conductor, where the flow of charge occurs by movement of electrons. Here the electrons are flowing to the left; if we could sit by any section of this conductor and count the electrons going by, we would count 6.25×10^{18} every second. Although the electrons are flowing to the left, we say that charge is flowing to the right at a rate of one coulomb per second because the electrons have negative charge. The current, equaling one ampere, is also to the right. This may seem confusing to you after extensively studying electron flow in physics, but engineers normally discuss electricity in terms of current and charge. Many practicing engineers (not, it is hoped, electrical engineers) have no doubt forgotten which atomic particles are actually moving.

An electric current is produced in a body by application of a voltage across the body. This voltage, also called a potential difference, provides the electromotive force which "pushes" the electrons or charge through the body. The ability of a particular body to resist this flow of charge is known as the resistance of the body. Resistance depends on the body's material as well as its size and shape. The property of a material which resists the flow of current is called its resistivity. Materials which are good electric insulators have high values of resistivity; materials which are good conductors of electricity have low values. The symbol for resistivity is usually ρ, with typical units of ohm-centimeters. Table 5.1 is a list of different materials' resistivities. Note that copper has the lowest value for common materials. This is one reason for its

TABLE 5·1
RESISTIVITIES OF VARIOUS MATERIALS

Material	Typical Resistivity, $\Omega \cdot cm$
Polystyrene	1.0×10^{18}
Mica	1.0×10^{15}
Polyvinyl chloride	1.0×10^{14}
Glass	1.0×10^{13}
Carbon	4.0×10^{-3}
Lead	2.1×10^{-7}
Steel	1.3×10^{-7}
Aluminum	2.7×10^{-6}
Gold	2.3×10^{-6}
Copper	1.7×10^{-6}
Silver	1.6×10^{-6}

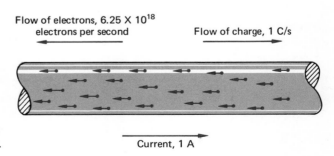

FIGURE 5·1
Flow of current in a conductor.

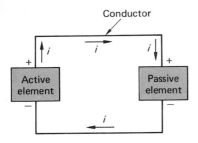

FIGURE 5·2
Current flowing in a circuit.

widespread use in electric wiring, but there are others, as you will see in Sec. 8.3.

If a body has a uniform cross section with an area A centimeters squared and the length of the body is l centimeters, its resistance to current flow, in ohms, is given by

$$R = \frac{\rho l}{A} \tag{5.1}$$

By inspecting this equation, we can see that for longer bodies the resistance is higher, because the charge must flow farther. Bodies of larger cross section have more room for the charge to flow, so they have lower resistance; in which case, a conductor is similar to a pipe, which carries a liquid. As an example of the use of Eq. (5.1), consider a copper conductor 100 ft long (3050 cm) and $\frac{1}{64}$ in in diameter (0.0397 cm). Using ρ from Table 5.1, Eq. (5.1) yields a resistance of 4.2 Ω.

5·2 SIMPLE CIRCUITS

The most basic form of electric circuit is shown in Fig. 5.2. An active element applies a voltage to a passive element in a closed electric path. Charge flows from the positive terminal of the active component through the conductor to the positive terminal of the passive element and through the passive element, finally returning to the negative terminal of the active element. We think of the current i as flowing in the same direction as the charge, even if the electrons are actually moving in the opposite direction. In passive elements, there is a voltage drop in the direction of current flow; in active elements, there is a voltage rise. If there is an open circuit anywhere, perhaps by a loose connection or broken conductor, the voltage is no longer applied to the passive element and current flow stops.

Figure 5.3 represents an ideal flashlight circuit. Here, two 1.5-V batteries are the active elements connected in series—the positive terminal of one to the negative terminal of the other. They each provide a voltage increase of 1.5 V in the direction of current flow and together apply a total of 3.0 V across the bulb, or the passive element. Since there are no alternative paths for current flow, the current must be the same at every point in this circuit: through the bulb, in the conductors connecting the bulb to the batteries, and also in the conductor between batteries. A switch such as S, shown in the figure, can be inserted anywhere in the loop to open the circuit and turn off the light.

Note that Fig. 5.3 uses the standard symbol for batteries of two

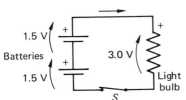

FIGURE 5·3
A flashlight circuit.

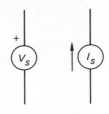

FIGURE 5·4
Symbols for voltage and current sources.

parallel lines, one longer than the other. The shorter line, like a minus sign, represents the negative battery terminal. The standard symbol for resistance has also been used in Fig. 5.3 to represent the bulb. Other common symbols, for a general voltage source and a current source, are shown in Fig. 5.4. The plus sign on the voltage source indicates its positive terminal; the arrow on the current source indicates the direction of the current it produces.

Figure 5.3 is an idealized model of a real electric circuit. Symbols are used to represent each of the electric components of interest, so that the character of the circuit can be examined and equations can be written to calculate the behavior of the circuit. Schematic drawings such as this are only *idealized* models: that is, the conductors do not have resistance, the batteries are good and each provides 1.5 V, and the switch and battery connections are clean and properly made so that they do not have resistance. If any aspects of this model of the circuit are not correct, results calculated from the model will not be correct. If some idealized items are not correct, adjustments must be made so that the model adequately represents the real circuit before calculations are performed. Even in this simple circuit additional resistance might be included if poor battery contacts are suspected or the voltages of the batteries may be reduced if they are old and weak.

Let's now consider Fig. 5.5a, where a resistor is connected to a current source. The current source is an electronic power supply very common in electrical laboratories, so we can use it to produce an adjustable current i through the resistor. If we select a series of current settings and measure and record the voltage across the resistor at each current setting, we can plot a graph of voltage vs. current, as in Fig. 5.5b. Here the voltage drop across the resistor depends linearly on the current through it. This is true for simple resistance elements; the slope of the line is the resistance R of the element. We can thus say that

FIGURE 5·5
Measurement of resistance.

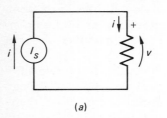

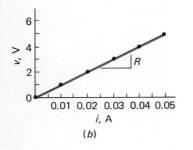

$$v = Ri \tag{5.2}$$

which is called Ohm's law. The units of R are actually volts per ampere, but this ratio has been given the name ohm (Ω). For the resistance described by Fig. 5.5b, $R = 100\ \Omega$.

VOLTAGE AND CURRENT LAWS

In the analysis of some simple circuits, there are two important laws which we apply. Kirchhoff's voltage law states that the sum of the voltage around any closed loop in a circuit must be zero. Kirchhoff's current law is applied at the nodes, or junctions of two or more conductors, of a circuit. To have equilibrium conditions at any node, charge cannot be accumulating, so the summation of currents at a node must be zero. These summations of voltages and currents are made with proper regard for sign, as were our previous summations of forces and moments; they may be stated mathematically in a similar way as

$$\sum_{j=1}^{n} v_j = 0 \tag{5.3}$$

to sum the voltage over n elements in a loop, or as

$$\sum_{k=1}^{m} i_k = 0 \tag{5.4}$$

to sum the currents in m conductors joining at a node.

Let us now apply Ohm's law and Khirchhoff's voltage and current laws to the circuit of Fig. 5.6, which you will note is essentially like the ideal flashlight circuit we mentioned earlier. We stated then that the current was the same throughout this circuit, but let's apply Eq. (5.4) at node A to take a closer look. We consider i_1, the current *into* node A, as positive; and i_2, the current *out of* node A, as negative. The summation of Eq. (5.4) is then

$$i_1 - i_2 = 0$$

and we see that

$$i_1 = i_2$$

for this simple case. This tells us that the current in path EA of Fig. 5.6 is the same as in path AB. Similarly, we could, of course, show that the currents in all the paths are the same.

We can also apply Khirchhoff's voltage law to Fig. 5.6, since the circuit $ABCDEA$ forms a closed loop. To apply it, we proceed in the direction of the current, recording voltage drops for the passive elements and voltage rises for the active elements. Thus, the summation of Eq. (5.3) yields

$$-v_R + 0 + 1.5 + 1.5 + 0 = 0 \tag{5.5}$$

where we started around the loop beginning with path AB and recording zeros for paths BC and EA. Solid lines without elements are always considered to be ideal conductors and no voltage change occurs across them. In fact, ideal conductors such as EA and BC may be considered to be simply large nodes. Rearranging Eq. (5.5), of course, results in

$$v_R = 3.0 \text{ V}$$

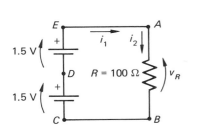

FIGURE 5·6
Analyzing the flashlight circuit.

as the voltage drop across the resistor. From Ohm's law we then find

$$v_R = Ri_2$$
$$3.0 = 100i_2$$
and
$$i_2 = 0.030 \text{ A}$$

is the current throughout this circuit.

Symbols for voltage and current

SERIES RESISTORS

In Fig. 5.6, we considered two batteries in series driving a resistive load. We now analyze the circuit of Fig. 5.7, where we have one battery driving two resistors in series. The current is i through both resistors and the entire circuit. We need apply only Khirchhoff's voltage law to find

$$-v_{R_1} - v_{R_2} + 1.5 = 0$$

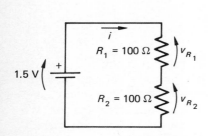

FIGURE 5·7
Series resistors.

or

$$v_{R_1} + v_{R_2} = 1.5 \tag{5.6}$$

Now we write Ohm's law for each resistor so that

$$v_{R_1} = R_1 i \tag{5.7}$$

and

$$v_{R_2} = R_2 i \tag{5.8}$$

If we substitute these two equations for v_{R_1} and v_{R_2} into Eq. (5.6), we obtain

$$R_1 i + R_2 i = 1.5$$

Knowing that $R_1 = R_2 = 100 \ \Omega$ for this circuit, we then have

$$(100 + 100)i = 1.5$$

and

$$i = 0.0075 \text{ A}$$

Returning this result to Eqs. (5.7) and (5.8), we then find

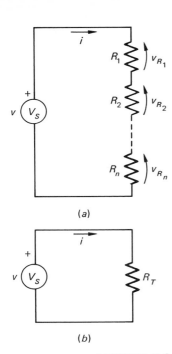

FIGURE 5·8
The equivalent total resistance.

$$v_{R_1} = R_1 i = 100(0.0075) = 0.75 \text{ V}$$

and also

$$v_{R_2} = R_2 i = 100(0.0075) = 0.75 \text{ V}$$

In this case with two equal resistors, half the applied voltage is found across each resistor.

When analyzing circuits containing resistors in series, it is usually convenient to utilize the equivalent total resistance of the series connection, which is simply the sum of the individual resistances. This can be seen by inspecting Fig. 5.8a, where a voltage v is applied across a number of resistances in series. The current through each is i because of the series connection, and summing voltages around the loop yields

$$-v_{R_1} - v_{R_2} - \cdots - v_{R_n} + v = 0$$

or

$$v_{R_1} + v_{R_2} + \cdots + v_{R_n} = v$$

Realizing that Ohm's law applies to each of these resistances and that the current is the same through each, we can rewrite Eq. (5.9) as

$$R_1 i + R_2 i + \cdots + R_n i = v$$

and then

$$v = (R_1 + R_2 + \cdots + R_n)i \quad (5.10)$$

If we think of the total resistance of these series elements as

$$R_T = R_1 + R_2 + \cdots + R_n$$

Eq. (5.10) becomes

$$v = R_T i \quad (5.11)$$

Since this equation looks exactly like Ohm's law, we say that Ohm's law can be applied directly for series elements by using the sum of the individual resistances as the total resistance. The equivalent circuit is simply as shown in Fig. 5.8b. The voltage drop across any single resistor in the connection can be found, once the current is known, by applying Ohm's law for that individual resistor.

PARALLEL RESISTORS

We now consider the parallel resistance circuit of Fig. 5.9a and the development of an equivalent resistance R_E. In the simple circuit of Fig. 5.9b, this equivalent resistance represents the overall voltage and current relationship for the original circuit. To obtain R_E, we begin by realizing that each resistance has the voltage v across it in Fig. 5.9a. This can be concluded by inspection of the circuit or by summing voltages around each loop. For example, from the loop shown as (1), v_{R_1} is the voltage across resistor R_1 and

$$-v_{R_1} + v = 0$$

so that

$$v_{R_1} = v$$

Then, from loop (2),

$$-v_{R_2} + v_{R_1} = 0$$

and

$$v_{R_2} = v_{R_1} = v$$

We could also show, by continuing to consider each loop, that $v_{R_n} = v$.

Let's now sum the currents at the large node where the conductors going to each of the resistors come together with the conductor from the voltage source. At this node,

$$-i_1 - i_2 - \cdots - i_n + i = 0$$

or

$$i_1 + i_2 + \cdots + i_n = i \tag{5.12}$$

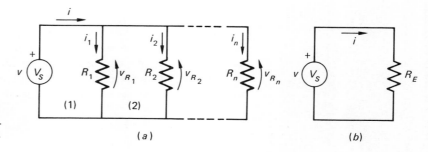

FIGURE 5·9
The equivalent parallel resistance.

Each of the resistors must, of course, follow Ohm's law, so $i_1 = v/R_1$, $i_2 = v/R_2$, ..., $i_n = v/R_n$. Substituting these expressions in Eq. (5.12), we find that

$$\frac{v}{R_1} + \frac{v}{R_2} + \cdots + \frac{v}{R_n} = i$$

and rearranging,

$$v = \frac{1}{\frac{1}{R_1} + \frac{1}{R_2} + \cdots + \frac{1}{R_n}} i$$

Now, if we let

$$R_E = \frac{1}{\frac{1}{R_1} + \frac{1}{R_2} + \cdots + \frac{1}{R_n}} \quad (5.13)$$

we have defined an equivalent resistance. This can be used in Fig. 5.9b to describe the voltage and current relationship for the overall circuit. The expression $v = R_E i$ is an equivalent statement of Ohm's law.

Let's consider a parallel circuit with only two resistors, as in Fig. 5.10. Both resistors are equal in magnitude, so $R_1 = R_2 = R$. From our definition of the equivalent resistance in Eq. (5.13),

$$R_E = \frac{1}{\frac{1}{R} + \frac{1}{R}} = \frac{1}{\frac{2}{R}} = \frac{1}{2} R$$

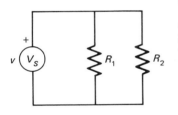

FIGURE 5·10
Two parallel resistors.

Thus with two equal resistors in parallel, the equivalent resistance is one-half the resistance of each. Let's now consider one resistance to be much larger than the other. If $R_1 = R$, let $R_2 = 10R$ so that the equivalent resistance

$$R_E = \frac{1}{\frac{1}{R} + \frac{1}{10R}} = \frac{1}{\frac{1.1}{R}} = 0.91\,R$$

Note that this equivalent resistance of $0.91\,R$ is smaller than both $R_1 = R$ and $R_2 = 10R$. An equivalent resistance is always less than the smallest of two parallel resistors. Resistance is always decreased by adding a parallel resistor.

Examples 5.1 and 5.2 further illustrate the analysis of simple circuits. The concepts of equivalent parallel and series resistances are utilized to simplify the solutions. In these examples, there are some important points which should be emphasized and can be seen quite well in Figs. 5.11c and 5.12c. For resistors connected in series, the current through the resistances is the same while the voltage drop across each is normally different. The largest voltage drop occurs across the largest resistance. If there are equal resistances, they show equal voltage drops. When there are resistors in parallel, the voltage drop across each is the same but the currents are normally different. The largest current passes

Example 5·1

Analyze the circuit of Fig. 5.11a to determine the current from the 100-V source and through each resistor and also the voltage drop across each resistor. Let $R_1 = 10\ \Omega$, $R_2 = 5\Omega$, $R_3 = 10\ \Omega$, $R_4 = 20\ \Omega$.

Solution

We realize first that R_1 and R_2 are in series with total resistance

$$R_{12} = R_1 + R_2 = 15\ \Omega$$

and, similarly, R_3 and R_4 are in series with total resistance $R_{34} = 30\ \Omega$. The simplified equivalent circuit is as shown in Fig. 5.11b. We now determine the current $i_1 = i_2$ which flows through the series resistors R_1 and R_2 by applying the full 100 V across the combination R_{12}.

$$i_1 = \frac{v}{R} = \frac{100}{R_{12}} = \frac{100}{15} = 6.67\ \text{A}$$

The full 100 V is also applied across R_{34} and thus the current $i_3 = i_4$ through the R_3 and R_4 resistances is

$$i_3 = \frac{v}{R} = \frac{100}{R_{34}} = \frac{100}{30} = 3.33\ \text{A}$$

The total current from the source is obtained by the summation of currents at node A in Fig. 5.11b.

$$-i_1 - i_3 + i = 0$$
$$i = i_1 + i_3 = 6.67 + 3.33 = 10.0\ \text{A}$$

Finally, the voltage drop across each resistor in Fig. 5.11a is determined by applying Ohm's law to each resistor,

$$v_1 = R_1 i_1 = 66.7\ \text{V}$$
$$v_2 = R_2 i_2 = 33.3\ \text{V}$$
$$v_3 = R_3 i_3 = 33.3\ \text{V}$$
$$v_4 = R_4 i_4 = 66.7\ \text{V}$$

The results are summarized in Fig. 5.11c.

through the smaller resistance and equal resistances will have equal currents.

In wiring a home or an automobile or any common machine, the circuits are normally connected as parallel circuits. Each device—light bulb, radio, washing machine, etc.—has the same voltage applied to it and each device draws current related to its electric resistance. The product of current and voltage is power, and the device drawing the most current, of course, draws the most power.

Example 5·1 (continued)

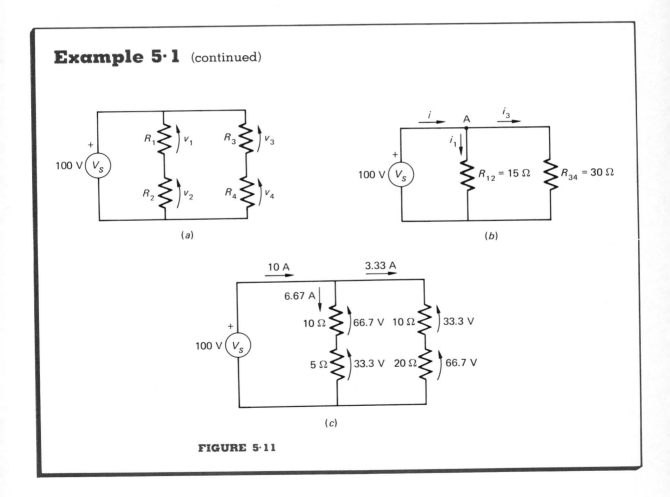

FIGURE 5·11

Example 5·2

For the circuit of Fig. 5.12a, determine the current through each resistor and the total current drawn from the 6-V source as well as the voltage drop across each resistor. Here

$$R_1 = R_2 = 200 \ \Omega \quad \text{and} \quad R_3 = R_4 = 100 \ \Omega$$

Solution

Since the resistors R_1, R_2, and R_3 are all in parallel, the corresponding voltage drops are equal and $v_1 = v_2 = v_3$. We calculate the equivalent resistance for this group as

$$R_E = \frac{1}{\frac{1}{R_1} + \frac{1}{R_2} + \frac{1}{R_3}}$$
$$= \frac{1}{\frac{1}{200} + \frac{1}{200} + \frac{1}{100}}$$
$$= 50 \ \Omega$$

The equivalent circuit is shown in Fig. 5.12b; there are now two resistances in series with total resistance $R_T = 50 + 100 = 150 \ \Omega$. The current from the source by Ohm's law is

$$i = \frac{v}{R} = \frac{6}{R_T} = \frac{6}{150} = 0.040 \text{ A}$$

and thus the voltage drops in Fig. 5.12b are

$$v_1 = Ri = 50(0.040) = 2.0 \text{ V}$$
$$v_2 = Ri = 100(0.040) = 4.0 \text{ V}$$

The individual currents can be calculated as

$$i_1 = \frac{v_1}{R_1} = \frac{2.0}{200} = 0.010 \text{ A}$$
$$i_2 = \frac{v_2}{R_2} = \frac{2.0}{200} = 0.010 \text{ A}$$
$$i_3 = \frac{v_3}{R_3} = \frac{2.0}{100} = 0.020 \text{ A}$$

and, of course, $i_4 = i = 0.040$ A. A summary of the results is shown in Fig. 5.12c.

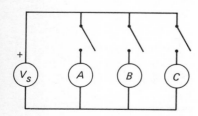

FIGURE 5·13
Parallel connection.

Standard parallel circuitry has the built-in advantage of permitting devices to be switched on and off easily. For example, in Fig. 5.13 the switches which supply power to devices A, B, and C are in parallel and can be operated completely independent of one another. Whenever one switch is closed, the source voltage will be immediately applied to the device it controls. However, in the series connection of Fig. 5.14, we see that all the switches must be closed before any of the devices operate and then all will be operating simultaneously. A parallel circuit is also convenient since the full source voltage is always applied directly to a device

Example 5·2 (continued)

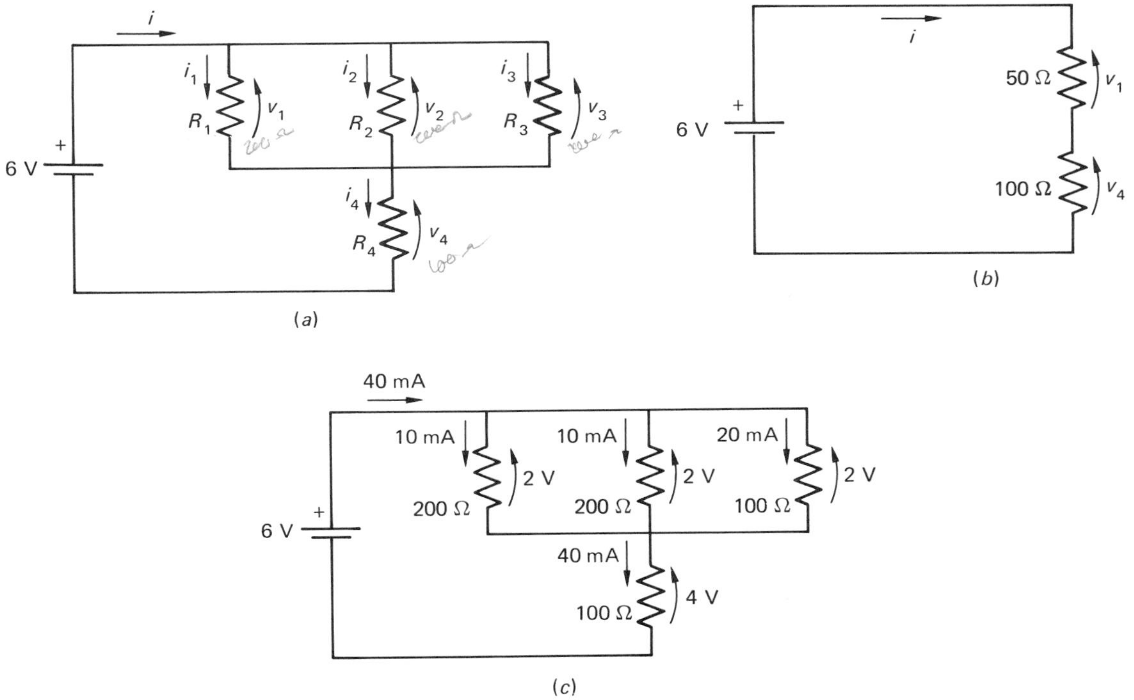

FIGURE 5·12

FIGURE 5·14
Series connection.

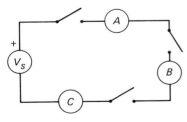

when it is turned on, so that each device can be designed for operation at that particular voltage. Of course, series connections are frequently made within particular devices for many reasons. However, the overall circuitry of a vehicle, building, or machine is generally connected in parallel fashion. Multicolored lights for holiday decorations have frequently been wired in series to hold costs down. The common difficulty with such lights is that the failure of one lamp opens the circuit and shuts off the entire string of lights.

5.2 SIMPLE CIRCUITS

PROBLEMS

5.1 Calculate the unknown current or voltage shown in each circuit in Fig. P5.1:

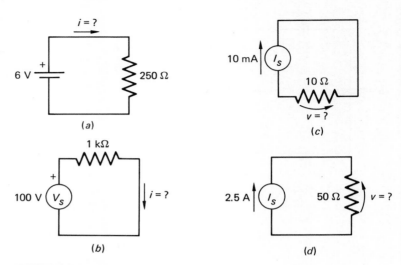

FIGURE P5·1

5.2 For the circuit in Fig. P5.2 determine the current i if
(a) $R_1 = 10\ \Omega$, $R_2 = 100\ \Omega$, $R_3 = 1000\ \Omega$
(b) $R_1 = 110\ \Omega$, $R_2 = 1000\ \Omega$, $R_3 = 1010\ \Omega$
(c) $R_1 = 10\ \Omega$, $R_2 = 0\ \Omega$, $R_3 = 15\ \Omega$

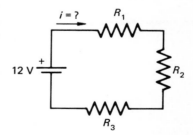

FIGURE P5·2

5.3 For the circuit in Fig. P5.3 determine the equivalent resistance of the parallel resistors if
(a) $R_1 = 100\ \Omega$, $R_2 = 100\ \Omega$, $R_3 = 100\ \Omega$
(b) $R_1 = 1000\ \Omega$, $R_2 = 1000\ \Omega$, $R_3 = 1000\ \Omega$
(c) $R_1 = 100\ \Omega$, $R_2 = 100\ \Omega$, $R_3 = \infty$
(d) $R_1 = 100\ \Omega$, $R_2 = 100\ \Omega$, $R_3 = 0$

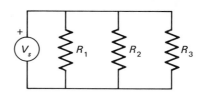

FIGURE P5·3

5.4 In the circuits in Fig. P5.4, solve for the current through and voltage across each of the resistances.

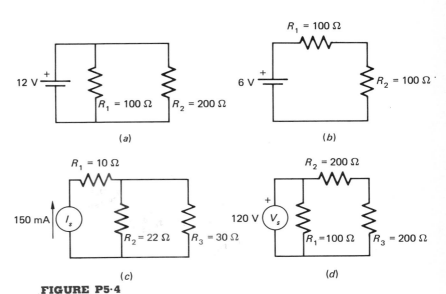

FIGURE P5·4

5.5 A diode is a special nonlinear resistance. It has the symbol ▸◂. When used in a circuit, it ideally has a resistance of $R_D = 0$ for current flow in the direction of the arrow in the symbol. It has a resistance $R_D = \infty$ for current in the opposite direction. In Fig. P5.5, determine the voltage v_{out} for the case where $v_{in} = +10$ V. Also, find v_{out} if $v_{in} = -10$ V.

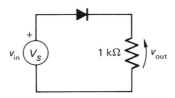

FIGURE P5·5

5.6 In the simple flashlight circuit shown earlier in Fig. 5.6, corrosion has recently occurred at the battery contact marked D so that a resistance $R_c = 10{,}000\ \Omega$ is present at that contact. Under this condition, what is the voltage v_R across the light bulb? Do you think that this will affect the operation of the flashlight?

5·3 APPLICATIONS OF VARIABLE RESISTANCE

An important method for obtaining or sensing information with electric circuitry is to utilize changes in resistance. Based on this approach, a complex electronic circuit can be given instructions or adjusted; measurements of temperature, force, pressure, etc., can be continuously made and displayed; and electronic circuits can be used to control processes or objects where direct control by a human operator would be too tedious or inadequate.

A potentiometer is a simple but important example of the use of adjustable resistance. Figure 5.15a shows a schematic drawing of a basic potentiometer circuit. We see a battery applying a voltage v to a resistance R_T and an arrow pointing to a particular location on the resistance. This is intended to represent the operation of many actual potentiometers where the overall resistance is a tight coil of resistive wire (as indicated by the zig-zag line which represents resistance). A sliding contact moves along the coil, making electric contact at any selected point in the same way that you could imagine the arrow in Fig. 5.15a moving up and down on the resistor. There are three electric contacts on a potentiometer, one at each end of the resistor and the third at the sliding contact which we have represented with the arrow.

In Fig. 5.15a the potentiometer is producing an output voltage v_o which will vary from 0 to v depending on the position of the slider. The equivalent circuit of Fig. 5.15b illustrates this in a way that we can analyze. We pretend that the potentiometer is broken into two separate resistances R_1 and R_2 corresponding to the portions above and below the sliding contact. Of course, the total resistance $R_T = R_1 + R_2$ and the output voltage from this circuit is the voltage drop across the resistor R_2. We assume now that the current i is the same through R_1 and R_2 and that no significant amount of current flows out of the circuit through the sliding contact. The current is computed based on the total potentiometer resistance as

$$i = \frac{v}{R_T}$$

The voltage drop across R_2 is then

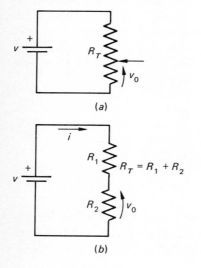

FIGURE 5·15
Operation of a potentiometer.

$$v_2 = R_2 i = R_2 \frac{v}{R_T}$$

and since the output voltage $v_o = v_2$,

$$v_o = \frac{v}{R_T} R_2 \tag{5.14}$$

Because the battery voltage v and total resistance R_T are fixed for this circuit, the output is simply proportional to R_2, as indicated by Eq. (5.14). This resistance R_2 changes with the position of the sliding contact. For rotational potentiometers as on many radios, for example, this resistance change occurs with the turn of a knob. For rectilinear potentiometers which have become more popular on electronic devices, the sliding contact actually moves in a straight line much as is indicated by the drawing of the potentiometer in Fig. 5.15a.

If a potentiometer has been wound uniformly or has been otherwise constructed to have a linear relation between slide-wire position and resistance, a simple circuit such as Fig. 5.15a can be used to provide relatively accurate measurement of position. For example, consider our potentiometer to be 6 in long and wound uniformly, with a 6-V battery supplying the circuit. Positioning the sliding contact at the end of the resistor where the negative battery connection is made will produce $v_o = 0$. Moving in 1 in from the end, we find $v_o = 1$ V; moving in 2 in, $v_o = 2$ V; at 3.2 in, $v_o = 3.2$ V; and so on until at 6 in, $v_o = 6$ V. This type of device is frequently used for the electronic control of position where the voltage v_o can be used to tell an electronic circuit the position of a machine tool or ship rudder, etc. The same circuit may also be used in scientific or engineering experiments where the voltage proportional to position can be read on a voltmeter or by computer or can be recorded electronically much more easily than an experimenter could get out a ruler and measure the position of interest.

In audio equipment, a nonlinear relationship between slide-wire position and resistance is frequently used in potentiometers. Called an *audio taper*, it is introduced to correct for the nonlinearity of the human ear. As you adjust the volume of a radio, you are rotating a knob connected to a potentiometer, which normally has an audio taper. With this type of potentiometer, the nonlinearity of the potentiometer is designed to cancel the nonlinearity of the human ear, so that the volume which we hear is more linearly related to the rotation of the knob.

Many other devices utilize the variation of resistance in their operation. The carbon microphone is an example of small pressure variations in air produced by audible sounds being con-

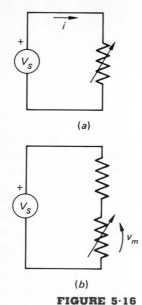

FIGURE 5·16
Simple microphone circuits.

verted into electric signals by variation of resistance. This is accomplished by either of the circuits of Fig. 5.16, where the varying resistance is shown with an arrow through it. The carbon microphone contains packed carbon particles; its electric resistance depends on pressure. As acoustic pressure variations occur because of sound waves and the resistance of the packed carbon fluctuates, the circuit of Fig. 5.16a produces current variations, since a constant voltage is supplied. The circuit of Fig. 5.16b with a constant voltage and a fixed series resistance produces fluctuations in the voltage v_m as the microphone resistance changes. Either the fluctuating current or voltage can be processed by additional electric systems to transmit and eventually reproduce the original sound. Generating the electric signal by resistance variation, however, can be the first step in the process.

Additional examples of devices that depend on variable resistance for their operation include thermistors. With a resistance dependent on temperature, they can be used in circuits similar to that shown in Fig. 5.16 to provide a current or voltage proportional to temperature. Also, the photoresistive cell commonly used today in the exposure meters of quality cameras has a resistance which depends on incident light. A circuit similar to that in Fig. 5.16a may be used where the small battery inserted into the camera provides the voltage source. A current meter is included directly in the circuit and its reading is proportional to the incident light. To guide the user in making the camera settings or to make them automatically, the speed of the film, the lens opening, and the exposure time are all considered in combination with the incident light through additional electric circuitry or mechanical components. Some cameras with light meters may not require a battery for their operation, because they contain photovoltaic cells which generate a voltage proportional to incident light; however, the photoresistive type is more useful at lower light levels and more compact.

PROBLEMS

5.7 In Fig. P5.7, $R_T = 10$ kΩ is the total resistance of a potentiometer, R_1 is the resistance between point A and the sliding contact, and R_2 is the resistance between the sliding contact and point B. A load resistance R_L is shown which is infinite in an ideal potentiometer circuit. Analyze the circuit and perform calculations so that you can plot a graph of v_o vs. R_2 for a range of R_2 from zero to R_T. Consider two cases: (a) $R_L = \infty$ and (b) $R_L = 10$ kΩ. Compare your results with Eq. (5.14).

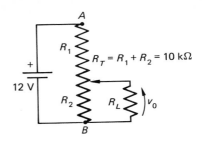

FIGURE P5·7

5.8 A potentiometer similar to that shown in Fig. P5.7 is being used to measure position electrically. The distance from A to B is 5 in and the potentiometer operates ideally ($R_L = \infty$). What value of battery voltage is needed to have v_o register 2 V/in of slider motion?

5.9 Consider the simple camera-light-meter circuit in Fig. P5.9. The ammeter measures the circuit current i and its needle is seen through the viewfinder on the camera. In a bright sun the resistance R_{pc} of the photoresistive cell is 500 Ω and in shade the resistance $R_{pc} = 5000$ Ω. Calculate the current i in both cases.

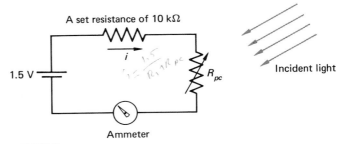

FIGURE P5·9

5·4 ELECTROMECHANICAL DEVICES

The electrical production of mechanical motion and the mechanical production of electric current form the basis for much of our modern technology. These capabilities result directly from electromagnetic principles and make possible electric generators, motors, relays, and the many machines and tools of our society which are based on these devices. There are three electromagnetic principles of primary importance in understanding the operation of electromechanical devices. The principles are summarized very briefly below. (Keep in mind that these are complex topics about which chapters and books have been written.)

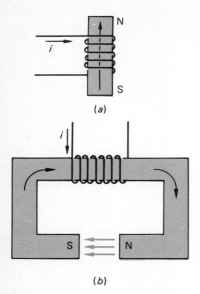

FIGURE 5·17
Electric currents create magnetic fields.

1. *Current flowing through any electrical conductor creates a magnetic field.* The magnetic field may be enhanced by winding or coiling the conductor around a core of ferromagnetic material, such as iron, cobalt, or nickle, which is particularly susceptible to magnetization. This is normally done in electromagnets, two typical configurations of which are shown in Fig. 5.17. The direction of the magnetic field produced by the current in the coil is given by the "right-hand rule." By coiling the fingers of your right hand around the magnetic core in the direction of current flow, your thumb points in the direction of the resulting magnetic field or flux and toward the north pole of the magnet. For the configuration of Fig. 5.17b, the magnetic flux in the air gap is directed from the north pole to the south pole. The strength of the magnetic field is proportional to the current, the number of turns in the coil, and that property of the core called the magnetic permeability.

2. *When an electrical conductor placed across a magnetic field is moved through the magnetic field, a voltage difference is produced between the ends of the conductor.* The voltage is proportional to the strength of the magnetic field and the magnitude of the component of velocity across the field. Velocity exactly in the direction of the magnetic flux does not produce voltage. For the conductor shown in Fig. 5.18, the voltage produced is

$$v = BlV \sin \theta \qquad (5.15)$$

where B = density of magnetic flux, Wb/m²
l = length of conductor in magnetic field, m
V = velocity of the conductor, m/s
θ = angle between directions of velocity and magnetic flux

The direction of the generated voltage may be established by another right-hand rule. Position the right hand now so that your fingers coil from the arrow representing velocity toward the direction of the arrows representing magnetic flux. Your thumb will point toward the end of the conductor, which will be positive if the generated voltage difference is measured.

3. *When an electric current passes through a conductor placed across a magnetic field, a force is produced which acts on the conductor.* The magnitude of this force is given by

$$F = Bli \qquad (5.16)$$

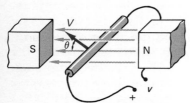

FIGURE 5·18
Motion of a conductor in a magnetic field.

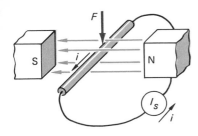

FIGURE 5·19
Force produced by a magnetic field.

where F = force, N
B = flux density, Wb/m^2
l = length of conductor in field, m
i = current, A

The direction of the force, perpendicular to both the conductor and the flux, is again given by a right-hand rule. This time the fingers are coiled from the direction of the current toward the direction of the flux and the thumb points in the direction of the produced force. Figure 5.19 shows an example where a current source is applied to a conductor in a magnetic field. Of course, we do not have equilibrium as shown; the produced force will immediately drive the conductor out of the field.

Remember now that the descriptions above are brief and that further details on these effects and why they occur should be learned from a more complete study of the physics (see, for example, Refs. 1 and 2). However, by keeping these three principles in mind, we can explore the operation of several important devices—galvanometers, ammeters, voltmeters, and electric motors.

THE GALVANOMETER

We will consider first the galvanometer, an important instrument which is the basis of operation for most ammeters and voltmeters used in general-purpose applications. An overview of a galvanometer is given in Fig. 5.20, where we see that the important components are a permanent magnet frame, a rotating coil, and a torsional spring. The coil is driven by a current whose magnitude we would like to measure. A force acts on the coil because it is a conductor carrying current while in a magnetic field. The coil and the pointer fastened to it rotate under the action of the induced force until an equilibrium position is reached where the spring provides a moment which balances that produced by the current. The position of the pointer is then a measure of the current. The scale shown in the figure can be labeled accordingly to indicate the magnitude of the current being measured.

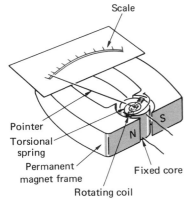

FIGURE 5·20
A galvanometer.

Let's take a closer look at the operation of the galvanometer, making use of Fig. 5.21, which shows a top and front view of the operating parts. A fixed iron core is used in the galvanometer to shape the magnetic field, making it radial (with respect to the center of rotation of the coil) in the gap between the fixed core and the permanent magnet. This makes the electromagnetic force produced on the coil always act tangentially so that the moment created about the coil's center of rotation will not depend on coil position in the range of operation. The electromagnetic force is

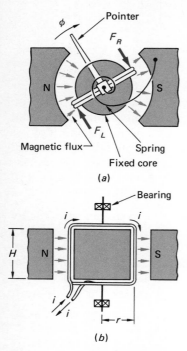

FIGURE 5·21
Top and front views of the galvanometer.

produced mostly on the vertical portions of the coil shown in Fig. 5.21b. Consider first the left side of the coil where current is flowing vertically upward. Applying the right-hand rule tells us that the force acting on this portion of the coil acts into the surface of the paper and is as shown by F_L in Fig. 5.21a. For the vertical portion of the coil to the right in Fig. 5.21b, the current is flowing downward, and application of the right-hand rule tells us that the force here acts out of the surface of the paper. This force is shown in Fig. 5.21a as F_R.

The magnitudes of the forces F_R and F_L are equal, since the current and magnetic field are the same on both sides of the core. Each force produces a moment about the axis of rotation of the coil. This axis has been carefully mounted in bearings to rotate easily. The total moment produced by these electromagnetic effects is

$$M_{em} = rF_L + rF_R = 2\,rF \tag{5.17}$$

where r is the radius of the coil and $F = F_L = F_R$ is the magnitude of each of the forces. If this moment is not counteracted by an equal and opposite moment, the coil will not be in equilibrium. The spring that provides this moment is called a torsional spring because it provides a moment, or "torque," about a particular axis of rotation. Like all springs, it inherently produces a reaction directly proportional to its deformation. The reaction here is a torque or moment about the axis of rotation of the coil and the deformation is the angle of rotation of the coil ϕ. The moment produced by the spring is ideally

$$M_s = k\,\phi \tag{5.18}$$

where k is a constant and is referred to as the spring stiffness. Typical units for k might be newton-meters per radian or foot-pounds per degree.

The requirement for the coil to be in mechanical equilibrium is that the sum of the moments about the axis of rotation be zero. Thus, with the electromagnetic moment acting clockwise and the spring moment counterclockwise

$$M_s - M_{em} = 0$$

is required where the minus sign is introduced because M_{em} is clockwise while M_s is counterclockwise. Substituting Eqs. (5.17) and (5.18) above, we find

$$k\,\phi - 2\,rF = 0$$

and remembering Eq. (5.16), we obtain

$$k\phi - 2rBli = 0 \qquad (5.19)$$

The length l of the conductor within the field depends on the magnet height H in Fig. 5.21 and the number of wrappings n on the coil, so that $l = nH$. We substitute this into Eq. (5.19) and rearrange the equation to find the position of the pointer as

$$\phi = \frac{2\,rBnH}{k}\,i \qquad (5.20)$$

After the galvanometer has been constructed, coil radius r, number of wrappings n, height H, magnetic field B, and spring stiffness k are all fixed. Only the current to the galvanometer changes. If we let a constant $C = 2rBnH/k$, we can write Eq. (5.20) as

$$\phi = Ci \qquad (5.21)$$

and we can easily see the linear proportionality between the applied current and the resulting angular position of the pointer in a working galvanometer. This is an important aspect of this instrument as well as most scientific instruments, since scales can be conveniently labeled and calibrated, and motion of the pointer corresponds directly to changes in the current being measured. The designer of the galvanometer has the task of providing smooth, sturdy bearings to support the coil; a rugged, flexible spring; a compact size; and values of B, n, H, and k that produce a proportionality constant C which is useful for the intended application.

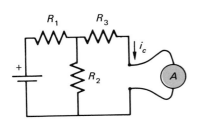

FIGURE 5·22

Measuring current with an ammeter.

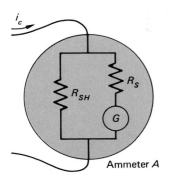

FIGURE 5·23

An ammeter contains a galvanometer and well-chosen resistors.

Ammeter A

AMMETERS

An ammeter is used to measure electric current. Since a galvanometer inherently measures electric current, as we have just discussed, it could be used directly as an ammeter. However, it would have limitations, since it works properly only for currents of a particular magnitude which produce a convenient deflection according to Eq. (5.21). It is also desirable for an ammeter to have a very low resistance, since it must be inserted in series in a circuit, as shown in Fig. 5.22, in order to measure a particular current. If the ammeter has a large resistance, it will change the character of the circuit when installed in series and distort the current reading it makes. The resistances shown in Fig. 5.23 can be helpful in making a galvanometer a more useful ammeter. The resistance

R_{SH} is called a shunt resistance, it causes some current in the circuit being measured to bypass the galvanometer (G) and controls the overall resistance of the ammeter. Since R_{SH} is in parallel with R_S and the galvanometer, the apparent resistance of the ammeter is always at least as low as R_{SH}. Normally R_{SH} can be made sufficiently low so that the ammeter does not disturb the operation of circuits when current measurements are being made.

With a little analysis of the circuit in Fig. 5.23, we can find that the current through the galvanometer is

$$i = \frac{R_{SH}}{R_S + R_G + R_{SH}} i_c \qquad (5.22)$$

where i_c is the current being measured in the circuit and R_G is the resistance of the galvanometer itself. Recalling Eq. (5.21) and substituting the expression above for i, we find the pointer angle

$$\phi = C \frac{R_{SH}}{R_S + R_G + R_{SH}} i_c \qquad (5.23)$$

in terms of the actual circuit current. Note now that adjustments in R_S and R_{SH} will change the amount of pointer motion for a given circuit current. This is the way changes in scale can be made on an ammeter.

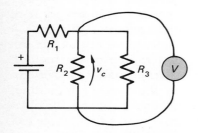

FIGURE 5·24
Making a voltage measurement.

VOLTMETERS

Let's now look at the way a galvanometer may be used in the measurement of voltage. Figure 5.24 shows a voltmeter being used to measure the voltage across resistors R_2 and R_3 connected in parallel. A closer look at the voltmeter in Fig. 5.25 shows it to be a galvanometer with a series resistance R_S. This resistance is kept as high as possible because the voltmeter must be inserted in parallel with circuit elements to make a voltage measurement as shown in Fig. 5.24. With too low a resitance, it may disturb the circuit and distort the measurement. A simple analysis of Fig. 5.25 shows us that the current through the galvanometer is

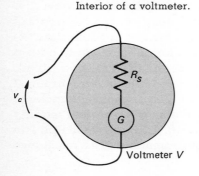

FIGURE 5·25
Interior of a voltmeter.

$$i = \frac{1}{R_S + R_G} v_c \qquad (5.24)$$

where v_c is the circuit voltage being measured and R_G is again the galvanometer resistance, normally much less than R_S. The pointer deflection will be

$$\phi = C \frac{1}{R_S + R_G} v_c \qquad (5.25)$$

which is proportional to the voltage being measured. The amount of deflection for a given voltage can be increased by reducing R_S. However, with lower values for R_S the voltmeter may tend to disturb the circuit it is measuring. The quality of a voltmeter is frequently indicated by a statement of resistance per volt which tells us the value of R_S when the meter is fully deflected with a 1-V input. The more sensitive the galvanometer is originally, the higher the resistance per volt for the voltmeter.

By adding electronic components and supplying a source of electric power, special voltmeters can be constructed to have a very high resistance. These are frequently used for applications in electronic circuits. It is also possible with the addition of a power source to construct a galvanometer circuit, called an *ohmmeter*, which directly measures electric resistance. In describing the measurement of voltage and current, we have focused on dc (direct-current) measurements where the voltage or current of interest is steady and constant as from a battery. In ac, or alternating-current, circuits the current continually changes direction; for commercial electric power in the United States, this change of direction occurs 60 times per second. Galvanometers can be applied in the measurement of ac voltage or current by adding an electronic component called a rectifier to the basic ammeter and voltmeter circuits.

ELECTRIC MOTORS

Electric motors are devices which we all have contact with on a daily basis for a variety of applications. Motors are electromechanical and convert electric power into mechanical power. We will now discuss the operation of a simple dc motor based on the three electromagnetic principles described earlier.

Consider Fig. 5.26a, which shows a single conducting coil supported by bearings so that it can rotate freely. A magnetic field formed by either an electromagnet or a permanent magnet is positioned so that the coil, called an *armature*, is within the magnetic field and the axis of rotation is perpendicular to the field. A battery may be used to produce a current i in the conducting coil, as shown in Fig. 5.26a. Because we have a conductor carrying current within a magnetic field, forces act on the coil. These forces are shown in Fig. 5.26a as given by the right-hand rule and, for this orientation and current, they produce a clockwise moment which will result in clockwise rotation of the coil or armature. As the coil rotates, it will reach the position shown in Fig. 5.26b where, in addition to entangling the wires, there is a severe

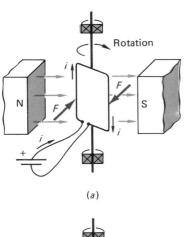
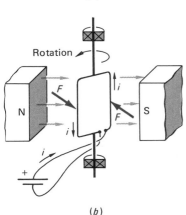

FIGURE 5·26
Principles of a simple motor.

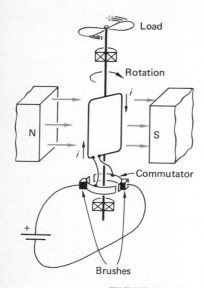

FIGURE 5·27
Proper rotation results with added brushes and commutator.

problem. At this point, application of the right-hand rule shows that the forces acting on the coil are now in the wrong direction. They will not continue the clockwise rotation which just began but instead produce a counterclockwise moment. It is necessary to switch the direction of the current in the coil in order for the motor to operate properly. Figure 5.27 shows an arrangement which automatically provides this switching function as the shaft rotates. The leads from the coil are attached to separate conductors mounted on the shaft but electrically insulated from the shaft and from each other in an assembly called a *commutator*. The motor "brushes" ride on the commutator as the shaft rotates and electrically connect the leads from the voltage source to the appropriate portion of the coil. The direction of the current in the coil is switched by the commutator during rotation to keep the electromagnetic torque acting in the direction of rotation. In Fig. 5.27, clockwise rotation will be maintained. In real motors, many coils are normally used and correspondingly the commutator consists of many separate segments. This provides a relatively uniform torque during rotation and a smooth operation results.

Let's consider the question of what brings an operating dc motor, driving a load, to a steady-state, or equilibrium, speed of rotation. Briefly, there are two basic reasons: the torque or moment available from the motor decreases as the speed increases and also the torque required to rotate the load increases as the speed increases. The steady-state rotational speed is where the torque available from the motor equals the torque needed to drive the load. The characteristics of both a motor and load are normally presented as torque-speed curves, as shown in Fig. 5.28. Here torque T is plotted against rotational speed symbolized by ω (the lowercase form of the Greek letter "omega"). In Fig. 5.28a, we have a typical curve for a motor. At zero speed, maximum torque is available, corresponding to the current produced with full applied voltage acting across the armature. At higher speeds, the

FIGURE 5·28
Motor and load characteristics.

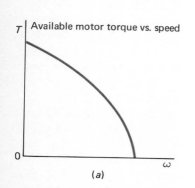

(a)

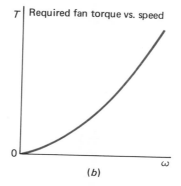

(b)

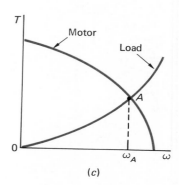

(c)

torque produced by electromagnetic action is reduced because the motion of the armature coils within the magnetic field creates a voltage opposing the applied voltage. This reduces the current in the armature coils and, in turn, the torque. Also, since the armature is spinning faster with higher shaft speed, more torque is required to keep it moving against friction with the air surrounding it and this subtracts from the torque available to drive a load. Figure 5.27 shows a fan blade attached to the motor shaft, which represents a load on our motor. If we measure the torque required to rotate this fan at various speeds, we obtain a plot similar to Fig. 5.28b. Here, at zero speed, no torque is required, but more torque is required at an increasing rate as the speed is increased. To find the actual equilibrium speed in a motor and load combination, we plot both characteristic curves on the same set of axes, as shown in Fig. 5.28c. Here, equilibrium is located at point A, where the curves intersect and the motor and load torques are equal. The equilibrium speed is ω_A. Note the inherent balance present. If the speed were to become slightly greater than ω_A, the load torque would be greater than the motor torque and would tend to slow down the speed. If the speed decreases to less than ω_A, the motor torque becomes greater than the load torque and will increase the speed toward ω_A.

PROBLEMS

5.10 In a certain galvanometer, a current of 0.10 A is known to produce 1.0 rad of needle motion. What is the sensitivity constant C in Eq. (5.21)?

5.11 A galvanometer has a magnet height $H = 1.0$ cm, a coil radius $r = 1.0$ cm, and 25 turns on the coil. The torsional spring has a stiffness $k = 0.025$ N·m/rad. If this galvanometer has a sensitivity constant $C = 10$ rad/A, what is the strength B of the magnetic flux in Wb/m²?

5.12 A parallel electric circuit is shown in Fig. P5.12a. Determine the current i_2 passing through R_2. In Fig. P5.12b an ammeter

FIGURE P5·12

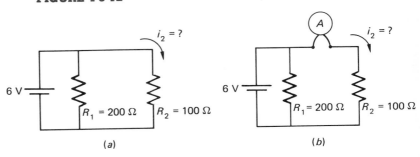

has been inserted to measure i_2. The ammeter has the internal arrangement of Fig. 5.23 with $R_{SH} = 1.0 \; \Omega$, $R_S = 0$, and $R_G = 10 \; \Omega$. What is the current i_2 now passing through R_2 and the ammeter? Any change is caused by the presence of the ammeter in the circuit.

5.13 Figure P5.13a shows the torque-speed curve for a certain small electric motor. If this motor is used to drive a blower which has a torque-speed curve as shown in Fig. P5.13b, what will be the operating speed when the two are connected together?

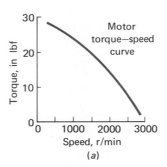
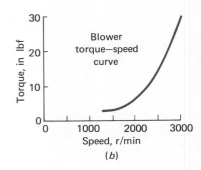

FIGURE P5.13

5.5 REMARKS

The topics covered in this chapter are very basic subjects for the practicing electrical engineer. However, they are also important on a daily basis to others such as mechanical, chemical, and civil engineers who happen to be involved with instrumentation or moving parts operated by electric motors or controlled by electronic circuits.

In the study of electric circuits and electromagnetic devices, the next step beyond the material presented here is physics for a more detailed look at the nature of current flow and electromagnetics. References 1 and 2 are representative of college-level physics books. For study of more complex electric circuits, where complicated arrangements of resistors, capacitors, and inductors are analyzed using special mathematical tools based on algebra and calculus, see Refs. 3 and 4. Then discussion of electronic devices and electronic circuits can be found in some typical textbooks such as Refs. 5 and 6, which will give you an idea of coursework in electronics. The operation and characteristics of electric motors and generators are covered as a separate subject normally known as electrical machinery and involving all types of ac and dc electric equipment (see Refs. 7 and 8).

REFERENCES

1. Resnick, R. and D. Halliday, *Physics*, Wiley, New York, 1977.
2. Bueche, F. J., *Introduction to Physics for Scientists and Engineers*, McGraw-Hill, New York, 1980.
3. Churchman, L. W., *Circuits: A First Course*, Holt, New York, 1976.
4. Budak, A., *Circuit Theory Fundamentals and Applications*, Prentice-Hall, Englewood Cliffs, NJ, 1978.
5. Ryder, J. D., *Electronic Fundamentals and Applications: Integrated and Discrete Systems*, Prentice-Hall, Englewood Cliffs, NJ, 1976.
6. Millman, J., and C. Halkias, *Electronic Fundamentals and Applications for Engineers and Scientists*, McGraw-Hill, NY, 1975.
7. Fitzgerald, A. E., C. Kingsley, and A. Kusko, *Electrical Machinery*, McGraw-Hill, New York, 1971.
8. DelToro, V., *Electromechanical Devices for Energy Conversion and Control Systems*, Prentice-Hall, Englewood Cliffs, NJ, 1968.

6 THE THERMAL-FLUID SCIENCES

The area of engineering study frequently called the thermal-fluid sciences includes the many aspects of fluid mechanics, thermodynamics, and heat transfer. These subjects are especially significant today because they provide the basis for understanding the energy problems and environmental problems of our modern world.

Fluid mechanics, thermodynamics, and heat transfer are generally of more importance to practicing engineers in the fields of aerospace, mechanical, and chemical engineering, because they are concerned with flows around vehicles and in pipes and tubes as well as the thermal processes found in engines, chemical plants, refineries, and electric power plants. Other types of engineers also involved with the thermal-fluid sciences include petroleum and textile engineers and also civil engineers, for whom drainage, channel flows, dams, and waste-disposal systems are important applications. Electrical engineers do not often develop detailed expertise in the thermal-fluid areas but are frequently involved with engineering teams in studying such problems, because the operation of electric heating systems, the cooling of electric components, and the actual production of electric power itself depend importantly on thermal-fluid phenomena.

We will discuss here a few topics in the thermal-fluid sciences to introduce you to the nature of the subject matter and to some of its more significant concepts.

6·1 MECHANICS OF FLUIDS

The term *fluid* is used in engineering to describe both liquids and gases. A major difference between a fluid and solid is that a fluid will always deform to take on a shape corresponding to its container, whereas a solid has its own shape. A gas will usually expand to occupy a containing vessel completely, whereas a liquid will not expand but will occupy only the lower portion of an oversized container in normal gravitational conditions.

Particles of fluid must behave according to the same physical laws of mechanics, including statics, that affect solid objects. The application of these laws, and others, to the study of fluids at equilibrium and in motion is known as *fluid mechanics*. In fluid mechanics the nature of fluid motion, as well as the pressures and forces acting on the fluid or produced by the fluid, is of interest.

Looking at Fig. 6.1a, we see a fluid contained in a piston-cylinder assembly with a force F applied to the piston. All the requirements of static equilibrium apply to this problem. In order for the piston to be in equilibrium, the sum of the vertical, or y-direction, forces acting on it must be zero. In Fig. 6.1b, the force F_f is the fluid force acting on the piston, so we easily sum forces to find $F_f - F = 0$ or $F_f = F$. Thus the fluid must supply a force F_f equal and opposite to the applied force F. The pressure of the fluid being compressed supplies this force, which is not actually applied at a point but is distributed over the entire fluid side of the piston, as shown in Fig. 6.1c. The pressure P_f is the force applied by the fluid per unit area with typical units of pounds-force per square foot, pounds-force per square inch, or newtons per square meter (the pascal). Of course, the total force applied by the fluid to the piston is $F_f = P_f A$, where A is the piston area. For the piston to be in equilibrium, we must have $P_f A = F$, so the fluid pressure will be $P_f = F/A$.

We have computed the pressure P_f without reference to the surrounding atmosphere; this result is called a *gage pressure*, since it is the pressure which would be most directly measured by a pressure gage connected to the cylinder. To find the absolute fluid pressure P_{fa} inside the cylinder we must now consider the atmospheric pressure P_{at} acting on the outer side of the piston. The free-body diagram of Fig. 6.1d actually shows the complete story. There, summing the vertical forces, we find

$$P_{fa}A - F - P_{at}A = 0 \qquad (6.1)$$

and solving this equation for the absolute fluid pressure yields

$$P_{fa} = \frac{F}{A} + P_{at}$$

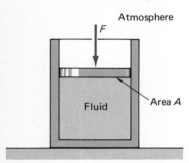
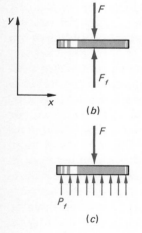
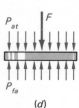

FIGURE 6·1 Pressure in a cylinder.

or

$$P_{fa} = P_f + P_{at}$$

Thus, the atmospheric pressure simply adds to the original result, so the gage pressure is simply

$$P_f = P_{fa} - P_{at} \qquad (6.2)$$

It is this difference between absolute and atmospheric pressure that is conveniently measured by pressure gages.

As a numerical example, consider the force on the piston $F = 1000$ lbf, the area of the piston $A = 10$ in^2, and atmospheric pressure $P_{at} = 14.7$ lbf/in^2 absolute. The resulting pressure of the fluid is then $P_f = 100$ lbf/in^2 gage or, in absolute terms, it is $P_{fa} = 114.7$ lbf/in^2 absolute.

PRESSURE IN FLUID COLUMNS

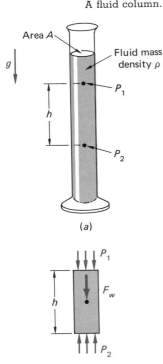

FIGURE 6·2
A fluid column.

In the previous discussion, we did not consider the density or mass of the fluid and the effect that it may have on pressure. This is typical in piston-cylinder problems, since the pressures are normally high and pressure variations due to fluid density are insignificant. However, there are many situations where existing pressure is created predominantly by fluid density. This is the case in the earth's atmosphere, in lakes, and in oceans. We can examine the pressure changes caused by fluid density by inspection of the column of liquid shown in Fig. 6.2a. The mass density of the fluid is ρ and the column is standing upright with gravity g. We would like to find the relationship between P_1 and P_2, the pressures at two levels separated by the vertical distance h. In Fig. 6.2b we have isolated the column of fluid between levels 1 and 2 and have indicated the vertical forces which act on that fluid. The forces must sum to zero for the fluid column to be in equilibrium. The pressure force at the top is P_1A and at the bottom it is P_2A. Note that pressure does not depend on direction; it may produce a force in any direction. The force of interest at the top of Fig. 6.2b thus results from P_1 acting downward and the force of interest at the bottom results from P_2 acting upward. The fluid mass is ρAh and the body force or weight of the fluid in the small column is $F_w = \rho g A h$. We can now sum the vertical forces to obtain

$$P_2 A - \rho g A h - P_1 A = 0$$

Area A may be canceled from this equation, so we may rearrange it to obtain

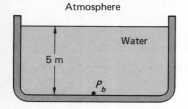

FIGURE 6·3
Pressure in a water tank.

$$P_2 - P_1 = \rho g h \qquad (6.3)$$

This is a very important and useful expression for the pressure difference between two fluid levels. The only important restriction is that ρ be the same at all points between the two levels. This often prevents us from using Eq. (6.3) for gases where density depends considerably on temperature and pressure. For example, Eq. (6.3) cannot be applied directly over substantial changes of altitude in the atmosphere. It works well, however, for nearly all liquid applications.

Consider the pool of water shown in Fig. 6.3. We would like to determine the pressure P_b at the bottom of this 5-m-deep pool. The pressure at the surface of the water is $P_{at} = 1.01 \times 10^5$ Pa absolute. Using Eq. (6.3) with $P_2 = P_b$, $P_1 = P_{at}$, $g = 9.81$ m/s², and a water density $\rho = 1000$ kg/m³, we find that

$$P_b - P_{at} = (1000)(9.81)(5)$$
$$P_b = P_{at} + 49{,}000$$
$$= 150{,}000 \text{ Pa absolute}$$

which is the absolute pressure at the pool bottom. A pressure increase of nearly 50 percent thus occurs between the atmosphere and the bottom of the tank. This is, of course, the reason you feel pressure in your ears when you dive into deep water.

MANOMETERS

A manometer is a convenient and accurate way for measuring pressure. Manometers are very often used in engineering laboratories and engineering applications. In many European service stations, a type of mercury manometer is the normal instrument for measuring tire pressure. Figure 6.4a shows a representation of

FIGURE 6·4
Operation of a manometer.

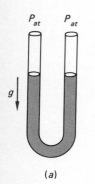
(a)

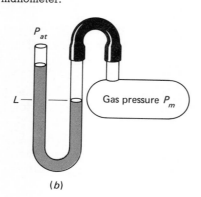
(b)

(c)

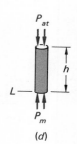
(d)

230 THE THERMAL-FLUID SCIENCES

a U-tube manometer where both legs are simply opened to the air of the atmosphere. The liquid levels in each of the legs are at even heights and the pressure above each liquid is atmospheric P_{at}. Now we fit a tube over the right leg of the U, connecting it to a gas at pressure P_m, which is the pressure to be measured. This applies P_m to the fluid on the right-hand side of the manometer, forcing it downward, and the fluid on the left-hand side is forced upward, as we see in Fig. 6.4b.

Figure 6.4c focuses on the lower section of the manometer below the level L indicated in Fig. 6.4b and at the level where the fluid surface in the right leg is now located. The manometer is at equilibrium, so the pressure within the liquid at level L on the left-hand side of the manometer must become equal to P_m as applied on the right, as shown in Figure 6.4c. Otherwise, the fluid will continue to rise on the left side of the tube. The increased pressure at level L, necessary for equilibrium, is created by the column of liquid above this level on the left side. Figure 6.4d shows the forces on this upper portion of the liquid. Pressure P_m is shown pushing upward here, and Eq. (6.3) can be applied to determine its magnitude, since the height h in Eq. (6.3) is easily measured. This equation shows

$$P_m - P_{at} = \rho g h \qquad (6.4)$$

If the fluid is mercury with $\rho = 26.2$ slugs/ft^3, gravity is standard at $g = 32.2$ ft/s^2 and a 3-in difference occurs in the fluid levels, $h = 3.000/12 = 0.250$ ft. Then

$$P_m - P_{at} = (26.2)(32.2)(0.250)$$
$$= 211 \text{ lbf/ft}^2$$

or
$$P_m - P_{at} = 1.47 \text{ lbf/in}^2$$

As written above, the pressure we measure is 1.47 lbf/in^2 gage. If atmospheric pressure is 14.70 lbf/in^2 absolute, the pressure $P_m = 16.17$ lbf/in^2 absolute.

Note that manometers are used so commonly that often the conversion to normal pressure units is not applied and the pressure is referred to only in terms of the liquid height h. For example, inches of water or centimeters of mercury are frequently used as convenient measures of pressure. These can, of course, be quickly converted to normal units by multiplying h times ρg.

LIQUID FLOW IN PIPES

The flow of liquids in pipes or tubes is something we experience every day, the most common example, of course, being when we

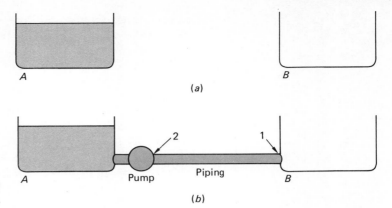

FIGURE 6·5
Pumping a liquid.

turn on a faucet to let water flow from the tap. We will try now to give you some insight into the way that liquid flows through pipes and also into the types of calculations that can be applied to describe the flow. Our discussion uses concepts of dimensional analysis given in Sec. 2.2.

Let's begin with the basic problem shown in Fig. 6.5a. We have some liquid in container A and we would like to move it to container B. It may be easy to suggest picking up container A and pouring the liquid into B, but this is often not possible, particularly if the container is 100 ft in diameter and holds a few million gallons of oil, for example. So, to move the liquid from A to B we install a pump and piping between the two containers. In order to answer questions related to the size of the pump, the pipe diameter, and the power for the pumping process, we must understand the general nature of the liquid flow within the pipe. A main point to be remembered is that because of friction between the liquid, which is moving, and the pipe, which is fixed, the liquid must be forced through the pipe. In Fig. 6.5b, the pressure at 2, the outlet of the pump, is larger than the pressure at 1, the inlet to tank B. This difference in pressure ($\Delta P = P_2 - P_1$) produces the force which pushes the liquid through the pipe. The liquid will reach an equilibrium, or steady, flow rate when the force required to move the liquid through the pipe exactly equals the force produced by the pressure difference. We would now like to know how large this pressure difference must be.

Consider first the various quantities which might influence the required pressure difference ΔP. The length of the pipe l, as well as the pipe's inside diameter, is certainly important. The average velocity at which the fluid moves through the pipe, as well as the properties of the fluid (the mass density ρ and the viscosity μ which measures the "thickness," or "gooiness," of the fluid), is significant. Densities and viscosities of various fluids are shown in

Table 6.1. We will assume that our pipe is smooth on the inside so that surface roughness need not be considered.

It is possible to develop an equation for pressure difference in terms of the quantities mentioned above by applying methods of mathematics and physics. That is, an algebraic description of $\Delta P = f(l, d, V, \rho, \mu)$ can be derived. However, this equation has only limited use. The most important pipe-flow results are experimental, based on studying pressure differences for different velocities of fluid flow, in different pipes and with different fluids. One experimental result, which simplifies the problem substantially, is that for long pipes the pressure difference is easily found to depend linearly on the pipe length. For example, with a flowing fluid, pressure measurements across 100 m of a pipe show twice the pressure difference as measured across the first 50 m of the pipe. Since we know that pressure changes linearly with pipe length, only the rate of change or slope $s = \Delta P/l$ is an important quantity and we are no longer interested in either ΔP or l alone. Experiments based on dimensional analysis have been used to find $s = f(d, V, \rho, \mu)$.

TABLE 6·1
FLUID DENSITIES AND VISCOSITIES

Fluid	Mass Density, kg/m³	Viscosity, N·s/m²
Mercury:		
0°C	1.36×10^4	1.7×10^{-3}
40°C	1.35×10^4	1.5×10^{-3}
Water:		
0°C	1.00×10^3	1.9×10^{-3}
40°C	9.90×10^2	6.5×10^{-4}
Oil (SAE 30):		
0°C	8.80×10^2	2.0
40°C	8.70×10^2	0.10
Carbon dioxide:		
Atmospheric pressure:		
0°C	1.96	1.4×10^{-5}
40°C	1.71	1.5×10^{-5}
Air:		
Atmospheric pressure:		
0°C	1.29	1.7×10^{-5}
40°C	1.12	1.8×10^{-5}
Helium:		
Atmospheric pressure:		
0°C	0.18	1.8×10^{-5}
40°C	0.15	1.9×10^{-5}

Note: 1 kg/m³ = 1.94×10^{-3} slugs/ft³
1 N·s/m² = 2.09×10^{-2} lbf·s/ft²
1 N·s/m² = 1 kg/m·s
1 lbf·s/ft² = 1 slug/ft·s

To understand this procedure, recall our previous work in dimensional analysis. For pipeflow, the variables are as follows:

$s = \Delta P/l$ has dimension F/L^3 or M/L^2T^2
d has dimension L
V has dimension L/T
ρ has dimension M/L^3
μ has dimension FT/L^2, or M/LT

Since we have $n = 5$ variables containing $k = 3$ fundamental dimensions and $n - k = 2$ dimensionless groups, the eventual result we expect is $\pi_1 = f(\pi_2)$. The first dimensionless group is referred to as the friction factor and is given the symbol f. It is defined as

$$f = \pi_1 = \frac{sd}{\frac{1}{2}\rho V^2} = \frac{\Delta P d}{\frac{1}{2}\rho V^2 l} \tag{6.5}$$

The second group is the Reynold's number, important for many different fluid flows and defined here as

$$\text{Re} = \pi_2 = \frac{V d \rho}{\mu} \tag{6.6}$$

You should check to see that these are actually dimensionless variables. In f, the one-half is inserted into the denominator because the expression $\frac{1}{2}\rho V^2$ has particular significance in fluid mechanics, representing the kinetic energy of the flowing fluid. The experimental relationship for $f = f(\text{Re})$ is shown graphically in Fig. 6.7. This graph has been verified by many experimenters over the past century using many different pipe sizes and liquids. The graph also applies to gases if the variation in density ρ is not significant within the pipe.

Note, of course, that the graph in Fig. 6.7 is not simply a single smooth curve. Instead, it contains an ill-defined area called the transition region, which separates two distinct curves. At low velocities and therefore low values for the Reynolds number Re, fluid flow, referred to as *laminar*, is smooth and well-behaved. The straight line plotted in Fig. 6.7 applies. At higher velocities and therefore higher values for Re, fluid flow is *turbulent*—not smooth but disordered and random. Turbulent flows are difficult to describe mathematically; however, the experimental curve shown in Fig. 6.7 does apply for a smooth pipe. In the transition region between laminar and turbulent flow in the figure, it is not clear which curve applies, because the flow may be either turbulent or laminar.

Example 6·1

The pump shown in Fig. 6.6 is pumping water through a 6-in-diameter pipe which is 2000 ft long. Assuming a smooth pipe, what must the pressure P_2 be if the flow rate is 2 ft³/s? The viscosity of the water $\mu = 2.52 \times 10^{-5}$ lbf·s/ft² and the mass density $\rho = 1.94$ slugs/ft³.

Solution

First we must determine the average velocity

$$V = \frac{q}{A}$$

for the pipe. With $d = 6$ in $= 0.5$ ft,

$$A = \frac{\pi d^2}{4} = \frac{\pi (0.5)^2}{4} = 0.196 \text{ ft}^2$$

and

$$V = \frac{q}{A} = \frac{2}{0.196} = 10.2 \text{ ft/s}$$

Now, the Reynolds number can be computed as

$$Re = \frac{Vd\rho}{\mu} = \frac{(10.2)(0.5)(1.94)}{2.52 \times 10^{-5}} = 3.93 \times 10^5$$

From Fig. 6.7 we can find for this value of Re that

$$f = \frac{\Delta P d}{\frac{1}{2}\rho V^2 l} = 0.014$$

or

$$\Delta P = 0.014 \frac{1}{2} \frac{\rho V^2 l}{d}$$

$$= 0.014 \frac{1}{2} \frac{(1.94)(10.2)^2(2000)}{0.5}$$

$$= 5700 \text{ lbf/ft}^2 = 39 \text{ lbf/in}^2$$

So the required pressure difference $\Delta P = P_2 - P_1 = 39$ lbf/in². Since the end of the pipe is discharging to the atmosphere $P_1 = P_{atm}$, the pressure $P_2 = 39$ lbf/in² gage or $P_2 = 54$ lbf/in² absolute.

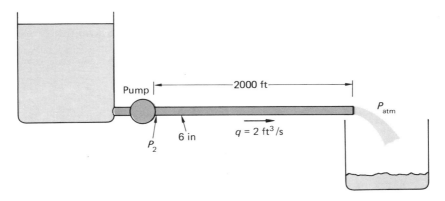

FIGURE 6.6

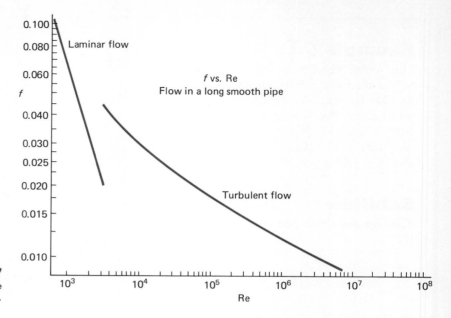

FIGURE 6·7
Friction relationship for pipe flow.

You can get a fairly good feeling for the concepts of laminar and turbulent flow by operating your bathroom faucet. Turn the tap until water flow just begins. It should be smooth and the stream should stay together. This is laminar flow. Increase the flow rate substantially now. The stream will become less smooth and more erratic and tend to disperse. This is turbulent.

We have included Example 6.1 to show how Fig. 6.7 can be applied to the solution of pipe-flow problems. It is important to keep all the units for the various quantities in good order. It is best to simply express every variable in fundamental units, either USCS or SI, and then perform the calculations. Also, note in this example that the pipe's length is long compared with its diameter. Strictly speaking, this is a requirement for the application of Fig. 6.7.

PROBLEMS

6.1 Calculate the gage pressure at the bottom of the tank shown in Fig. P6.1 for each of the cases below. Express your result in appropriate units.
  (a) $\rho = 1000$ kg/m³, $g = 9.81$ m/s², $h = 10$ m
  (b) $\rho = 1.94$ slug/ft³, $g = 32.2$ ft/s², $h = 32.8$ ft
  (c) $\rho = 850$ kg/m³, $g = 32.2$ ft/s², $h = 300$ in
  (d) $\rho = 50$ lbm/ft³, $g = 6.50$ ft/s², $h = 0.99$ m

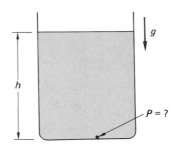

FIGURE P6·1

6.2 A mercury manometer is being used to measure the pressure in an automobile tire. If the manometer reads 115 cm of mercury, what is the tire pressure (gage)? Under normal atmospheric conditions, what is the absolute pressure in the tire?

6.3 It has been suggested that a water manometer could be used to accurately measure tire pressure. Is this practical? What manometer size would be needed to measure a gage pressure of 30 lbf/in²?

6.4 Using Eq. (6.6), calculate the Re for these conditions:
    (a) $\rho = 1000$ kg/m³   (b) $\rho = 1.94$ slug/ft³
        $d = 0.152$ m $d = 0.50$ ft
        $V = 0.317$ m/s $V = 1.04$ ft/s
        $\mu = 1.21 \times 10^{-3}$ N·s/m² $\mu = 2.52 \times 10^{-5}$ lbf/ft²
    (c) $\rho = 500$ kg/m³   (d) $\rho = 62.4$ lbm/ft³
        $d = 30.4$ cm $d = 6.0$ in
        $V = 0.634$ m/s $V = 12.5$ in/s
        $\mu = 2.42 \times 10^{-3}$ N·s/m² $\mu = 2.52 \times 10^{-5}$ lbf·s/ft²

6.5 Water is flowing through a garden hose at a rate of 1 gal/min. The inside diameter of the hose is $\frac{3}{8}$ in. Determine the average velocity V of the water flowing in the hose (note that 1 ft³ = 7.48 gal). If the hose is 100 ft long and $\mu = 2.1 \times 10^{-5}$ lbf·s/ft² what is the pressure drop ΔP from the faucet to the nozzle of the hose?

6.6 If a hose with an inside diameter of $\frac{1}{2}$ in were used in Prob. 6.5, what would be the average velocity and the pressure drop?

6·2 CONSERVATION OF MATTER

That matter is not created and does not vanish in physical processes is very often a factor in the solution of engineering problems. Of course, there are situations where the relationship between mass and energy can be significant ($E = mc^2$ is Einstein's famous equation for the relationship). In those cases some variation in mass may result, but such cases occur only rarely and are mostly of interest to atomic physicists. You will very likely

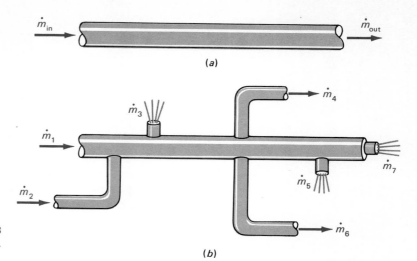

FIGURE 6·8
Pipe flows.

spend your entire career confident that the amount of matter or mass at the end of a chemical reaction is the same as the amount before the reaction. That is, the number of atoms of each element will not change during a reaction, even though the compounds in the products of the reaction differ from the initial reactants.

From another viewpoint, if we consider the flow of matter or mass into a container, we know that only two things can happen to it. It can either stay in the container or flow out of the container. In an equilibrium, or steady-flow, situation, storage is not taking place; and the rate at which mass is flowing into the container must be directly balanced by an equal flow of mass out of the container.

Flow in a pipe is the simplest example of this steady-flow situation. Figure 6.8a is a pipe with a mass flow rate $\dot{m}_{in}$ (read this as "em dot in") going into and $\dot{m}_{out}$ leaving the pipe. Mass flow rate is the amount of mass passing a point per unit time, with dimension M/T. Typical units are kilograms per second, slugs per second, pounds-mass per second, etc. When flow into a pipe begins, as in Fig. 6.8a, some time will be required for the pipe to fill, so some mass storage must initially take place. However, when steady flow or equilibrium is reached, the pipe is filled and $\dot{m}_{in} = \dot{m}_{out}$ because there is no longer storage occurring—"what goes in must go out." This same principle applies for more-complicated flow arrangements like Fig. 6.8b. Here the inlet and outlet flow rates are the sum of the various sprays and pipe flows shown, so that

$$\dot{m}_1 + \dot{m}_2 = \dot{m}_3 + \dot{m}_4 + \dot{m}_5 + \dot{m}_6 + \dot{m}_7 \qquad (6.7)$$

is the equation for the conservation of matter. This type of equation is also frequently referred to as a conservation of mass equation, or sometimes as a continuity equation.

As an example of the way we are often able to obtain important information from application of the conservation of mass, or continuity, principle, let's consider Fig. 6.9, where a piston-cylinder assembly is shown moving a load at the constant velocity V. There are flow connections to the cylinder on both sides of the piston. In order for the piston in the cylinder to be moving to the right, there must be a flow into the left side $\dot{m}_1$ and a flow out of the right side $\dot{m}_2$. Our concept of conservation of mass, or continuity, tells us that $\dot{m}_1 = \dot{m}_2$ when we have steady flow, as we must if the piston velocity is constant. In addition to this requirement, we also have

$$\dot{m}_1 = \dot{m}_2 = \rho V A_p \tag{6.8}$$

because the product VA_p describes the volume flow rate of the fluid moved by the piston. If the piston moves V meters per second, it must displace VA_p cubic meters per second of fluid. Multiplication by density ρ then completes Eq. (6.8), defining the mass flow rate through the cylinder. Rearranging Eq. (6.8) yields

$$V = \frac{\dot{m}_1}{\rho A_p}$$

for the piston velocity in terms of flow rate, density, and piston area. This is one of the most basic equations applied in the study of hydraulic systems, where pressurized hydraulic oil is used to develop very large forces and power levels in machinery that is actually quite compact. Modern construction equipment like the backhoe is a good example of such machinery.

An approach to studying the behavior of hydraulic systems and piping networks that is often utilized is based on the equivalent

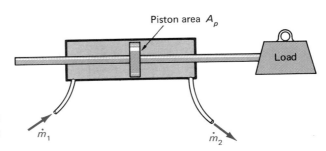

FIGURE 6·9
A hydraulic cylinder moving a load.

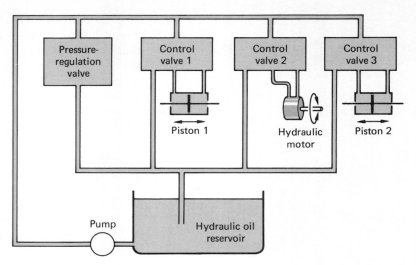

FIGURE 6·10
Circuit diagram for a hydraulic system.

electrical representation of the hydraulic system. This can be done because the flow of current in wires has many similarities to the flow of fluids in pipes and tubes. For example, the conservation of mass equation for pipe flow as in Eq. (6.7) takes on the same summation form as the summation of current in an electric circuit. In addition, the pressure developed by pumps forces fluid through pipes and restrictions just as voltage from a battery or generator forces current through wires and resistors. The phenomena are quite similar, although electric circuitry tends to be more well behaved and more linear in nature. Figure 6.10 shows a typical hydraulic circuit. Representing various aspects of this circuit by an equivalent electric circuit is often a substantial aid in understanding its behavior.

A COMBUSTION PROBLEM

Figure 6.11 shows 500 cm^3 ($\frac{1}{2}$ L) of air at standard atmospheric pressure and temperature (20 °C). We are to add gasoline to form a mixture and explode it with a spark. We would like to determine how much gasoline to add and what mixture of combustion products results from the detonation. You will see that the principle of conservation of matter plays an important role in solving problems of combustion and chemical reaction such as this.

Gasoline is actually a mixture of many different compounds of carbon and hydrogen known as hydrocarbons. It is convenient, however, as a basis for the analysis of combustion, to consider it to be a single compound, typically octane C_8H_{18}. The combustion of this fuel and other hydrocarbons in a mixture with air oxidizes both the carbon and the hydrogen to form water

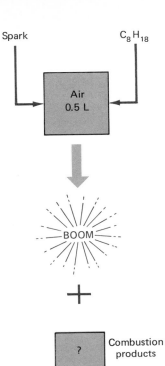

FIGURE 6·11
Combustion of air and octane.

$$2 H_2 + O_2 \rightarrow 2 H_2O$$

and carbon dioxide

$$C + O_2 \rightarrow CO_2$$

When the supply of oxygen is insufficient for the reaction with carbon to form carbon dioxide

$$2 C + O_2 \rightarrow 2 CO$$

also occurs and carbon monoxide forms. These reactions are accompanied by the release of heat, which characterizes and is, of course, the useful purpose of combustion.

Air is composed of only 21% oxygen. Neglecting other lesser components, we have 79% nitrogen present during combustion which can also play a role in the combustion reaction. There are $79/21 = 3.76$ times as many molecules of N_2 in air as compared with O_2, so based on one molecule of C_8H_{18}, the combustion equation is

$$C_8H_{18} + x\ O_2 + 3.76(x)\ N_2 \rightarrow 8\ CO_2 + 9\ H_2O + 3.76(x)\ N_2$$

We want to find x, the amount of air required for complete combustion. Since we want complete combustion, CO_2 is shown on the right side of the equation along with the other products of combustion. The conservation of matter has been applied to establish the amount of each product indicated. That is, since we began with eight atoms of carbon on the left side of the equation, we must find eight on the right. With 18 atoms of hydrogen on the left, they must also be on the right. All the atoms present before combustion must also be present afterward. We add the number of oxygen atoms appearing in the products of combustion on the right now and find $x = [(8)(2) + 9]/2 = 12.5$ to be the amount of oxygen necessary on the left. The equation for complete combustion including the required amount of oxygen and nitrogen is thus

$$C_8H_{18} + 12.5\ O_2 + 47.0\ N_2 \rightarrow 8\ CO_2 + 9\ H_2O + 47.0\ N_2 \quad (6.9)$$

No reaction has been shown for the nitrogen; it appears on the right as a product of combustion in uncombined form. The water in the combustion products may be in the form of vapor or liquid, depending on the temperature and pressure.

To determine the amount of fuel required for a theoretical combustion of our 500 cm³ of air at standard atmospheric temperature and pressure, we make use of Eq. (6.9) to realize that for every mole of fuel we should have 12.5 moles of O_2. A mole of a given compound is a number of grams of that compound equal to its molecular number. It contains 6.03×10^{23} molecules of the compound. Also a mole of any gas at standard atmospheric pressure and temperature of 20°C occupies a volume of 24.0×10^3 cm³.

The 500 cm³ of air contains

$$\frac{0.21(500) \text{ cm}^3 \text{ O}_2}{24.0 \times 10^3 \text{ cm}^3 \text{ O}_2/\text{mol}} = 4.38 \times 10^{-3} \text{ mol O}_2$$

We calculate the required fuel as

$$\frac{1 \text{ mol C}_8\text{H}_{18}}{12.5 \text{ mol O}_2} 4.38 \times 10^{-3} \text{ mol O}_2 = 3.50 \times 10^{-4} \text{ mol C}_8\text{H}_{18}$$

Since one mole of octane weighs 114 g, we can say that

$$3.50 \times 10^{-4} \text{ mol C}_8\text{H}_{18} \frac{0.114 \text{ kg}}{\text{mol C}_8\text{H}_{18}} = 3.99 \times 10^{-5} \text{ kg C}_8\text{H}_{18}$$

This is the mass of octane fuel which should be added to the 500 cm³ of air to obtain complete combustion. By a similar analysis the mass of oxygen, nitrogen, and each product of combustion can also be determined.

Since we have determined the mass of fuel involved, we can now determine the energy released during the combustion. We utilize the results of standardized experiments which show that the combustion of octane releases energy in the form of heat of approximately 46 million joules per kilogram of fuel. Our idealized combustion of octane with 500 cm³ of air will therefore release

$$3.99 \times 10^{-5} \text{ kg C}_8\text{H}_{18} \frac{46 \times 10^6 \text{ J}}{\text{kg C}_8\text{H}_{18}} = 1840 \text{ J}$$

As a result of the calculations performed above and similar calculations for the combustion products, we are now able to completely describe the before-and-after picture of our theoretical combustion, as shown in Fig. 6.12. Note that both before and after the combustion, we have a total mass of 6.42×10^{-4} kg. The mass ratio of air to fuel in the initial mixture is

$$\text{AF} = \frac{6.02 \times 10^{-4}}{3.99 \times 10^{-5}} = 15.1$$

This theoretical air-fuel ratio indicates the amount of air that should be combined with octane for its complete combustion.

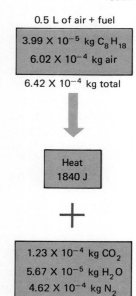

FIGURE 6·12 Details of the combustion process.

COMBUSTION IN PRACTICE

The preceding example considered the combustion of gasoline with 500 cm^3 of air. This is the size of a typical cylinder in an automotive engine, so our analysis represents one typical firing of the cylinder. The amount of gasoline used here and the amounts of combustion products may seem small and insignificant. But in every operating engine, thousands of combustions like this one take place per minute and there are millions of operating engines. The total result is a consumption of fuel and a generation of exhaust products that threatens our economy and environment.

We will look a little more closely at combustion now by discussing some aspects of an actual ignition as compared with the theoretical ignition of an air-fuel mixture described above. An important difference can be caused by dissimilarities between the composition of the theoretical air-fuel mixture and an actual air-fuel mixture in real combustion. In our simple theory, we assumed that each molecule of fuel would be sufficiently close to the right number of oxygen molecules in order to react properly and that forming the air and fuel mixture with the proper air-fuel ratio then theoretically produces a perfect combustion. In reality, this theoretical ratio may be difficult to achieve or may actually not be the most desirable ratio. As a consequence, any given mixture may have too little fuel (be *lean*) or too much fuel (be *rich*).

We can determine the effect of lean and rich mixtures on combustion by writing the chemical equations for situations with too little or too much fuel. For a lean case we may consider only 95 percent of the required octane that we had in Eq. (6.9) and rewrite the equation for 0.95 mol of octane, obtaining

$$0.95\ C_8H_{18} + 12.5\ O_2 + 47.0\ N_2 \rightarrow$$
$$7.6\ CO_2 + 8.55\ H_2O + 47.0\ N_2 + 0.625\ O_2 \quad (6.10)$$

Note in this equation that the reduced amount of fuel results in some oxygen's being left over after the reaction. In this situation, at the high temperatures which occur during combustion, there is a tendency for the leftover oxygen to react with the always available nitrogen and form nitric oxide (NO). Later, in the atmosphere, nitric oxide from combustion products can react with additional oxygen to form nitrogen dioxide (NO_2). This is reddish brown in color and is a pollutant, a primary cause of haze and smog.

Now consider the result of a rich mixture. First, for one with slightly too much fuel, we obtain the following combustion equation by considering Eq. (6.9) with an extra 5% octane:

$$1.05\ C_8H_{18} + 12.5\ O_2 + 47.0\ N_2 \rightarrow$$
$$1.25\ CO + 7.15\ CO_2 + 9.45\ H_2O + 47.0\ N_2 \quad (6.11)$$

Not enough oxygen is available here for complete reaction with the carbon and some carbon monoxide is formed in addition to carbon dioxide. Since the reaction to form carbon monoxide produces less energy than the carbon dioxide reaction, the efficiency of combustion is reduced. In addition, carbon monoxide is a poisonous gas and a significant cause of pollution.

In the case of a highly rich mixture, the situation is more severe and the following form of combustion equation results:

$$1.5\ C_8H_{18} + 12.5\ O_2 + 47.0\ N_2 \rightarrow$$
$$0.029\ C_8H_{18} + 11.8\ CO + 13.2\ H_2O + 47.0\ N_2 \quad (6.12)$$

Here all the carbon is oxidized to carbon monoxide and still there is insufficient oxygen for all the fuel. Some of the C_8H_{18} cannot react, so it appears as a product of combustion. It is clearly inefficient for fuel to go unburned during combustion, since it is completely wasted. In addition, the exhaust of such unburned hydrocarbons into the atmosphere creates pollution and, in turn, eye irritation. With some particular forms of hydrocarbons, there may be a danger that the hydrocarbons are carcinogenic.

In actual combustion, the mixture is not completely uniform. There are regions near walls and valves, for example, where low temperatures are found and incomplete combustion occurs. Some traces of the pollutants nitric oxide, carbon monoxide, and unburned hydrocarbons always remain. For mixtures which are rich overall in fuel, the problems of carbon monoxide and unburned hydrocarbons are predominant. Inefficient use of fuel, reflected by poor gas mileage in automotive use, results. In some cases, though, when the most energy possible is desired from a given volume of air-fuel mixture (the case of full throttle in an automotive engine), a mixture somewhat rich overall that will produce maximum energy will be used. For starting a cold engine, a rich mixture is also desired, since it results in dependable ignition. In general, however, a mixture of air and fuel which is somewhat lean overall yields the most economical performance in combustion, since it is more certain to burn the available fuel efficiently. In overall lean mixtures, the problem of nitric oxide production is troublesome with respect to pollution. In addition to this, mixtures which are very lean do not ignite dependably and may misfire, wasting fuel and producing unburned hydrocarbons.

It should be clear from our discussion of combustion that the conservation of matter is an important tool for understanding the combustion process in both theory and practice.

PROBLEMS

6.7 For the arrangement of flows and nozzles shown in Fig. P6.7, determine the unknown in each case below. Assume that steady flow exists.
(a) $\dot{m}_1 = ?$, $\dot{m}_2 = 2$ kg/s, $\dot{m}_3 = 0$, $\dot{m}_4 = 5$ kg/s, $\dot{m}_5 = 0$
(b) $\dot{m}_5 = ?$, $\dot{m}_1 = 10$ kg/s, $\dot{m}_2 = 5$ kg/s, $\dot{m}_3 = 3$ kg/s, $\dot{m}_4 = 0$
(c) $\dot{m}_2 = ?$, $\dot{m}_1 = 1$ kg/s, $\dot{m}_3 = 5$ kg/s, $\dot{m}_4 = 5$ kg/s, $\dot{m}_5 = 6$ kg/s
(d) $\dot{m}_4 = ?$, $\dot{m}_1 = 10$ kg/s, $\dot{m}_2 = 5$ kg/s, $\dot{m}_3 = 10$ kg/s, $\dot{m}_5 = 10$ kg/s

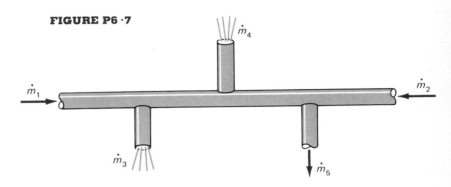

FIGURE P6·7

6.8 In Fig. P6.8, the fluid has a mass density $\rho = 850$ kg/m³ and the piston has an area of 5 cm². If $\dot{m}_1 = 1.0$ kg/s and $\dot{m}_2 = 0.10$ kg/s, calculate the piston velocity V assuming steady-state conditions.

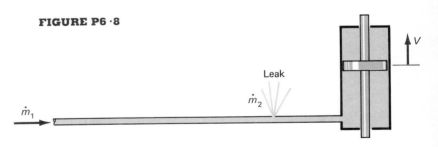

FIGURE P6·8

6.9 A certain one-cylinder engine has a cylinder with a volume of 500 cm³. One ideal firing of the cylinder at full throttle corresponds to filling the cylinder with air and fuel and burning it. The resulting combustion is as shown in Fig. 6.12. When this engine runs at a speed of 2000 r/min, combustions are taking place in the cylinder 1000 times per minute. If the ideal process of Fig. 6.12 is followed, what must be the mass-flow rate of fuel into the engine? What are the mass-flow rates of H_2O and CO_2 out of the engine? What is the rate at which heat is released by the combustion process (in watts)?

6.10 For the idealized rich combustion described by Eq. (6.11), determine the mass of octane fuel which would be burned with 500 cm³ of air. Determine the mass of carbon monoxide resulting from the combustion.

6·3 CONSERVATION OF ENERGY

The basis of the conservation of energy is that like mass, energy is neither created nor destroyed in physical processes, so after a physical event, the amount of energy is the same as before the event; only its form may have changed. The exception to this concept is in atomic physics, where energy-mass transformations may occur, related by $E = mc^2$. However, as mentioned in the conservation of matter discussion, this type of transformation is only rarely significant for us.

The first law of thermodynamics is the name given to the conservation of energy. This law is a foundation of the science of thermodynamics, which deals with heat and work and properties of substances related to heat and work. Within the science of thermodynamics, we find the basis for the operation of steam power plants, automotive engines, refrigerators, air conditioners, jet engines, and many other machines which are an important part of our modern society. It is often said that the first law of thermodynamics simply states that "you can't get something for nothing." It is this fact which requires us to deplete our energy resources when we warm our buildings in winter or cool them in summer, or when we drive to the supermarket or generate electricity. The second law of thermodynamics, which we will not try to discuss in any depth here, is often thought of as modifying our simplified description of the first law to state that "you can't get something for nothing—in fact, you can't break even." It is because of second-law requirements that the conversion of the energy in fuel to a useful form is never 100 percent efficient. Some energy must always be wasted. An example of this waste is the requirement for cooling water in steam power plants used for generating electricity. This water actually carries away heat produced by the power-plant fuel, but it is a necessary part of power-plant operation. The resulting warm-water discharge is a by-product of the generation of electricity. In some European communities, this warm water is used for heating homes, but it is generally discharged into a lake or river, creating artificially warm areas. To use our energy resources more effectively, much engineering effort is being directed toward improving the efficiency of energy-conversion processes and also utilizing waste heat.

AN EQUATION FOR CONSERVATION OF ENERGY

An equation for the conservation of energy during a physical event can be written for the matter or mass involved in the event. We call this matter or mass a thermodynamic system; in words our equation is

Energy after = energy before + net heat added to the system − net work done by the system

We usually use the algebraic symbols Q = heat added and W = work done. The energy we consider is the sum of the internal energy U of the matter or mass, its kinetic energy KE, and its potential energy PE. In algebraic form, the conservation of energy equation is thus

$$U_f + \text{KE}_f + \text{PE}_f = U_i + \text{KE}_i + \text{PE}_i + Q - W \qquad (6.13)$$

where the subscripts i and f indicate the initial and final conditions, respectively.

In Eq. (6.13) there are three different kinds of energy with which we must be concerned. Internal energy U is a result primarily of the molecular energy of the particles composing the material of the thermodynamic system and is directly related to its temperature. It is not easy to calculate the exact internal energy of a material. However, unless there is a change in phase (from liquid to vapor, for example), we can normally compute a change in internal energy as the product of the specific heat of the material, its mass, and its change in temperature. The specific heats for various fluids and solids are shown in Table 6.2. Kinetic energy is the energy possessed by a mass due to its velocity, where KE = $\frac{1}{2}mV^2$ and m is the amount of mass moving at velocity V. Potential energy is the energy of a mass which results from an elevated position of the mass within a gravitational field, PE = mgh, where g is acceleration due to gravity and h is the height of the mass above some reference level. In all these cases, energy may be interpreted as an ability to do work. If a thermodynamic system does work, there is a decrease in energy. This accounts for the minus sign in Eq. (6.13). Of course, if work is done on a thermodynamic system, W is negative and thus there is an increase in energy. The transfer of heat Q into a system increases its ability to do work and thus its energy level, as reflected by an increase in its internal energy and temperature level.

TABLE 6·2
SPECIFIC HEATS OF VARIOUS MATERIALS

Material	Specific Heat, J/kg·K
Polyester	2×10^3
Mica	8×10^2
Carbon	7.5×10^2
Steel	5.0×10^2
Copper	4.2×10^2
Silver	2.3×10^2
Gold	1.3×10^2
Mercury	1.4×10^2
Water	4.18×10^3
Oil	1.8×10^3
Carbon dioxide*	8.4×10^2
Air*	1.0×10^3
Helium*	5.2×10^3

* Constant-pressure specific heats at atmospheric pressure and 20°C.

Note:
1 J/kg·K = 5.98 ft·lbf/(slug)(°R)
= 2.39 × 10⁻⁴ Btu/(lbm)(°R)

The dimension of all the energy variables, as well as heat and work, is force times length (FL), as was mentioned in the section on dimensions. Typical units are foot-pounds-force or joules (where 1 J = 1 N·m). The British thermal unit is frequently used as a unit for heat; it was originally defined to be the amount of energy needed to raise the temperature of one pound-mass of water one degree Fahrenheit. The Btu is equal to 778.3 ft·lbf, or 1055 J. In Eq. (6.13), as with all general equations, it is best to have each of the variables in its most fundamental units before trying to apply the equation.

APPLYING THE CONSERVATION OF ENERGY EQUATION

As a first example of the use of the conservation of energy equation, suppose that you have thrown a ball into the air straight up and want to estimate the initial velocity you were able to give the ball as you threw it. A friend is on the ninth floor of your dormitory and says that at its peak, the ball just reached his eye level. If we allow 10 ft per floor, we can say that your throw went 90 ft in the air. When we consider the ball to be a thermodynamic system, we apply Eq. (6.13) by using the instant at which the ball is released to correspond to state i and the instant when the ball reaches a peak height of 90 ft to be state f. We now consider the various terms in Eq. (6.13). The internal energy $U_f = U_i$, since no significant change in temperature occurs between the time the ball is thrown and when it reaches peak height. The kinetic energy $KE_f = 0$, since the ball has zero velocity as it is reaching its peak height and starting downward. However, as the ball is thrown, $KE_i = \frac{1}{2}mV_i^2$ is the initial kinetic energy and $PE_i = 0$, since we take the height at this point to be $h_i = 0$. At the peak, $PE_f = mgh_f$, where $h_f = 90$ ft because the ball has reached approximately 90 ft above the release point. No important heat generation occurs at the low speed at which the ball is moving, so $Q = 0$. We assume that the friction of air acting on the ball produces a negligible force, so no work is done by the ball as it rises—that is, $W = 0$. Making all these substitutions into Eq. (6.13) reduces it to

$$mgh_f = \tfrac{1}{2}mV_i^2$$

so that

$$V_i = \sqrt{2gh_f}$$

Substituting the numbers,

$$V_i = \sqrt{2(32.2)(90)} = 76 \text{ ft/s}$$

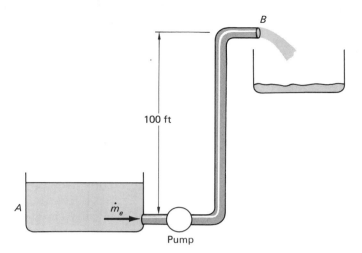

FIGURE 6·18
Pumping water to an elevated tank.

script o refers to outward flow. The mass-flow rate $\dot{m}_e = \dot{m}_o$ for steady flow.

As a first example of the application of Eq. (6.15), consider Fig. 6.18, where we show a pump sending 1000 gal/h of water 100 ft in the air. We would like to calculate the power requirement of the pump. We begin by selecting the connecting pipe and pump as being our thermodynamic system. We note that the velocity of the fluid in tank A is zero and the velocity at pipe outlet B is small enough, so it is reasonable to say that $V_e = V_o = 0$. The pressure at tank A and at the outlet B is only atmospheric, so that $P_e = P_o = P_{atm}$. No heat transfer occurs, so $\dot{Q} = 0$ and we must simply calculate $\dot{W}$ to obtain the pump power. First the mass-flow rate is computed for $\rho = 1.94$ slug/ft^3 and 1 ft^3 = 7.48 gal:

$$\dot{m} = \dot{m}_e = \dot{m}_o = (1.94)(1000) \frac{1 \text{ ft}^3}{7.48 \text{ gal}} \frac{1 \text{ h}}{3600 \text{ s}}$$

$$= 0.0720 \text{ slugs/s}$$

Equation (6.15) with $h_e = 0$ and $h_o = 100$ ft may now be written as

$$(0.0720)\left(0 + 0 + u_e + \frac{P_{atm}}{1.94}\right) + 0$$

$$= (0.0720)\left[0 + (32.2)(100) + u_o + \frac{P_{atm}}{1.94}\right] + W \quad (6.16)$$

Now we assume that no temperature rise of the water occurs between inlet and outlet, so $u_o = u_e$ and we simplify Eq. (6.16) to find

Example 6·3

Hydraulic turbines operate with flowing fluid turning the blades of the turbine similar to the way wind turns a windmill. A certain hydraulic turbine is to be driven by water from a reservoir 75 m above it. If 1000 kW is the power expected from the turbine, what is the required water-flow rate? Assume that the reservoir pressure and turbine exit pressure are essentially atmospheric and that the water velocity is negligible as it leaves the turbine and as it leaves the reservoir. With no significant temperature rise $u_e = u_o$.

Solution

For this steady-flow situation with $\rho = 1000$ kg/m³, Eq. (6.15) becomes

$$\dot{m}\left[0 + (9.81)(75) + u_e + \frac{P_{atm}}{1000}\right] + \dot{Q}$$
$$= \dot{m}\left(0 + 0 + u_o + \frac{P_{atm}}{1000}\right) + \dot{W}$$

No heat addition is considered and outlet height $h_o = 0$ and $h_e = 75$ m. The equation simplifies to

$$736\dot{m} = \dot{W}$$

Since the required rate of work is 1000 kW, we write

$$\dot{m} = \frac{1,000,000}{736} = 1360 \text{ kg/s}$$

This is, of course, the minimum conceivable flow rate which could be required. We have made idealizing assumptions. Pressure drops, inefficiency, etc., mean that more flow than this will be actually needed to produce the desired power.

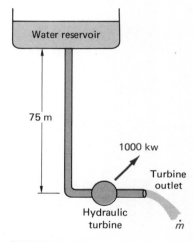

FIGURE 6.19

$$\dot{W} = (-0.0720)(32.2)(100)$$
$$= -232 \text{ ft·lbf/s}$$
$$= -0.422 \text{ hp}$$

The minus sign here means that work is done on our water and not the reverse. The 232 ft·lbf/s is the minimum power that can lift 1000 gal/h to a height of 100 ft in the air. We can expect losses due to pressure drop in the pipes, friction in the pump parts, and simple inefficiency in the pumping process. If we were to determine the actual power input to the pump by measuring the torque and speed and multiplying the two numbers, we might actually find that $\dot{W}_p = 307$ ft·lbf/s. This would be quite typical of real processes where more energy input is required than the useful energy obtained. Efficiency is an indication of what fraction of a machine's energy is useful. In percent,

$$\varepsilon = \frac{\text{Useful energy}}{\text{Energy supplied}} \times 100 \tag{6.17}$$

In this case, we would find

$$\varepsilon = \frac{232}{307} \times 100 = 76 \text{ percent}$$

and only 76 percent of the energy supplied performs the useful function. The rest goes into perhaps heating the bearings of the pump by friction, doing work against the friction of pipes, heating the water a bit by inefficient pumping, or spraying the water from the outlet with a substantial velocity. We did not try to include these items in our application of Eq. (6.15) because we focused on only the lifting of the water, which is the useful goal. Example 6.3 is the reverse of this one, in which falling water is used to do work. Example 6.4 describes a problem of heating flowing water.

One consequence of the laws of thermodynamics is to show that it is not possible to build machines which will run indefinitely without a power source (you can't get something for nothing). Such machines are called perpetual motion machines, for which many people have sought patents. The energy crisis of recent years has encouraged many "backyard" inventors to continue to pursue the concept of perpetual motion, either consciously or unconsciously. It is important to recognize machines which perform useful work without an energy input and to realize that they are not possible. As an example of this type of machine, see Fig. 6.20a, which is an electric generator which produces electric power as its output when its input shaft is rotated by mechanical power. It has been suggested that if we connected an electric motor to the input shaft of the generator, the motor could turn the generator to produce electricity. Part of this generated electricity would be needed to provide an input for the electric motor and the other portion could then be used to light your house or operate your television.

We see difficulty with this approach to producing electricity when we draw a box around it as in Fig. 6.20b and call it a thermodynamic system. We see electric power (work) coming out of the thermodynamic system and no energy entering. The machine is

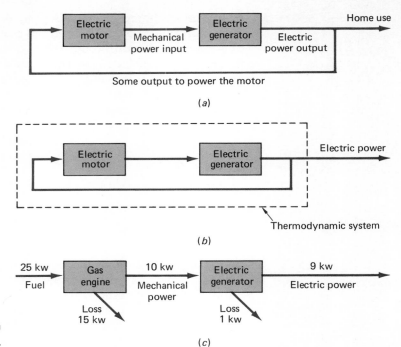

FIGURE 6·20
Generating electric power.

attempting to produce something for nothing. Let's look at Fig. 6.20c now to see the actual problem. We have in this figure the typical configuration of a small generating plant. A gasoline engine is driving an electric generator. Energy in the form of a flow of combustible fuel is transformed into mechanical power with a large loss due to inefficiency. The resulting mechanical power drives the generator, but because of its inefficiencies, the electric power it produces is less than the mechanical power which drives it. This is exactly the problem with the scheme of Fig. 6.20a. The electric power produced by the generator is not enough to power the motor which drives the generator. Not only is there no electric power left over for home use, as shown in Fig. 6.20a, but there is not enough to make the arrangement operate at all.

PROBLEMS

6.11 A ball has been dropped from the roof of a small building. Just before hitting the ground, it is known to be traveling at 10 m/s. Assuming that air friction may be neglected, calculate the height of the building using the conservation of energy equation, Eq. (6.13).

Example 6·4

A water heater is known to be 70 percent efficient. It is used to heat water continuously from a temperature of 50 to 110°F with a mass-flow rate of 2 gal/min. At what rate must heat be supplied? If a fuel like propane with a heating value of 5.00×10^8 ft·lbf/slug is used, what is the rate at which fuel must be supplied? The fluid velocities in and out of the heater are the same as are the pressures and there is no change in height.

Solution

Equation (6.15) for this steady-flow process is

$$\dot{m}\left(\frac{1}{2}V_e^2 + gh_e + u_e + \frac{P_e}{\rho_e}\right) + \dot{Q}$$
$$= \dot{m}\left(\frac{1}{2}V_o^2 + gh_o + u_o + \frac{P_o}{\rho_o}\right) + \dot{W}$$

With $V_e = V_o$, $h_e = h_o$, $P_e = P_o$, and $\rho_e = \rho_o$; and knowing that $\dot{W} = 0$ for the heater, we can rewrite this equation as

$$\dot{Q} = \dot{m}(u_o - u_e)$$

For a flow of 2 gal/min and a water density of 1.94 slugs/ft³, and with 1 ft³ = 7.48 gal, we have

$$\dot{m} = \rho q$$
$$= (1.94)(2)\frac{1\ ft^3}{7.48\ gal}\frac{1\ min}{60\ s}$$
$$= 8.65 \times 10^{-3}\ slugs/s$$

For water, the specific heat $C = 1$ Btu/(lbm)(°F) $= 2.51 \times 10^4$ ft·lbf/(slug)(°F). The change in internal energy

$$u_o - u_e = C(T_o - T_e)$$
$$= (2.51 \times 10^4)(110 - 50)$$
$$= 1.51 \times 10^6\ ft·lbf/slug$$

and then

$$\dot{Q} = (8.65 \times 10^{-3})(1.51 \times 10^6)$$
$$= 1.31 \times 10^4\ ft·lbf/s$$

is the useful heat-flow rate. At an efficiency $\varepsilon = 70$ percent using Eq. (6.17), we find

$$\dot{Q}_{actual} = \frac{\dot{Q}_{useful}}{0.70}$$
$$= 1.87 \times 10^4\ ft·lbf/s$$

as the required heat from the fuel. The necessary fuel flow is

$$\dot{m}_{fuel} = \frac{\dot{Q}_{actual}}{heating\ value}$$
$$= \frac{1.87 \times 10^4}{5.00 \times 10^8}$$
$$= 3.74 \times 10^{-5}\ slug/s$$
$$= 4.34\ lbm/h$$

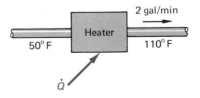

FIGURE 6·21

6.12 A projectile, shown in Fig. P6.12, is fired from a "cannon." The projectile has a mass of 2 kg and is propelled by an explosion known to release 90,000 J of heat energy. If all this energy were used to accelerate the projectile, what would be its exit velocity?

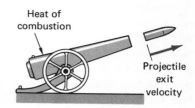

FIGURE P6·12

6.13 In Prob. 6.12, the exit velocity of the projectile has been measured at 200 m/s. How much energy is actually found in the exiting projectile? What has happened to the other energy? If the missing energy heats the 10-kg mass of steel of which the firing chamber is constructed, what would be the temperature rise produced by one firing of the cannon?

6.14 Electricity is to be produced by a hydraulic turbine driving an electric generator. The hydraulic turbine is powered by water from an elevated reservoir, as shown in Fig. P6.14. The mass-flow rate $\dot{m}$ = 100 slugs/s, the height h = 150 ft, and the outlet velocity v_o = 10 ft/s. Determine the power output ideally possible from the turbine. If the overall efficiency is 75 percent for the system, what will be the electric output power?

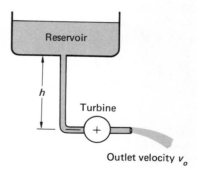

FIGURE P6·14

6.15 Consider Prob. 6.14 with $\dot{m}$ = 1000 kg/s, h = 50 m, v_o = 3 m/s. Now what is the electric output power expected? For a reservoir depth of 2 m, what surface area would be necessary to maintain this output for a 3-hour period?

6.16 The waste-heat-recovery system shown in Fig. P6.16 warms a new fluid which is needed for a chemical process by cooling a warm fluid being discarded. This reduces the amount of

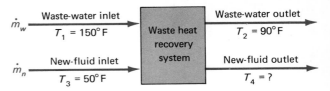

FIGURE P6·16

fuel needed to heat the new fluid to the temperature required.

In Fig. P6.16, a waste-water-flow rate $\dot{m}_w = 1$ slug/s is being cooled [specific heat $C_w = 2.51 \times 10^4$ ft·lbf/(slug)(°F)]. The new fluid being warmed is a light oil that has flow rate $\dot{m}_n = 4$ slug/s and a specific heat $C_n = 1.50 \times 10^4$ ft·lbf/(slug)(°F).

Consider the cooling of the waste water as shown. How much heat will be obtained? If the recovery system is 70 percent efficient, how much of this heat can be added to the new fluid? What outlet temperature T_4 will this produce?

6·4 HEAT TRANSFER

In our discussions of thermodynamic systems in the previous section, we found that heat flow played an important role in our study of the flow of energy in and out of a system. It is also the phenomenon that defines the temperature of all objects. The study of heat transfer is the study of ways to predict and control the flow of heat and to compute temperatures resulting from heat flow. The emphasis may be at times on enhancing the flow of heat and at other times on reducing heat flow, but always an understanding of the heat-transfer process is involved.

Heat transfer occurs wherever there is a temperature difference, by means of conduction, convection, and radiation. These three modes are normally involved simultaneously in real problems, but particular modes can be identified as predominant in certain situations. We will begin our discussion of heat transfer by briefly describing the three modes and cases where each is singularly important.

Conduction is the primary mode of heat transfer within a given material or between materials in direct physical contact. It is the dominant mode of heat transfer in solids, resulting directly from temperature differences within the material. An example of conduction is when the outer surface of an oven becomes hot as a result of heat conducted through the oven wall from the hotter inner surface. Another example is a roast beef inside the oven, with the temperature of the outer surface of the beef high and the temperature of the interior lower. Heat is conducted inward from the outside of the beef, thereby cooking it. The same type of heat conduction process can be important in distributing heat from localized sources to avoid damaging high-temperature buildup. This is often the case in electronic components and automotive engines.

Convection is the mode of heat transfer wherein fluid motion plays a role. In a typical convection process, heat is conducted from a solid surface to neighboring fluid particles that are then swept away by motion of the fluid. The fluid motion may be a "forced" motion provided by wind or a pump or fan. It may also be a "natural" motion created in an otherwise still fluid by the rising of warmer fluid particles near the solid surface. Increasing the amount of fluid motion makes it easier for convective heat transfer to occur by reducing the thickness of the relatively immobile fluid layer which lies next to the solid surface. An important example of convective heat transfer is the heat loss from the outer surfaces of buildings to the atmosphere during the winter months. Also, the transfer of heat from the air in an oven to the outer surface of a cooking roast beef is by convection. The fan blades in an automobile force air through the radiator, where heat is transferred to the air by convection from the outer surfaces of the radiator. (Note that a radiator could more appropriately be called a convector, since most of its heat transfer is by convection.)

Radiation heat transfer occurs because all bodies naturally emit a *thermal radiation,* which is an electromagnetic radiation normally emitted at wavelengths longer than visible light in the infrared portion of the electromagnetic spectrum. The exact nature and amount of emitted radiation is dependent on the temperature of the emitting body. Very high temperature bodies, such as the sun and light-bulb filaments, for example, emit a substantial amount of radiant energy in the visible region which we can see as light. The heat that warms our planet earth and governs its temperature is the radiant energy emitted by the sun and absorbed by the earth after transmission through millions of miles of nearly empty space. Radiation heat transfer becomes the predominant mode of heat transfer whenever there is little atmosphere or when large temperature differences exist between neighboring bodies. An equilibrium temperature for a spacecraft results when the radiation energy emitted by the spacecraft exactly balances the radiation energy received from the sun. The bright reddish-orange coloring that we associate with the temperatures of flames and molten metals is an indication of high-temperature thermal radiation. Heat transfer by radiation plays an important role in high-temperature processes, including combustion in furnaces, boilers, and combustion engines, and is also important in steel production and metal working.

The mathematical techniques for dealing with heat-transfer

problems are introduced in the material below. We have focused here on the conduction and convection modes of heat transfer, but related concepts are also applied in radiation heat transfer.

CONDUCTION

The simplest equation describing the process of heat conduction is

$$\dot{Q} = -kA \frac{d\tau}{dx} \qquad (6.18)$$

where $\dot{Q}$ = rate of heat transfer (with dimension FL/T),[1] watts or Btu/h
k = material property called thermal conductivity (dimension $FL/TL\theta$), W/(m)(K) or Btu/(h)(ft)(°F)
A = cross-sectional area of interest (dimension L^2), m² or ft²
τ = temperature (dimension θ), K or °F
x = linear position (dimension L), m or ft

Equation (6.18) is valid for heat conduction along a single direction in steady state. The derivative term $d\tau/dx$ is the slope or gradient of the temperature in the x direction. The minus sign appears because heat flows in the direction of decreasing temperature. That is, heat flows in the positive x direction when $d\tau/dx$ is negative, and vice versa.

Figure 6.22 is an example of simple heat conduction. We have a wall with the inner surface temperature τ_i at 60°F and the outer surface temperature τ_o at 30°F. Heat is flowing through the wall from the higher-temperature inner surface to the lower-temperature outer surface. At steady state, the heat flow must be the same any place within the wall. Thus the slope $d\tau/dx$ must be constant and the temperature variation is linear inside the wall, as indicated by the plot of τ versus x shown in Fig. 6.22. If the wall is 4 in, or 0.33 ft, thick, the temperature gradient in the wall is

$$\frac{d\tau}{dx} = \frac{30 - 60}{0.33} = -90°F/ft$$

This is negative because the temperature is decreasing in the x direction. If the wall consists of a material whose thermal conductivity $k = 0.05$ Btu/(h)(ft)(°F), then the heat transfer through the wall, using Eq. (6.18), will be

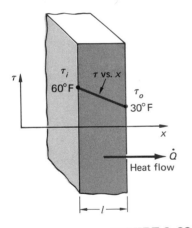

FIGURE 6·22
Simple heat conduction.

[1] Remember that T is the symbol for the dimension of time and θ is the symbol for the dimension of temperature.

$$\dot{Q} = -(0.05)A(-90)$$

or

$$\frac{\dot{Q}}{A} = 4.5 \text{ Btu/(h)(ft}^2)$$

expressed per unit area of the wall. If we have 1000 ft² of this wall, the total heat loss is $\dot{Q}$ = 4500 Btu/h under these conditions. Note that the calculated $\dot{Q}$ is positive so that heat flow is in the $+x$ direction and from the high temperature side to the low temperature side as expected. The value of k = 0.05 Btu/(h)(ft)(°F), which we just considered above is typical of insulating materials. Other thermal conductivities for a variety of materials, including some liquids and gases, are shown in Table 6.3.

The steady-state temperature variation in one-dimensional heat transfer (τ versus x) is normally linear through a thickness of uniform solid material, as we saw in Fig. 6.22. In such cases, the

TABLE 6·3
TYPICAL THERMAL CONDUCTIVITIES

Material*	Thermal Conductivity, W/(m)(K)†
Asbestos	0.16
Cork	0.043
Glass wool	0.043
Rock wool	0.035
Wood, pine	
Perpendicular to grain	0.10
Parallel to grain	0.24
Steel	45
Copper	380
Silver	415
Gold	295
Mercury	8.3
Oil, light	0.13
Water	0.60
Carbon dioxide	
At atmospheric pressure, 0°C	0.014
At atmospheric pressure, 40°C	0.017
Air	
At atmospheric pressure, 0°C	0.024
At atmospheric pressure, 40°C	0.027
Helium	
At atmospheric pressure, 0°C	0.14
At atmospheric pressure, 40°C	0.16

* At temperature of 20°C unless otherwise indicated.
† 1 W/(m)(K) = 0.5777 Btu/(h)(ft)(°F) = 0.1248 ft·lbf/(s)(ft)(°F).

temperature gradient in Eq. (6.18) can be calculated by dividing the temperature difference by the material thickness so that

$$\frac{d\tau}{dx} = -\frac{\tau_H - \tau_L}{l} \quad (6.19)$$

where τ_H = higher surface temperature
τ_L = lower surface temperature
l = material thickness

Substituting this into Eq. (6.18), we find that

$$\dot{Q} = \frac{kA}{l}(\tau_H - \tau_L) \quad (6.20)$$

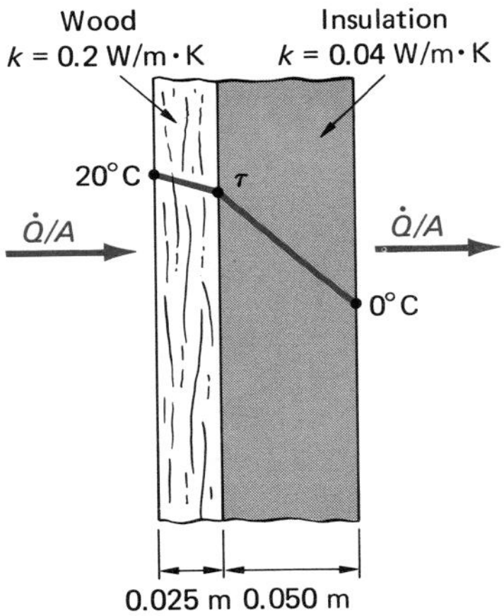

is the rate of heat transfer through the material from the hotter surface to the cooler surface. This equation is in a very convenient form, showing how the rate of heat flow depends directly on the temperature difference across a thickness of material. If there is no temperature difference, there is no heat flow.

As another example of heat-conduction calculations, consider Fig. 6.23, where we have a 2.5-cm thickness of wood with $k = 0.2$ W/(m)(K) and a 5-cm layer of insulation with $k = 0.04$ W/(m)(K). The inner-surface temperature is 20°C and that of the outer surface is at 0°C. We would like to determine the heat flow through this combination of material and also the temperature τ at the interface between the materials. We will assume that there is very solid contact at the interface. To handle this problem, we will work per square meter of surface area, rearranging Eq. (6.20) to be

$$\frac{\dot{Q}}{A} = \frac{k}{l}(\tau_H - \tau_L)$$

We then write this equation for each of the materials in our wall. For the wood, the warmer surface is at 20°C and the cooler surface is simply τ, so that

$$\frac{\dot{Q}}{A} = \frac{0.20}{0.025}(20 - \tau) \quad (6.21)$$

and similarly for the insulation,

$$\frac{\dot{Q}}{A} = \frac{0.04}{0.050}(\tau - 0) \quad (6.22)$$

In steady state the heat flow through the wood must equal the heat flow through the insulation, since the materials are essentially in

FIGURE 6·23
Heat conduction through two materials.

"series" and heat entering the wood must leave through the insulation. Thus the $\dot{Q}/A$ in Eqs. (6.21) and (6.22) must be the same, so we can equate them:

$$\frac{0.20}{0.025}(20 - \tau) = \frac{0.04}{0.05}(\tau - 0)$$

We then solve for τ, obtaining

$$(8)(20) - 8\tau = 0.8\tau$$
$$8.8\tau = 160$$
$$\tau = 18.2°C$$

Now that we have this temperature, we return to Eq. (6.21) to compute the heat flow as

$$\frac{\dot{Q}}{A} = \frac{0.20}{0.025}(20 - 18.2)$$
$$= 14.4 \text{ W/m}^2$$

This is the heat flow per unit area and 100 m² of such a wall would allow 1.44 kW of heat flow with this temperature difference. In Fig. 6.23, we have shown the temperature plot within the wall. We see the linear variation of temperature in each material but with different slope. For the same heat flow, the better insulating material shows a steeper temperature gradient and most of the temperature drop takes place across the insulation.

CONVECTION

A solid body whose surface temperature is τ_s is shown in Fig. 6.24. It is in contact with a liquid or gas (in this case, air) which has a temperature τ_∞ at a distance removed from the solid body. We assume that τ_s is greater than the surrounding temperature τ_∞, so heat transfer from the surface to the fluid is taking place by convection. The rate of convective heat transfer is given by

$$\dot{Q} = hA(\tau_s - \tau_\infty) \quad (6.23)$$

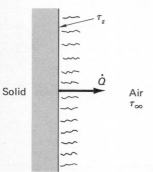

FIGURE 6·24
Convection from a solid surface.

where $\dot{Q}$ = heat-transfer rate (dimension FL/T)
A = surface area (dimension L^2)
τ_s = surface temperature (dimension θ)
τ_∞ = fluid temperature (dimension θ)
h = heat-transfer coefficient, also referred to as the film coefficient or convective heat-transfer coefficient (dimension $FL/TL^2\theta$), W/(m²)(K) or Btu/(h)(ft²)(°F).

TABLE 6·4
REPRESENTATIVE VALUES FOR THE CONVECTION HEAT TRANSFER COEFFICIENT

Situation	Typical h, $W/(m^2)(K)$*
Air, free convection	5–30
Air, forced convection	30–300
Oil, forced convection	60–1800
Water, forced convection	300–6000
Boiling water	3000–60,000
Condensing steam	6000–120,000

* $1\ W/(m^2)(K) = 0.1764\ Btu/(h)(ft^2)(°F)$
   $= 0.03812\ ft·lbf/(s)(ft^2)(°F)$.

This equation for convection is often called Newton's law of cooling. It is deceptively simple but does show directly how the heat-transfer rate is proportional to the surface area and the temperature difference between surface and fluid. The colder the fluid, the faster heat will leave the surface. Of course, if the fluid is hotter than the surface, heat transfer will be in the opposite direction, into the surface. One difficulty with Eq. (6.23) is that temperature often varies across the surface of interest and it may become necessary to define an average temperature or to divide the surface into several small areas which can each be considered to have uniform temperature. It is the heat-transfer coefficient h, however, which is the most difficult aspect of Eq. (6.23). Its value depends on the shape of the surface as well as the flow of the surrounding fluid and often varies from point to point on the surface. Determining the value of h for particular situations has furnished employment for many chemical and mechanical engineers over the past decades and the work is still continuing.

It is possible to give some general indication of the values of h in different fluids and flow situations, as you see in Table 6.4. However, these are simply typical ranges, useful only for the most approximate calculations. You may notice in this table that still air in free convection shows the lowest values for h and that higher values are found for moving air (forced convection) and for liquids. The heat-transfer coefficients with boiling or condensing liquids can be extremely high, with a great deal of heat being transferred in a small area. This effect has been utilized in difficult cooling problems, including that of rocket motors.

EXPERIMENTAL DETERMINATION OF h

For accurate heat-transfer calculations, specific values of the heat-transfer coefficient h must be available. In some situations,

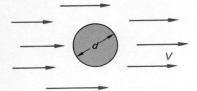

FIGURE 6·25
Fluid flow across a cylinder.

accurate values of h can be calculated based on physics and analysis, but the results of careful experiments based on methods of dimensional analysis are normally utilized. As an example of the experimental case, we will consider the situation of Fig. 6.25—a cylindrical tube in a fluid cross flow. We want to determine the heat-transfer coefficient between the outer surface of the tube and the flowing fluid.

As in all dimensional analysis problems, we begin by listing the various quantities on which h may depend. In this case, it has been found important to include

d = outside diameter of cylinder, dimension L
V = velocity of the fluid, dimension L/T
μ = viscosity of the fluid, dimension FT/L^2 or M/LT
ρ = mass density of the fluid, dimension M/L^3
k = conductivity of the fluid, dimension $FL/TL\theta$ or $ML/T^3\theta$
C_p = specific heat of the fluid (at constant pressure), dimension $FL/M\theta$ or $L^2/T^2\theta$

Also, recall that the dimension of h is $FL/TL^2\theta$ or $M/T^3\theta$. In analyzing the dimensions of these variables, both force and mass cannot both be considered fundamental dimensions. We have considered only mass to be fundamental and have altered the dimensions of the variables k, C_p, and h accordingly, replacing the force dimension F by ML/T^2. Based on the dimensional analysis procedure described in Chap. 2, the seven variables containing the four fundamental dimensions of M, L, T, and θ can be reduced to three dimensionless groups. The three groups normally selected are the following.

The Nusselt number:

$$\pi_1 = \frac{hd}{k}$$
$$= \mathrm{Nu} \tag{6.24}$$

The Prandtl number:

$$\pi_2 = \frac{C_p \mu}{k}$$
$$= \mathrm{Pr} \tag{6.25}$$

And another version of the Reynolds number which we have seen previously, in Chap. 2 and in Sec. 6.1:

$$\pi_3 = \frac{V d \rho}{\mu}$$
$$= \mathrm{Re} \tag{6.26}$$

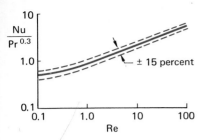

FIGURE 6·26
Experimental result for heat transfer from a cylinder.

The general approach to determining the heat-transfer coefficient for a cylinder in cross flow is to establish a relationship for Nu as a function of Pr and Re using a series of experiments with different diameter tubes, different fluids, and different velocities. The results of several experiments performed in the 1920s and 1930s have been correlated by Kreith [1] with a graph of

$$\frac{\text{Nu}}{\text{Pr}^{0.3}} = f(\text{Re}) \qquad (6.27)$$

as shown in Fig. 6.26. All the available data for oil, water, and air fall within the band of ±15 percent shown in the figure. The centerline of this band has been obtained by curve-fitting methods and is given adequately by the equation

$$\text{Nu} = 0.82\ \text{Pr}^{0.3}\ \text{Re}^{0.4} \qquad (6.28)$$

when Re is between 3 and 100. Remembering that $\text{Nu} = hd/k$, we can make use of Fig. 6.26 or Eq. (6.28) to determine the heat-transfer coefficient at the outside of the cylinder. Note that the resulting h is an average value representing the overall cylindrical surface. The results are useful only when fluid is flowing past the cylinder. For the case of a cylinder sitting in still air, free convection resulting from air currents created by the warming or cooling of air surrounding the cylinder produces a heat-transfer coefficient in a different manner and must be studied in different experiments.

As an example of the use of Eq. (6.28), consider that we have a wire in an electric assembly and a fan is blowing cooling air over the assembly. In the process of finding the current which can be safely carried by the wire, we need to know the heat-transfer coefficient. The wire diameter is $d = \frac{1}{16}$ in and the air blowing across it has a velocity of $V = 1$ ft/s. If the air near the wire has a temperature of about 100°F, we find the properties of the air to have the values $\rho = 2.2 \times 10^{-3}$ slugs/ft^3, $\mu = 3.8 \times 10^{-7}$ lbf·s/ft^2, $C_p = 6000$ ft·lbf/(slug)(°R), and $k = 3.4 \times 10^{-3}$ ft·lbf/(s)(ft)(°F). These may be seen in Tables 6.1 to 6.3. We can calculate the values for the dimensionless groups as follows:

$$\text{Pr} = \frac{C_p \mu}{k} = \frac{6000 \text{ ft·lbf/(slug)(°R)}\ 3.8 \times 10^{-7} \text{ lbf·s/ft}^2}{3.4 \times 10^{-3} \text{ ft·lbf/(s)(ft)(°F)}} = 0.67$$

and

$$\text{Re} = \frac{V d \rho}{\mu} = \frac{1 \text{ ft/s}\ \frac{1}{(16)(12)} \text{ ft}\ 2.2 \times 10^{-3} \text{ slugs/ft}^3}{3.8 \times 10^{-7} \text{ lbf·s/ft}^2} = 30$$

These values for Pr and Re can then be used in Eq. (6.28) to find

$$\text{Nu} = 0.82 \, \text{Pr}^{0.3} \, \text{Re}^{0.4} = (0.82)(0.67^{0.3})(30^{0.4}) = 2.8$$

Since Nu = hd/k, we now have

$$h = 2.8 \frac{k}{d} = 2.8 \frac{3.4 \times 10^{-3} \text{ ft·lbf/(s)(ft)(°F)}}{\frac{1}{(16)(12)} \text{ ft}}$$

$$= 1.8 \text{ ft·lbf/(s)(ft}^2\text{)(°F)}$$

or

$$h = 8.3 \frac{\text{Btu}}{\text{(h)(ft}^2\text{)(°F)}}$$

This is the average heat-transfer coefficient around the wire.

CALCULATING EQUILIBRIUM TEMPERATURE

The various equations of heat transfer are often used to predict the temperatures of objects or bodies. For example, consider Eq. (6.20) for conduction:

$$\dot{Q} = \frac{kA}{l}(\tau_H - \tau_L)$$

Or Eq. (6.23) for convection:

$$\dot{Q} = hA(\tau_s - \tau_\infty)$$

If the heat-flow rate is known, either equation can be used to calculate temperature. In the heat-conduction equation, when the temperature on one side of a wall is known, the other temperature can be found. If the environment's temperature τ_∞ is known in convection, the surface temperature τ_s can be calculated. In cases where combinations of convection and conduction occur, equilibrium temperatures can still be calculated by combining the appropriate equations as long as the heat-flow rate is defined.

We will look at the tent-heating problem of Fig. 6.27 for an example of temperature computation. Inside is a small heater producing heat at a rate of 1500 W. We assume that air for ventilation does not carry off a significant amount of this heat. $\dot{Q}$ = 1500 W must thus be transferred by convection from the air inside the tent to the tent walls and then from the tent walls to the outside air. The thin tent wall should not provide any significant

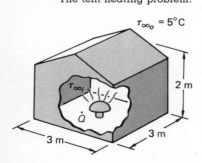

FIGURE 6.27
The tent heating problem.

insulation, so we will let the inside and outside surface temperatures be the same with $\tau_{si} = \tau_{so} = \tau_s$. An average value of $h_i = 7$ W/(m²)(K) would be a typical value for the convective heat-transfer coefficient between the inside air and the inside surface. An average heat-transfer coefficient between the outside air and outside surface of $h_o = 10$ W/(m²)(K) would be reasonable for a calm night. For these conditions and an outside air temperature of $\tau_{\infty o} = 5°C$, we would like to calculate $\tau_{\infty i}$, which is the inside air temperature. The convection equations for the inside and outside are, respectively,

$$\dot{Q} = h_i A(\tau_{\infty i} - \tau_s) \tag{6.29}$$

and

$$\dot{Q} = h_o A(\tau_s - \tau_{\infty 0}) \tag{6.30}$$

We find the total surface area of the tent to be approximately

$$A = 4(2 \times 3) + 3 \times 3 = 33 \text{ m}^2$$

Substituting the various numbers in Eqs. (6.29) and (6.30), we obtain

$$1500 = 7 \times 33(\tau_{\infty i} - \tau_s) \tag{6.31}$$
$$1500 = 10 \times 33(\tau_s - 5) \tag{6.32}$$

Equation (6.32) can be solved for the wall temperature:

$$\tau_s = 5 + \frac{1500}{330} = 9.5°C$$

and this is substituted into Eq. (6.31) to obtain the inside temperature

$$\tau_{\infty i} = \tau_s + \frac{1500}{7 \times 33} = 9.5 + \frac{1500}{231}$$
$$= 16°C$$

This inside temperature is, of course, an average value, which does not represent the temperature very close to the heater, for example. The average inside temperature will gradually rise to this value after the heater is ignited. The basic principle here is that the heat produced inside the tent must be dissipated by convection from the outer surface of the tent. The tent-interior temperature and tent-wall temperature rise until the heat given off by the tent equals the heat produced inside. After the equilibrium temperature is reached, if the heater output were to be decreased slightly, the tent-interior and -wall temperature would fall until the convected heat from the tent matched the new heater output.

If there were no internal heat generation, there would be no convective heat transfer. Equations (6.29) and (6.30) then tell us that there would be no temperature rise and the interior, wall, and exterior temperatures would all tend to be equal. It might be mentioned, however, that people give off heat. Therefore, even without a heater, the presence of people in our tent would create an increase in the interior temperature.

PROBLEMS

6.17 For the one-dimensional conduction problem in Fig. P6.17, τ_i and τ_o are the temperatures of the inner and outer surfaces. Determine the rate of conduction heat transfer $\dot{Q}$ when:
(a) $\tau_i = \tau_o$
(b) $\tau_i = 20°C$, $\tau_o = 10°C$, $k = 0.035$ W/(m)(K)
(c) $\tau_i = 20°C$, $\tau_o = 0°C$, $k = 0.035$ W/(m)(K)
(d) $\tau_i = 20°C$, $\tau_o = 0°C$, $k = 0.10$ W/(m)(K)

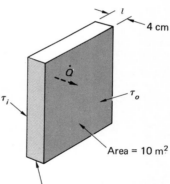

FIGURE P6.17   Slab with thermal conductivity k

6.18 The R value commonly used in home insulation has been defined so that $R = l/k$ (thickness divided by conductivity) in units of hours-square feet-degrees Fahrenheit per British thermal unit. A particular insulation has an R value of 20 with a thickness of 6 in. Determine the thermal conductivity of that insulation.

6.19 A homeowner has insulated a 1750-ft² attic with R30 insulation. On a certain day, the inner temperature of the insulation is 65°F and the outer temperature of the insulation is 30°F. What is the heat loss through the attic insulation?

which means that during a short time interval Δt in which a velocity change Δv occurs, the acceleration approximately equals $\Delta v/\Delta t$. Equation (7.6) becomes less of an approximation with smaller time intervals and becomes equivalent to Eq. (7.5) when the time interval becomes infinitesimal, approaching zero. The d's are used instead of Δ's to indicate this.

We now return to Eq. (7.4) but replace a by Eq. (7.5). This produces

$$m \frac{dv}{dt} + bv = F_p \qquad (7.7)$$

which is a differential equation for velocity v as a function of force F_p and time t. It is called a differential equation because it contains derivatives. After the equation is "solved" in the normal mathematical sense, an algebraic equation is obtained for v in terms of F_p and t. We can also produce a computer-based or "numerical" solution if we know the particular values for m and b and have a description of the force F_p. The algebraic solution for a differential equation is normally most desirable, but computer-based solutions can be obtained for difficult differential equations when algebraic solutions are not possible. These solutions are an important aspect of modern technology, including computations for automotive and electronic design, combustion and noise studies, as well as those for economic and weather forecasting. The computer-based numerical solutions are often referred to as "computer simulations." Although Eq. (7.7) is a simple differential equation, it is beyond our scope here to obtain its algebraic solution. We can, however, solve it readily by a computer simulation, as you will see below.

A SIMPLE NUMERICAL SOLUTION

Consider a case where the vehicle of Fig. 7.3 and Eq. (7.7) is initially at rest so that at the instant when time $t = 0$ and $v = 0$ and a constant propulsion force of $F_p = 250$ lbf is suddenly applied. The vehicle has a mass of $m = 100$ slugs (3220 lbm). Recall that $b = 5$ lbf·s/ft, as shown in Fig. 7.3d. We want to determine the vehicle velocity as a function of time. To begin, we return to Eq. (7.4) and make use of the approximate Eq. (7.6) to find

$$m \frac{\Delta v}{\Delta t} + bv = F_p \qquad (7.8)$$

This equation is then rearranged to compute the change in vehicle velocity during the time interval Δt. If v_1 and t_1 are the velocity and time at the beginning of the interval and v_2 and t_2 are the values at the end of the interval, we can rearrange Eq. (7.8) to compute the change in velocity as

$$\Delta v = \frac{1}{m}(F_p - bv_1)\Delta t \qquad (7.9)$$

so that the velocity v_2 and time t_2 at the end of the time interval are then

$$v_2 = v_1 + \Delta v \qquad (7.10)$$

and

$$t_2 = t_1 + \Delta t \qquad (7.11)$$

Note that in Eq. (7.9) we use v_1 on the right-hand side in computing Δv. This is necessary, of course, since we do not know v_2 until we have used the results of Eq. (7.9) in Eq. (7.10). By starting with a known velocity at a known time, we can use Eqs. (7.9) to (7.11) repeatedly to determine the transient response of the vehicle velocity. If the time step Δt is sufficiently small, the result will be accurate.

Let's now carry out these calculations for our sudden application of $F_p = 250$ lbf. With the defined values for m and b, Eq. (7.9) becomes

$$\Delta v = \tfrac{1}{100}(250 - 5v_1)\Delta t \qquad (7.12)$$

Our first attempt at the transient calculation is made with $\Delta t = 10$ s. Beginning the first time interval at $t = t_1 = 0$, where $v = v_1 = 0$, we calculate the first Δv as

$$\Delta v = \tfrac{1}{100}(250 - 0)10 = 25 \text{ ft/s}$$

Equation (7.10) then yields

$$v_2 = v_1 + \Delta v = 0 + 25$$
$$= 25 \text{ ft/s}$$

and Eq. (7.11) results in

$$t_2 = t_1 + \Delta t = 0 + 10$$
$$= 10 \text{ s}$$

Our procedure in the second time interval begins by updating the values of time and velocity, so that we now set $t_1 = 10$ s and $v_1 = 25$ ft/s. This results in

$$\Delta v = \tfrac{1}{100}[250 - (5)(25)](10)$$
$$= 12.5 \text{ ft/s}$$

and

$$v_2 = v_1 + \Delta v = 25 + 12.5$$
$$= 37.5 \text{ ft/s}$$

with

$$t_2 = t_1 + \Delta t = 10 + 10$$
$$= 20 \text{ s}$$

This defines the conditions at the end of the second time interval. We update again, setting the time and velocity at the beginning of the third interval to $t_1 = 20$ s and $v_1 = 37.5$ ft/s. The calculation then continues repetitively, in the manner outlined, as long as desired.

The results for this case are shown in Table 7.1, where we can see that a steady-state velocity is gradually approached. However, the solution in Table 7.1 is not an accurate one, since this Δt is not sufficiently small. Making Δt smaller requires many more time steps for calculation of the transient response. However, digital computers are natural tools for this type of repetitive calculation. Small time steps and accurate transient solutions can be obtained conveniently by digital computation. A FORTRAN com-

TABLE 7·1
TRANSIENT CALCULATIONS ($\Delta t = 10$ s)

Time Step	t_1, s	v_1, ft/s	Δv, ft/s
1	0	0.00	25.00
2	10	25.00	12.50
3	20	37.50	6.25
4	30	43.75	3.13
5	40	46.88	1.56
6	50	48.44	0.78
7	60	49.22	0.39
8	70	49.61	0.20
9	80	49.81	0.10
10	90	49.91	0.05
11	100	49.96	

FIGURE 7·4
FORTRAN program for the vehicle response.

```
      PROGRAM DYNAM
      XMASS = 100.                              Define the mass m
      B = 5.                                    and drag b.
      DELT = 2.                                 Set time interval Δt,
      NBOUT = 5                                 number of steps between printer
      TLIM = 100.                               outputs and desired limit on time.
      TONE = 0.                                 Set initial time t₁ and
      VONE = 0.                                 initial velocity v₁.
      KOUNT = NBOUT                             Initialize output counter.
  300 IF (KOUNT.NE.NBOUT) GO TO 100             Should we output now?
      WRITE(6,200) TONE, VONE                   Output
  200 FORMAT (1X, 2 F10.2)                      statements,
      KOUNT = 0                                 reset output counter.
  100 FP = 250.                                 Calculate the
      DELV = (FP − B*VONE)*DELT/XMASS           new values
      TONE = TONE + DELT                        for t₁
      VONE = VONE + DELV                        and v₁.
      KOUNT = KOUNT + 1                         Count the step.
      IF (TONE.LT.TLIM) GO TO 300               Are we finished?
      STOP
      END
```

TABLE 7·2
TRANSIENT CALCULATIONS
($\Delta t = 2$ s)

Time Step	t_1, s	v_1, ft/s
1	0	0
6	10	20.48
11	20	32.57
16	30	39.71
21	40	43.92
26	50	46.41
31	60	47.88
36	70	48.75
41	80	49.26
46	90	49.56
51	100	49.74

TABLE 7·3
TRANSIENT CALCULATIONS
($\Delta t = 0.010$ s)

t_1, s	v_1, ft/s
0	0
10	19.68
20	31.61
30	38.85
40	43.24
50	45.90
60	47.51
70	48.49
80	49.09
90	49.45
100	49.66

puter program for our current problem is presented in Fig. 7.4 along with some descriptive information. In looking through this program, note the changes from our algebraic symbols to comparable FORTRAN variable names. The nature of the solution and the time between outputs is controlled by the time interval $\Delta t =$ DELT and NBOUT, which is the number of calculations between outputs. With DELT = 10 and NBOUT = 1, the results of Table 7.1 would be obtained by executing this computer program. With DELT = 2 and NBOUT = 5 as shown in Fig. 7.4, the results in Table 7.2 are obtained. This value of $\Delta t = 2$ s is still relatively large and the results in Table 7.2 are not very accurate. However, using the computer we can easily obtain results with much smaller Δt. In Table 7.3 we find results obtained by the program with $\Delta t =$ DELT = 0.010 s and NBOUT = 1000. This value of Δt is adequately small for this problem and the results of Table 7.3 are quite accurate. The convenient way to check on the smallness of Δt is to reduce Δt further (normally cutting it in half). If no important change in results is observed, Δt is small enough. In this case, for example, if we set $\Delta t =$ DELT = 0.0050 s and NBOUT = 2000 and run the program, the identical values shown for t_1 and v_1 in Table 7.3 are again obtained. Often extra significant figures are included in a computer output to ensure sufficient accuracy in this comparison process. We have plotted the results of Tables 7.1 to 7.3 in Fig. 7.5. Here you can see the velocity gradually approaching steady state for the accurate solution ($\Delta t = 0.010$ s) and the inaccuracy during the transient when too large a Δt is used.

The topic we have been discussing here is known as the numerical solution of differential equations. The particular technique—a simple one—is called Euler's method after the

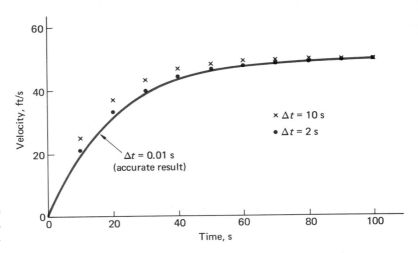

FIGURE 7·5
A plot of the transient calculations.

nineteenth century mathematician whose many discoveries are of great importance in modern technology. There are more complex methods available today for computer use which we will not try to discuss. These have been developed mostly to allow larger Δt during a solution while still producing an accurate result. This can lead to important savings in computer time for complex transient problems.

PROBLEMS

7.1 A camera and film for making a still photograph may be considered to be an engineering system. If we think of the output of this system as being the quality of a particular photograph, list the various things which should be thought of as inputs to the system while taking the photograph.

7.2 Consider the following common objects to be engineering systems. Define the things you might view as inputs to each system while it is in operation and the things you think are outputs. (a) A clothes washing machine; (b) a bicycle; (c) a television receiver; (d) a motorboat.

7.3 Keypunch the FORTRAN program of Fig. 7.4 and run it on an available computer system. Adjust DELT to a small value to obtain an accurate solution. Compare your result with Fig. 7.5. Cut your value of DELT in half and run the program again. If your two results are the same, you no doubt have an accurate solution.

7.4 For the transient-response problem of Fig. 7.3, the force F_p changes suddenly from 0 to 250 lbf at time $t = 0$. If F_p then changes back to zero at time $t = 50$ s, what is the complete response of v to this new input? Modify the FORTRAN program in Fig. 7.4 as necessary to obtain the complete and accurate solution. Plot a graph which shows the velocity v as a function of time from $t = 0$ to $t = 250$ s.

7.5 A certain small boat has a mass $m = 300$ kg. In the water it has drag-force characteristics approximately as shown in Fig. P7.5, which includes the change in drag as the boat begins to plane. At time $t = 0$, the boat is at rest ($v = 0$) when a force $F_p = 2000$ N suddenly begins to accelerate it. Modify the FORTRAN program of Fig. 7.4 to obtain an accurate solution to this problem. Plot a graph of velocity vs. time until a steady-state velocity is reached.

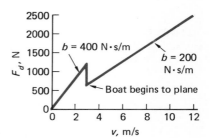

FIGURE P7·5

7·2 FIRST-ORDER SYSTEMS

The simple vehicle system and the transient response which we have described above are typical of first-order systems. These systems can be recognized because their differential equation has only one derivative term, as we saw in Eq. (7.7), which is repeated below

$$m \frac{dv}{dt} + bv = F_p \qquad (7.7)$$

The one derivative here is, of course, dv/dt. As we will see later, there are many different kinds of systems that can be represented by a differential equation with only one derivative. The general behavior of these systems can be quite predictable. It is characterized by the time constant of the system if the terms on the left side of the equation are the system output and its derivative multiplied by constant coefficients. This is the situation in Eq. (7.7), where the system output is v, its derivative is dv/dt, and the constant coefficients are m and b. To obtain the time constant, we divide the equation by the coefficient of the output variable. Equation (7.7) then becomes

$$\frac{m}{b} \frac{dv}{dt} + v = \frac{F_p}{b} \qquad (7.13)$$

and the time constant, which is symbolized by the Greek letter tau, is

$$\tau = \frac{m}{b} \tag{7.14}$$

This quantity will always have the dimension of time. For our example, it is in units of seconds, as can be seen by substituting units into Equation 7.14:

$$\tau = \frac{\text{slugs}}{\text{lbf·s/ft}}$$

Remembering that 1 lbf = 1 slug·ft/s²,

$$\tau = \frac{\cancel{\text{slugs}}}{\frac{(\cancel{\text{slug}})(\cancel{\text{ft}})}{\cancel{\text{s}^2}} \cdot \frac{\text{s}}{\cancel{\text{ft}}}} = \text{s}$$

and the units of τ are indeed seconds.

The importance of the time constant is that it provides an immediate indication of how long is required for the system to respond to a change in input. When a first-order system is at equilibrium and its input suddenly changes to a new constant value, the system output will make 63 percent of the change to a new equilibrium value one time constant (τ units of time) after the change. After three time constants (3τ units of time), 95 percent of the change is complete. Figure 7.6 shows a typical transient response of a first-order system to a sudden input change. This is, of course, the same general form as the transient we found in Fig. 7.5 where $\tau = m/b = 20$ s. The time constant is a convenient term for characterizing the dynamic behavior of a system. If may often be used in-

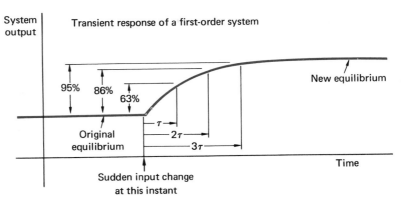

FIGURE 7·6 Response of a first-order system to a sudden input change.

7.2 FIRST-ORDER SYSTEMS

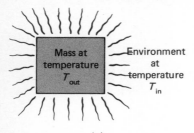

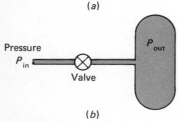

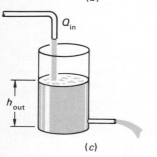

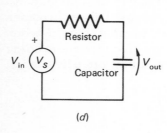

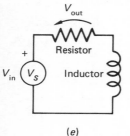

FIGURE 7·7
Common first-order systems.

formally to describe systems which display a transient response similar to Fig. 7.6, even if the equation for the system is unknown or does not have the form of Eq. (7.7).

Figure 7.7 contains several examples of systems which are essentially first order. Figure 7.7a shows a simple solid mass with a uniform temperature where the temperature of its surroundings T_{in} is like a system input and controls the mass temperature T_{out}. At equilibrium, of course, T_{out} must equal T_{in}. But if the surrounding temperature T_{in} suddenly changes, T_{out} will begin changing and gradually approach the new T_{in} with a response having the form of Fig. 7.6. Similarly, in Fig. 7.7b the tank air pressure P_{out} will respond to the applied pressure P_{in}. In Fig. 7.7c, the height h_{out} of fluid in the tank with outflow depends on the rate of input flow Q_{in}. For the electric circuit in Fig. 7.7d, the capacitor voltage V_{out} responds to applied voltage V_{in}. In the inductance circuit of Fig. 7.7e, the resistor voltage V_{out} has a first-order response to the voltage V_{in}. All these systems can be represented by a first-order differential equation similar to Eq. (7.7). Solutions to the equations can be obtained using the numerical technique we discussed earlier. The response of any of these systems to a sudden change in input will show the same general behavior as we found in Figs. 7.5 and 7.6; and a time constant can be defined for each system which characterizes the amount of time required for the response of that system.

CALCULATING EQUILIBRIUM VALUES AND RATE OF CHANGE

The differential equation which we found to describe the velocity of our simple vehicle model in terms of propulsion force was

$$m \frac{dv}{dt} + bv = F_p \qquad (7.7)$$

Like all differential equations, this is valid for the system at any instant of time. It can be used directly to make simple computations for any instant without completely solving the differential equation.

For example, to find the equilibrium velocity of the vehicle from Eq. (7.7), we let $dv/dt = 0$ and solve for v. We do this because we know that at equilibrium the velocity is not changing and its rate of change dv/dt must equal zero. Thus

$$m(0) + bv = F_p$$

at equilibrium and therefore at equilibrium

$$v = \frac{F_p}{b}$$

In our original problem, we have $F_p = 250$ lbf and $b = 5$ lbf·s/ft so that the equilibrium velocity

$$v = \frac{250}{5} = 50 \text{ ft/s}$$

This agrees with the results we obtained in Fig. 7.5 after extensive calculation. The response gradually approaches an equilibrium value of $v = 50$ ft/s.

We can use the approach above to also calculate rates of change at any instant of time assuming we know the velocity. Suppose that we want to know the rate of change of velocity (acceleration) of our vehicle at the instant F_p was applied when the velocity was zero. Equation (7.7) at that instant is

$$m\frac{dv}{dt} + b(0) = F_p$$

and we see that

$$\frac{dv}{dt} = \frac{F_p}{m}$$

For the values of $F_p = 250$ lbf and $m = 100$ slugs that we have considered, we find that when $v = 0$,

$$\frac{dv}{dt} = \frac{250}{100} = 2.5 \text{ ft/s}^2$$

A check on the slope of the curve in Fig. 7.5 at the instant where $t = 0$ and $v = 0$ shows the acceleration to be in agreement with this convenient calculation. In the same way, the rate of change dv/dt can be calculated from Eq. (7.7) for any specified velocity. Other differential equations can be treated similarly.

PROBLEMS

7.6 A pot of water is brought to a boil. The heat is removed and it is allowed to cool to room temperature of 20°C. While cooling, its temperature is measured at various times and the graph in Fig. P7.6 is plotted. From the graph, estimate the time constant τ during the cooling process.

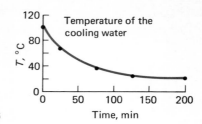

FIGURE P7·6

7.7 Consider the first-order differential equations below. Determine the time constant τ and the equilibrium, or steady-state, value for v_o when $v_i = 100$.

(a) $\dfrac{dv_o}{dt} + v_o = v_i$

(b) $5\dfrac{dv_o}{dt} + v_o = v_i$

(c) $5\dfrac{dv_o}{dt} + 2v_o = v_i$

(d) $5\dfrac{dv_o}{dt} + 2v_o = 5v_i$

7.8 The differential equation

$$10\frac{dT_{out}}{dt} + T_{out} = T_{in}$$

describes the temperature of the mass in Fig. 7.7a. The temperature T_{out} is initially 70°F. At time $t = 0$, the temperature T_{in} is suddenly changed to 100°F. Sketch a graph showing T_{out} versus time as it approaches 100°F.

7.9 A driver performs a transient-response test to estimate the drag characteristics of his car. While driving on a deserted road, he obtains coast-down data by putting the car in neutral and reading the speedometer at various times as it slows down. He is able to obtain the graph shown in Fig. P7.9. The weight of the car is 2415 lb. Determine the value of the viscous friction coefficient b indicated by these results. Your units for b should be pounds-force-seconds per foot.

FIGURE P7·9

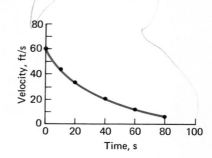

FIGURE 7·16
Symbol for an inductor.

ergy in electromagnetic form with the strength of the magnetic field dependent on the amount of electric current and on the nature of the conductor. Coiling the wire conductor and inserting an iron core enhances the magnetic field.

Inductance is the result of electromagnetic effects. Inductors are normally made of coiled wire and often have iron cores. In fact, the symbol for an inductor is a coil as shown in Fig. 7.16. When the inductor is part of an electric circuit, any change in the current i through the inductor creates a voltage v across it, so that

$$v = L \frac{di}{dt} \tag{7.38}$$

where L represents the amount of inductance. The derived unit for L is volt-seconds per ampere, which is called the henry (symbol H). As you will see below, Eq. (7.38) is used as the component equation for an inductor when developing circuit equations.

DYNAMICS IN ELECTRIC CIRCUITS

When we introduced the concept of transient response in Secs. 7.1 to 7.3, we used examples of mechanical systems. We developed differential equations by working with Newton's law and the component equations for the mass, the spring, and damping components of mechanical systems. We combined the component equations to form a differential equation for the system and we were then able to solve the differential equation and study the transient response. The same approach can also be used for dynamic electric systems.

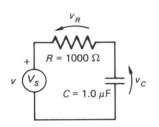

FIGURE 7·17
A resistor-capacitor circuit.

Consider the resistor-capacitor circuit of Fig. 7.17. Here we see a voltage source providing the input voltage v. We also see the voltage drop v_R across the resistance and the capacitor voltage v_C. The resistor and capacitor are in series, so we would like to establish a differential equation relating the voltage v_C (which will be the system output) to the voltage v (the system input). We begin by writing the component equation for the resistor:

$$v_R = Ri_R \tag{7.39}$$

and for the capacitor:

$$i_C = C \frac{dv_C}{dt} \tag{7.40}$$

We realize that since the resistor and capacitor are in series, the currents through them must be equal, so

$$i_R = i_C \tag{7.41}$$

We also realize that the sum of the voltages around the loop must be

$$v - v_R - v_C = 0$$

or

$$v_R + v_C = v \tag{7.42}$$

Since an equation with v_C and its derivatives on the left and v on the right is desired here to relate the system output and input, we should focus our attention on Eq. (7.42). Let's substitute Eq. (7.39) for v_R. We then have

$$Ri_R + v_C = v$$

Since $i_R = i_C$, we obtain

$$Ri_C + v_C = v$$

Now we use the capacitor relation in Eq. (7.40) to find

$$RC\frac{dv_C}{dt} + v_C = v \tag{7.43}$$

This is the desired differential equation for output v_C in terms of input v. Note that it is a first-order differential equation, since it has only a first derivative term. Thus, the circuit of Fig. 7.17 is a first-order system.

Refer back to Sec. 7.2 where we discussed first-order systems. There we said that in the form of Eq. (7.43), where unity is the coefficient of v_C, the system time constant τ is the coefficient of the derivative term. Here

$$\tau = RC$$

We can check the units for τ by recalling that ohms are volts per ampere and farads are ampere-seconds per volt. The units for the time constant are then

$$\tau = \frac{V}{A}\frac{A \cdot s}{V} = s$$

so τ is measured in seconds. For the circuit of Fig. 7.17, we have $R = 1000\ \Omega$ and $C = 1.0 \times 10^{-6}$ F; thus

$$\tau = RC = (1000)(1.0 \times 10^{-6})$$
$$= 0.001\ s$$

This is a time constant of one millisecond. The voltage v_C in Fig. 7.17 will respond to a sudden change in the voltage v, as shown in Fig. 7.6. After a change in v from one constant value to another, v_C will gradually change from its original equilibrium value to the new equilibrium value. In this case, "gradually" means that 63 percent of the change will be completed in 0.001 s and nearly all the change will be completed in 0.003 s.

The relationship between the equilibrium value for v_C and any constant value for v can be found by inspecting the system differential equation. For equilibrium $dv_C/dt = 0$; thus Eq. (7.43) shows us that

$$RC(0) + v_C = v \quad \text{and} \quad v_C = v$$

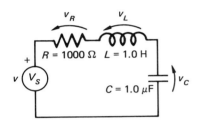

FIGURE 7·18
An R-L-C circuit.

The capacitor voltage at equilibrium is thus equal to the constant input voltage for our simple circuit.

Our next sample of a dynamic electric circuit is shown in Fig. 7.18. To form this circuit, we have introduced an inductor of one henry into the circuit of Fig. 7.17; we now have an R-L-C series circuit. With two nonresistance elements in the circuit (the inductor and the capacitor), you will see that this electric circuit is a second-order system.

As we analyze the circuit of Fig. 7.18, we will again consider the source voltage v to be our system input and the capacitor voltage v_C to be the system output. We will establish the differential equation, now relating this output v_C to the input v. The component equations for the resistor, capacitor, and inductor are, respectively,

$$v_R = Ri_R \tag{7.44}$$

$$i_C = C\frac{dv_C}{dt} \tag{7.45}$$

$$v_L = L\frac{di_L}{dt} \tag{7.46}$$

These must be combined with the circuit equations to obtain the differential equation. Since the elements are all in series, we know that the currents in each must be equal:

$$i_R = i_L = i_C$$

The voltages around the loop may be summed, so that

$$v - v_R - v_L - v_C = 0$$

which becomes

$$v_R + v_L + v_C = v \tag{7.47}$$

We want only v_C and its derivatives on the left, so we substitute the equations for the resistor and inductor into Eq. (7.47) to obtain

$$Ri_R + L\frac{di_L}{dt} + v_C = v$$

We realize that the currents are equal, so that

$$Ri_C + L\frac{di_C}{dt} + v_C = v$$

We then apply the capacitor component equation from Eq. (7.45) to find

$$RC\frac{dv_C}{dt} + L\frac{d}{dt}\left(C\frac{dv_C}{dt}\right) + v_C = v$$

The derivative operation is performed; we rearrange to obtain

$$LC\frac{d^2v_C}{dt^2} + RC\frac{dv_C}{dt} + v_C = v \qquad (7.48)$$

This is now a second-order differential equation which describes the circuit of Fig. 7.18. We could use our numbers for R, L, and C in this equation and carry out a computer simulation similar to the one shown in Fig. 7.9 for the mechanical system or we could use methods of differential equations[2] to solve the equation. However, in Sec. 7.3, we have already discussed the general behavior of second-order systems. Let's use the ideas we established there to help us understand the transient behavior of the circuit described by Eq. (7.48).

If we consider a sudden change in the input v from one constant value to another constant value, the system will follow the general behavior shown in Fig. 7.12. The equilibrium value of v_C can be seen directly from Eq. (7.48) because we know the derivatives must be zero at equilibrium, so

$$LC(0) + RC(0) + v_C = v$$

and thus

$$v_C = v$$

---

[2] These methods are still beyond our scope.

at equilibrium. If v changes suddenly from one constant value to another, v_C will be changing between the same two values. However, its transient is governed by the natural frequency f_n and damping ratio ζ corresponding to Eq. (7.48).

Refer back to Eq. (7.33), where we described the natural frequency and damping ratio. There we stated that with a unity output coefficient (on v_C here), the coefficient of the second derivative term defines the natural frequency. In Eq. (7.48), the coefficient of interest is LC and

$$\frac{1}{4\pi^2 f_n^2} = LC$$

This can be solved to obtain

$$f_n = \frac{1}{2\pi}\sqrt{\frac{1}{LC}}$$

Substituting $L = 1.0$ H and $C = 1.0$ μF, we find that

$$f_n = \frac{1}{2\pi}\sqrt{\frac{1}{1.0 \times 1.0 \times 10^{-6}}}$$
$$= 159 \text{ Hz}$$

This system will then have a tendency to oscillate at 159 Hz unless there is substantial damping. We check the damping ratio by returning again to our discussion of Eq. (7.33), where we found that the coefficient of the first derivative term defines the damping ratio. This means that

$$\frac{\zeta}{\pi f_n} = RC$$

for Eq. (7.48) with $R = 1000$ Ω and $C = 1.0 \times 10^{-6}$ μF, we have

$$\zeta = \pi(159)(1000)(1.0 \times 10^{-6}) = 0.50$$

This result tells us that the variation of v_C during the transient resulting from a sudden change in v will show an oscillation (ζ is less than 1). However, according to Figs. 7.11 and 7.12, v_C will quickly settle down to equilibrium, since $\zeta = 0.50$ is not very small. The change to the new equilibrium position should be expected to finish within the time needed for two natural oscillations, which is $2/f_n = 2/159 = 0.013$ s.

We have been able to make a description of the transient behavior by simply inspecting the differential equation of Eq.

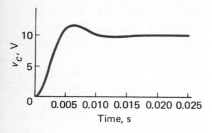

FIGURE 7·19
Calculated response for the R-L-C circuit.

(7.48) and determining the natural frequency and damping ratio. The expected response of v_C to a sudden change in v from 0 to 10 V has been carefully computed and the results are shown in Fig. 7.19. You can see that our discussion above has been generally accurate.

PROBLEMS

7.14 Consider Fig. 7.17 with the new values of $R = 10,000\ \Omega$ and $C = 10.0\ \mu\text{F}$. What is the differential equation now describing the relationship between v_C and v? Determine the time constant τ.

7.15 In Prob. 7.14, $v_C = v = 50$ V when suddenly, at $t = 0$, the input changes to $v = 100$ V. Sketch a graph of the output v_C versus t during the transient response to this change in v.

7.16 In Fig. 7.18, let $R = 100\ \Omega$ with $L = 10$ H and $C = 10\ \mu\text{F}$. Write the differential equation which relates the output v_C to the input v with these new values. What is the new natural frequency f_n and damping ratio ζ?

7.17 The system of Prob. 7.16 is at equilibrium with $v_C = v = 0$ V when, at $t = 0$, the input v changes suddenly to $v = 10$ V. Sketch a graph of the output v_C versus t during the response to this change in v.

7·5 FREQUENCY RESPONSE

In the previous sections of this chapter, we have given considerable attention to the transient response of dynamic systems. We have discussed specifically the transient response of a system occurring after a sudden change in the system input from one constant value to another constant value. This is an important type of transient condition that shows the basic character of system behavior. However, inputs and outputs in real systems are often much more general and complex than the type we have discussed. In this section, we will describe the concept of frequency response, which is the approach used commonly to understand system behavior under more general conditions.

Some examples of time histories of variables in realistic systems are shown in Fig. 7.20. Figure 7.20a shows the type of waviness that a typical road might have as seen by a car driving along it. This graph is a plot of road-surface vertical position as a function of time. The waviness shown serves as an input to the car's suspension system; the output of the suspension system might be the acceleration felt by a passenger seated in the vehicle. A representative plot of this acceleration as a function of time is depicted in Fig. 7.20b. For an audio system in a concert

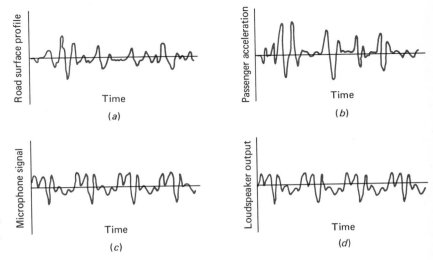

FIGURE 7·20
Typical plots of system inputs and outputs.

hall, the voltage produced by a microphone may take on the graphical form of Fig. 7.20c. This is the input to the public address system. The output of the system is the sound produced by the loudspeakers, which may appear as in Fig. 7.20d. The graphs shown in Fig. 7.20 are typical of time histories that may be found in real systems. It should be clear that these graphs represent complex behavior—much more complex than an input changing from one constant value to another.

BACKGROUND

The concept of frequency response is based on the most simple periodic function:

$$z = \sin(2\pi ft) \tag{7.49}$$

The sine in this equation is the same sine that you are familiar with from algebra and trigonometry. The quantity $2\pi ft$ is merely an angle, expressed in radians. The variable t stands for time, of course, and is normally in units of seconds. The variable f is frequency, in hertz (or cycles per second). The important property of the sine function here is that the sine of an angle repeats itself every 360°, or every 2π rad. For example, $\sin \pi/2 = \sin(\pi/2 + 2\pi) = \sin(\pi/2 + 4\pi) = 1$ and $\sin 3\pi/2 = \sin(3\pi/2 + 2\pi) = \sin(3\pi/2 + 4\pi) = -1$.

If we plot a graph of z versus t using Eq. (7.49) with $f = 2$ Hz, we obtain the result shown in Fig. 7.21. You should be able to check this graph by simply plugging f and any t into Eq. (7.49). Re-

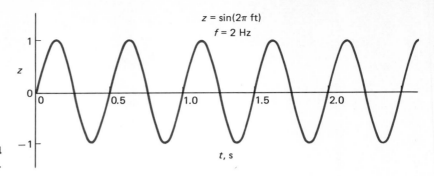

FIGURE 7·21
Plot of a sine wave.

member that the angle is in radians. Note that Fig. 7.21 shows two complete cycles of the sine "wave" every second, which is why we say the frequency is 2 cycles per second, or 2 Hz.

The sinusoidal function of Eq. (7.49) is called a periodic function because it repeats itself periodically. It is the most basic periodic function, one that is frequently found in natural phenomena—as in the vibrations of a tuning fork, the oscillations of a pendulum, and the generation of alternating electric current. The natural oscillations of our mass-spring system shown in Fig. 7.10 are also sinusoidal. Equation (7.49) allows us to describe this simple and common periodic behavior; but we are not yet able to describe complex waveforms of the type we saw in Fig. 7.20. Consider now the equation

$$z = A \sin (2\pi f t + \phi) \qquad (7.50)$$

The addition of the variables A and ϕ (Greek letter phi) now define the amplitude and phase of the sine wave. The amplitude, controlled by A, simply refers to the size of the wave. The phase, or phase angle, of the wave corresponds to ϕ, which is an angle measured in radians. It shifts the wave along the time axis. Some examples of Eq. (7.50) with different values of A and ϕ are shown in Fig. 7.22. A comparison of Fig. 7.22b and d in particular shows the effect of phase where $\phi = \pi$ radians (180°) completely reverses the behavior of the sine wave.

Complex input signals or waveforms of the type shown in Fig. 7.20 can be formed by adding together sine waves of different frequencies, different magnitudes, and different phase. This summation of a series of sine waves is referred to as a Fourier series after the French scientist who described the procedure in 1822. The mathematical form of a Fourier series can be written as

$$z = \sum_{i=1}^{n} A_i \sin (2\pi f_i t + \phi_i) \qquad (7.51)$$

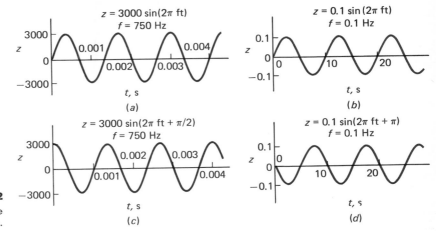

FIGURE 7·22
Sine waves of different amplitude and phase.

Here n is the number of terms to be included and the A_i, f_i, and ϕ_i are the various amplitudes, frequencies, and phases corresponding to the individual sine waves. By properly selecting these values and including enough individual sine waves, essentially any waveform can be duplicated. Some examples are shown in Fig. 7.23. Here you can see curves that look a great deal like the arbitrary and complex ones we saw in Fig. 7.20. You can also see that the rather uniform pattern of a "square" wave may be developed as a summation of sine waves.

We have now described the way very general system inputs may be considered to result from a combination of sine waves of different frequencies. Since this is true, we can learn a great deal about the response of a system subject to general inputs by study-

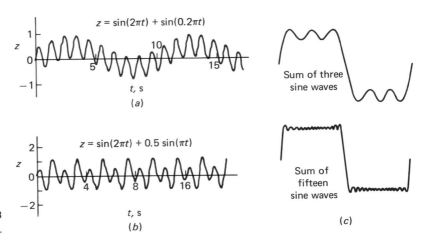

FIGURE 7·23
Summation of sine waves.

7.5 FREQUENCY RESPONSE

ing its response to individual sine waves at various frequencies. This, of course, is the reason why stereo enthusiasts are concerned about frequency response. By studying the behavior of a stereo component at various single sinusoidal frequencies, the behavior of the component can be evaluated while music composed of many sinusoidal frequencies is being played.

We will consider below the behavior of simple engineering systems when subject to sinusoidal inputs. We will focus on first- and second-order systems and will base our efforts on numerical solution of the differential equation describing the system.

FREQUENCY RESPONSE OF FIRST-ORDER SYSTEMS

Let's return to the first-order vehicle system we introduced in Sec. 7.1. A vehicle is being driven by the propulsion force F_p—the system input—and velocity v—the output. This is shown in Fig. 7.24 and the system differential equation is

$$m \frac{dv}{dt} + bv = F_p \qquad (7.7)$$

as we found in Eq. (7.7) or

$$\frac{m}{b} \frac{dv}{dt} + v = \frac{1}{b} F_p \qquad (7.52)$$

as we found in Eq. (7.13) with the time constant $\tau = m/b$.

Consider now that the propulsion force is a periodic function $F_p = 250 \sin(2\pi f t)$ lbf. This will correspond to applying a force in the forward direction when the sine term is positive and then in the reverse direction as the sine term becomes negative. We want to determine the reaction of the vehicle speed to this force, which is alternately pushing and pulling on the vehicle.

Recall the numerical solution we developed earlier for Eq. (7.7) and the FORTRAN program of Fig. 7.4 to carry out the computations. We can apply the same method here for the sinusoidal input but must make a few adjustments in the program. The new program is shown in Fig. 7.25 for the low-frequency input of $f = 0.0010$ Hz. The most important change in this program is at statement 100 where we now have introduced the propulsion force F_p as a function of time defined by the sine wave. This is in the portion of the program which contains the current values for time TONE and thus FP will always have the appropriate value. Other changes have been made in the program to output FP as well as VONE, to set the frequency at its desired value and to adjust DELT, NBOUT, and TLIM to suitable values. Note that with such a low

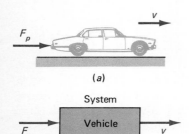

FIGURE 7·24
A simple vehicle system.

```
          PROGRAM FREQ1
          XMASS = 100.
          B = 5.
          PI = 3.141593
          DELT = .1
          NBOUT = 200
          TLIM = 4000.
          TONE = 0.
          VONE = 0.
          FREQ = .001
          KOUNT = NBOUT
          FP = 0.
      300 IF (KOUNT.NE.NBOUT) GO TO 100
          WRITE(6,200) TONE, FP, VONE
      200 FORMAT (1X, 3 F10.2)
          KOUNT = 0
      100 FP = 250.*SIN(2.*PI*FREQ*TONE)
          DELV = (FP − B*VONE)*DELT/XMASS
          TONE = TONE + DELT
          VONE = VONE + DELV
          KOUNT = KOUNT + 1
          IF (TONE.LT.TLIM) GO TO 300
          STOP
          END
```

FIGURE 7·25
FORTRAN program for the sinusoidal input.

value of FREQ = 0.001, a large value for TLIM is required to see several oscillations.

The results of the computation with $f = 0.0010$ Hz are shown in Fig. 7.26. Here we find that at this low frequency, the velocity, which is the system output, remains essentially in equilibrium with the input. The maximum velocity experienced during the oscillations is 50 ft/s in the forward direction. The velocity of -50 ft/s corresponds to the reverse direction. These are the same constant velocities that we would expect at equilibrium with a constant force of 250 lbf in the forward direction or with the force of -250 lbf applied in the reverse direction. The input frequency here is low enough so that the system dynamics are not playing a significant role in determining the behavior of the output.

We now increase the frequency in the FORTRAN program, being careful to pay close attention to making Δt small enough for accurate results. The results obtained at increased frequencies are shown in Fig. 7.27. In each case there is a brief period near $t = 0$ where the output is in a state of transition; however, after this period the output does follow a sine wave and may be described by

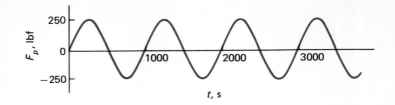
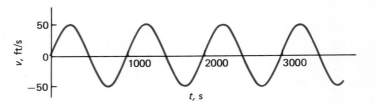

FIGURE 7·26
System input and output at $f = 0.001$ Hz.

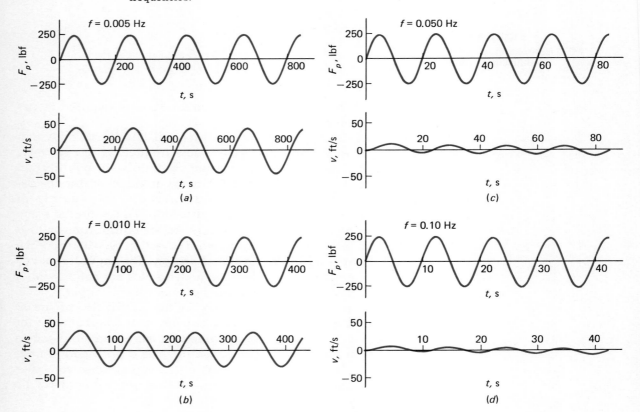

FIGURE 7·27
Inputs and outputs at higher frequencies.

312 DYNAMIC SYSTEMS

$$v = A \sin(2\pi ft - \phi) \tag{7.53}$$

In each part of Fig. 7.27, the input propulsion force is $F_p = 250 \sin 2\pi ft$, as shown in the individual graphs. However, the output oscillations decrease in size at higher frequency, so that the value of A in Eq. (7.53) becomes smaller. Also, the value of ϕ increases with frequency as the output oscillations lag behind the input.

The size of the output oscillations is of particular interest in many applications. The value of A can be obtained by careful measurements on graphs similar to those in Fig. 7.27. For example, in Fig. 7.27a, a value of $A = 42$ ft/s can be determined. Table 7.4 shows the values of A which occur at various input frequencies and a plot of A versus f is shown in Fig. 7.28.

This graph shows how the output oscillation size varies with frequency for the particular first-order system of Eq. (7.52) with $m = 100$ slugs, $b = 5$ lbf·s/ft, and $F_p = 250 \sin 2\pi ft$. However, a graph of much more general use can be obtained by a little manipulation.

We begin by defining A_o as the amplitude of the output oscillation at very low frequency. This is the size of the oscillation when it is slowly following and staying in equilibrium with the input as for the low-frequency case of Fig. 7.26, where $A = A_o = 50$ ft/s. With decreasing frequency the oscillation amplitude approaches this value as shown in Fig. 7.28. Also, at very low frequency, the derivative term in Eqs. (7.7) and (7.52) is essentially zero (there is only a slow change with respect to time). This means that A_o may be easily calculated by dividing the amplitude of the input oscillations in F_p by b to obtain the value of $A_o = 250/5 = 50$ ft/s.

With the definition of A_o we can replace Table 7.4 with the more general Table 7.5. Here the values of A have been divided by A_o and the frequency f has been multiplied by τ. When plotted, this produces the graph of Fig. 7.29, which may be used for any first-order system. Note that both A/A_o and $f\tau$ are dimensionless.

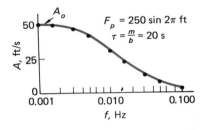

FIGURE 7·28
Frequency response plot of A versus f.

TABLE 7·4
AMPLITUDE A AT VARIOUS FREQUENCIES

f, Hz	A, ft/s
0.001	50
0.0015	49
0.003	47
0.005	42
0.010	31
0.015	23
0.030	13
0.050	7.9
0.100	4.0

TABLE 7·5
MODIFIED FREQUENCY AND AMPLITUDE DATA

f_τ	A/A_o
0.020	1.00
0.030	0.98
0.060	0.94
0.100	0.84
0.200	0.62
0.300	0.47
0.600	0.26
1.000	0.16
2.000	0.08

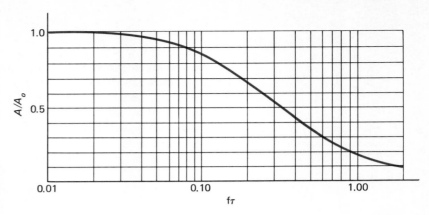

FIGURE 7·29
The generalized plot for first-order systems.

In Example 7.1, we apply Fig. 7.29 to obtain the oscillation amplitude in a first-order electric system.

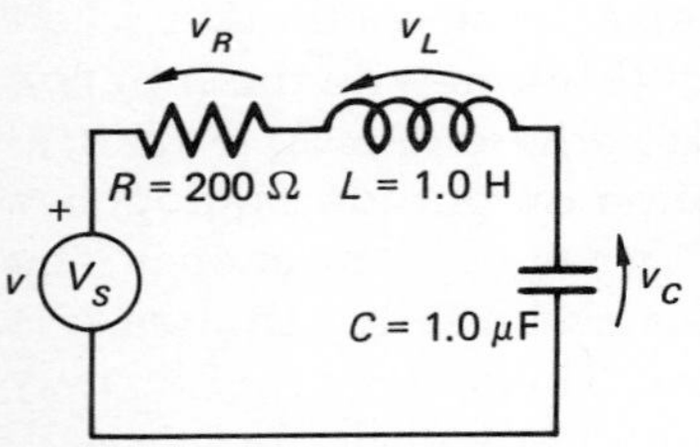

FREQUENCY RESPONSE OF SECOND-ORDER SYSTEMS

We will discuss the frequency response of second-order systems by considering the R-L-C electric circuit of Fig. 7.30. This circuit is similar to the one we considered earlier in Fig. 7.18, and its differential equation is

$$LC \frac{d^2 v_C}{dt^2} + RC \frac{dv_C}{dt} + v_C = v \tag{7.48}$$

FIGURE 7·30
A second-order system.

as we derived at that point. To consider this system subject to a sinusoidal input, let v be a sine wave:

$$v = 10 \sin 2\pi ft$$

The FORTRAN computer program in Fig. 7.31 can be used to solve for the resulting system output v_C at various frequencies, provided that care is used in keeping the time step DELT adequately small. The development of this program has been based on rearranging Eq. (7.48) to obtain

$$\frac{dr}{dt} = \frac{1}{LC}(v - v_C - RCr) \tag{7.54}$$

where we define

$$r = \frac{dv_C}{dt}$$

as the rate of change of the voltage v_C. Then, realizing that

$$\frac{\Delta r}{\Delta t} \approx \frac{dr}{dt}$$

Equation (7.54) can be written in approximate form as

$$\Delta r = \frac{\Delta t}{LC}(v - v_C - RCr) \tag{7.55}$$

Since

$$\frac{\Delta v_C}{\Delta t} \approx r$$

FIGURE 7·31
FORTRAN program for the second-order system with sinusoidal input.

```
      PROGRAM FREQ2
      R = 200.
      XL = 1.
      C = .000001
      PI = 3.141593
      DELT = .00001
      NBOUT = 100
      TLIM = .5
      TONE = 0.
      SRONE = 0.
      VCONE = 0.
      FREQ = 15.9
      KOUNT = NBOUT
      V = 0.
  300 IF(KOUNT.NE.NBOUT) GO TO 100
      WRITE(6,200)TONE, V, VCONE
  200 FORMAT (1X, 3 F10.2)
      KOUNT = 0
  100 V = 10.*SIN(2.*PI*FREQ*TONE)
      DELSR = (V - VCONE - R*C*SRONE)*DELT/XL/C
      DELVC = SRONE*DELT
      VCONE = VCONE + DELVC
      SRONE = SRONE + DELSR
      TONE = TONE + DELT
      KOUNT = KOUNT + 1
      IF (TONE.LT.TLIM) GO TO 300
      STOP
      END
```

we also have

$$\Delta v_C = r \, \Delta t \qquad (7.56)$$

Equations (7.55) and (7.56) form the basis for the FORTRAN program of Fig. 7.31. Note that the variable name SRONE has been used to represent r_1 in the program.

In the circuit of Fig. 7.30 we have $R = 200 \, \Omega$, $L = 1.0$ H, and $C = 1.0 \, \mu$F. Based on our work with Eq. (7.48), our second-order differential equation has the natural frequency

$$f_n = \frac{1}{2\pi} \sqrt{\frac{1}{LC}} = \frac{1}{2\pi} \sqrt{\frac{1}{1.0 \times 1.0 \times 10^{-6}}} = 159 \text{ Hz}$$

Also, considering the damping ratio,

$$\frac{\zeta}{\pi f_n} = RC$$

$$\zeta = \pi R C f_n = \pi 200 \times 1.0 \times 10^{-6} \times 159 = 0.10$$

Example 7·1

The electric circuit of Fig. 7.17 is given an input voltage $v = 10 \sin 2\pi(200)t$. What is the size of the oscillations in the output v_C?

Solution

Equation (7.43) was developed to describe this electric circuit:

$$RC \frac{dv_C}{dt} + v_C = v$$

In this case, the time constant $\tau = RC = (1000)(1 \times 10^{-6}) = 0.001$ s. Since the frequency $f = 200$ Hz, $f\tau = 0.200$. From Fig. 7.29, we now find $A/A_o = 0.62$. We obtain the value of A_o by considering the derivative term to be zero in the differential equation. For the given oscillation amplitude in v of 10 V, we also find $A_o = 10/1 = 10$ V, so that

$$A = 0.62 \, A_o = 6.2 \text{ V}$$

This means that the output oscillations in v_C will show an amplitude of 6.2 V at the frequency of 200 Hz.

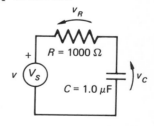

FIGURE 7·17
(Repeated)

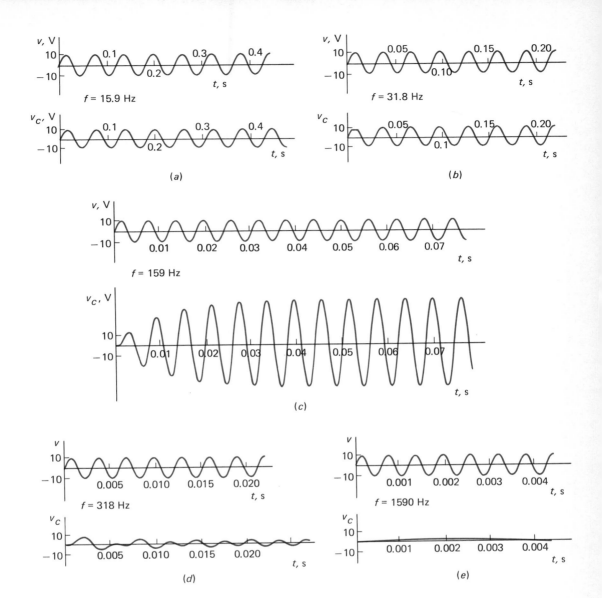

FIGURE 7·32
Response of a second-order system at various frequencies.

Thus the circuit of Fig. 7.30 has a natural frequency of 159 Hz and a damping ratio of 0.10. The frequency response of the circuit has been determined by applying the FORTRAN program for $f = \text{FREQ} = 15.9$ Hz, 31.8 Hz, 159 Hz, 318 Hz, and 1590 Hz. The results are presented in Fig. 7.32, which shows quite clearly the characteristic frequency response of a second-order system.

At very low frequency in Fig. 7.32a we can see that the output v_c is essentially identical to the input v. For the slightly higher frequency in Fig. 7.32b, the output v_C differs from v primarily near $t = 0$, but the steady oscillations are much the same. In Fig. 7.32c, the frequency of the input matches the natural frequency of the circuit and we see the effect of resonant behavior. The steady-output oscillations in v_C, achieved after a short transient, are larger in magnitude than the input oscillations. Then, at frequencies higher than the natural frequency, the size of the oscillations are reduced again, as shown in Fig. 7.32d. In Fig. 7.32e, at ten times the natural frequency, it is difficult to see any oscillations at all in v_C.

The amplitude of the output oscillations can be plotted for this system just as we did in Figs. 7.28 and 7.29 for the frequency response of the first-order system. In this case, however, we plot only the more general results in Fig. 7.33, which shows A/A_o as a function of f/f_n. The quantity A_o is again the output-oscillation amplitude at very low frequency. Figure 7.33 contains various curves for second-order systems with different values of damping ratio ζ. The curve for $\zeta = 0.1$ corresponds to the results shown earlier in Fig. 7.32. The large peak of $A/A_o = 5$ found at $f/f_n = 1$ indicates a resonant effect typical of second-order systems with low damping ratios. Input oscillations at the same frequency as the natural frequency produce large output oscillations. With larger damping ratios, such as $\zeta = 0.5$ shown in Fig. 7.33, this resonance is very much reduced. With still larger damping ratios, such as $\zeta = 1.0$ shown in Fig. 7.33, no resonance occurs at all.

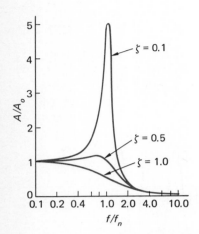

FIGURE 7·33
Frequency-response plot for second-order systems.

PROBLEMS

7.18 A sinusoidal input voltage $v_i = 100 \sin 2\pi ft$ is applied to an electric circuit. The input frequency $f = 100$ Hz. Determine the amplitude A of oscillations in the output v_o for each equation below:

(a) $0.01 \dfrac{dv_o}{dt} + v_o = v_i$

(b) $0.001 \dfrac{dv_o}{dt} + v_o = v_i$

(c) $0.01 \dfrac{dv_0}{dt} + 10 v_0 = v_i$

(d) $0.01 \dfrac{dv_0}{dt} + v_0 = 10 v_i$

7.19 The tank pressure P_o in Fig. P7.19 is related to P_i by the equation

$$10 \frac{dP_o}{dt} + P_o = P_i$$

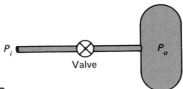

FIGURE P7·19

If the pressure $P_i = 2 \sin 2\pi t$ (lbf/in²), what is the size of the resulting oscillation in the pressure P_o?

7.20 A sinusoidal input is applied to a second-order electric circuit. The input $v_i = 10 \sin 10\pi t$. Determine the amplitude A of the output oscillations in v_o if the circuit is described by the equations below.

(a) $0.001 \dfrac{d^2 v_o}{dt^2} + 0.0064 \dfrac{dv_o}{dt} + v_o = v_i$

(b) $0.001 \dfrac{d^2 v_o}{dt^2} + 0.032 \dfrac{dv_o}{dt} + v_o = v_i$

(c) $\dfrac{d^2 v_0}{dt^2} + 32 \dfrac{dv_0}{dt} + 1000\, v_0 = 1000\, v_i$

(d) $\dfrac{d^2 v_0}{dt^2} + 320 \dfrac{dv_0}{dt} + 100{,}000\, v_0 = 100{,}000\, v_i$

7.21 The system shown in Fig. P7.21 is a simplified model of part of a vehicle suspension system. If the vehicle motion x is the output of this system and the wheel motion y is the input, the differential equation for this system is

$$m \frac{d^2 x}{dt^2} + b \frac{dx}{dt} + kx = b \frac{dy}{dt} + ky$$

Although this equation shows a derivative term on the right-hand side, the natural frequency f_n and damping ratio ζ may still be found in the normal way based on the left side of

FIGURE P7·21

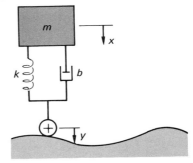

the equation. The mass supported here is $m = 25$ slugs and the stiffness $k = 1000$ lbf/in. To eliminate resonance and give good transient behavior, a damping ratio of $\zeta = 0.5$ is desired. What is the required value for b in units of pounds-force-seconds per foot?

7·6 SUMMARY

The purpose of this chapter has been to introduce you to the concept of dynamic systems—the study of transients and change in systems of interconnected components. It is a topic of interest in all branches of engineering and can be highly mathematical. The study of differential equations is necessary before you can develop a complete understanding of dynamic systems. However, the computer approach used here for solving differential equations is very often used in real engineering work.

It should be mentioned that dynamic behavior has both good and bad aspects. Resonance is an example of a dynamic characteristic which can be indispensable or disastrous depending on the application. The tuning circuit of a radio and the ticking of a mechanical watch both depend on resonant behavior. However, shock absorbers are installed in an automobile to prevent resonance, and engineering structures of many types have failed because of resonant behavior.

Systems analysis methods are studied in all fields of engineering. References 1 and 2 are typical of the books related to electrical and mechanical engineering systems. Chemical engineers might study Reference 3 as well as similar books. References 4 and 5 are examples of systems applications in civil engineering; Reference 6 deals with industrial engineering topics.

REFERENCES

1. Beachley, N. H. and H. L. Harrison, *Introduction to Dynamic System Analysis,* Harper & Row, New York, 1978.
2. Ogata, K., *System Dynamics,* Prentice Hall, Englewood Cliffs, NJ, 1978.
3. Luyben, W. L., *Process Modeling, Simulation, and Control for Chemical Engineers,* McGraw-Hill, New York, 1972.
4. Parker, H. W., *Wastewater Systems Engineering,* Prentice Hall, Englewood Cliffs, NJ, 1975.
5. Morlok, E. K., *Introduction to Transportation System Engineering and Planning,* McGraw-Hill, New York, 1978.
6. Turner, W. C., K. Case, and J. H. Mize, *Introduction to Industrial and Systems Engineering,* Prentice Hall, Englewood Cliffs, NJ, 1978.

PART FOUR

DESIGN AND APPLICATIONS

In our previous chapters, we have discussed engineering calculations and some of the physical principles used by engineers. Engineering is more than simply calculating, computing, and solving equations, however. It is an exacting and a creative profession, one that applies technology imaginatively and proficiently to meet the needs of society. In Part IV we will show how engineers get down to business—how they use engineering principles to cope with a wide range of actual problems. Chapter 8 presents an overview of the methods and ideas used in engineering design. Chapters 9 through 13 look more closely at engineering design, focusing on applying engineering theory to real situations.

We have selected a wide range of examples for discussion in an effort to appeal to people with varying interests in the different branches of engineering. We have tried to describe many of the details of engineering design and to illustrate the way engineering concepts and calculations are used as tools in the design process.

the proper equipment is available, and the cost and schedule of the testing are acceptable. The same is true in the development of manufacturing and construction techniques, production planning, airline scheduling, etc. These are not normally considered to be the problems of a design engineer, but design is definitely required. Physical hardware may be a part of these problems, but emphasis is placed on the design of a procedure or strategy to meet a particular need in an effective manner. The engineering profession is filled with problems of this type, where the process of engineering design is performed in a general way. The people involved are not called design engineers and are not designing a product; but they do apply the concepts of engineering design to satisfy the special needs which are an ordinary part of their work.

AN OUTLINE OF THE PROCESS

The process of engineering design is outlined in Fig. 8.1, which is a flow diagram that represents the most classic problem in engineering design, where the goal is to produce a marketable product for sale at a profit. As we have just discussed, engineering design is more general than this diagram suggests and not all engineering design problems fall into this category. However, producing a marketable product is a traditional design problem, whether applied to a small device like a better mousetrap or to a major system such as a nuclear reactor. The process may be somewhat modified for more general problems in engineering design which do not necessarily deal with physical things; nonetheless, the basic character of the process remains the same.

It is important to realize that the overall design process begins with recognition of an identifiable *need* that can be satisfied by the resulting design. The need may be an undeniable one with respect to society as a whole or simply a practical one with respect to a specific market for a proposed product. For example, a more fuel efficient automotive engine is urgently needed by a great many Americans; an electrically powered grass trimmer might have an adequate enough market to justify its manufacture.

The initial statement of need is nearly always made in vague and general terms. Real life is different from the normal classroom situation, where a poorly defined problem can be a cause for student complaints. In real life the initial statement of need is usually poorly defined. Part of the design process, as well as any problem-solving process, is to translate the initial general statement into a specific and well-defined problem. Only after this has been done can we establish goals and begin the actual design effort. In the case of an electric grass trimmer, the initial need may be vaguely described simply as a more-convenient way to cut grass near walls and around objects. This description must

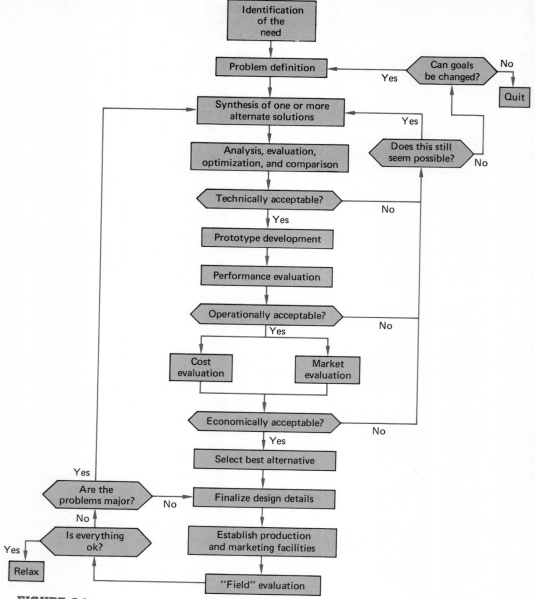

FIGURE 8·1

be translated into a problem definition which specifies clearly such things as the allowable weight of the trimmer and the types of grass and weeds it must be able to cut, as well as whether it is to be operated with one or two hands and whether it is to have a cord or to be cordless.

Upon carefully defining the problem, we see in Fig. 8.1 that the next step in the process is to begin the suggestion of alternative

328 ENGINEERING DESIGN: AN OVERVIEW

A control center for an operating chemical process. The process, which is computer-controlled, is monitored with several computer graphics displays. The computer programs that guide the operator must be designed as carefully as is the equipment. Design principles must be applied by the programmer even if he or she is not called a design engineer. (*Taylor Instrument Co.*)

solutions. This is the "inspiration phase," which offers special opportunity for creativity and innovation in engineering design—a point at which the individual inventor can make a unique and profitable contribution. In group efforts it is a time for mutual stimulation of ideas by freely interchanging thoughts and by making open and unrestricted suggestions. Such a group meeting, wherein the purpose is to accumulate provocative suggestions, is called a "brainstorming session." The group is given an outline of the design problem and asked to propose solutions. The creative process is indirectly encouraged by keeping criticism and study or analysis at a minimum since the only goal is to stimulate further suggestions and to arrive at a list of interesting possibilities. Impractical ideas cannot be stifled because some variation of a wild scheme may lead to an outstanding design. The use of nylon line as the cutting element in a grass trimmer is an example of the type of "wild scheme" with considerable potential that might be suppressed in a meeting with immediate critical evaluation.

The time for critical analysis and evaluation is after a list of alternative solutions has been developed, as we have indicated by the "analysis and evaluation" block of Fig. 8.1. At this point, the evaluation of each solution is normally hypothetical, with pencil, paper, computer, and most importantly, an inquiring mind being the available tools. Some experimentation may take place at this time, but talk and analysis are normally inexpensive compared with the actual construction of an operating device. Careful consideration and penetrating analysis in this phase can avoid the large expense of developing an unworkable idea. (Needless to say, there are many situations where a key concept can be evaluated more easily by a simple experiment, thereby avoiding a difficult analysis. For example, cutting grass with nylon line can be demonstrated by using a piece of fish line and a small electric drill. This is a much more direct and economical test of the concept than the mathematical analysis of a nylon filament impacting a blade of grass, which could take weeks to complete.)

During the evaluation of several alternatives, we must not only identify the workable solutions but also compare them, saving the more attractive ones and discarding the least attractive. At this stage, it is important that each alternative be considered fairly before comparing it with the other possibilities. We say that each alternative should be "optimized." In our own experience, we would never compare the quality of two automotive manufacturers by the result of a race in which a factory-tuned X mobile beat a Y mobile prepared by a backyard mechanic. The products of each must be adjusted and tuned to achieve their best performance before any comparison is made. The same is the case when comparing alternative solutions to an engineering design problem. Adjustments and tuning (i.e., optimization) of each alternative must be performed prior to any serious comparison.

After analysis and evaluation of each alternative, a decision must be made on the technical acceptability of the available solutions, as shown in Fig. 8.1. Are there one or more solutions that seem to work and that satisfy the problem definition? If no proposed solution is technically acceptable, the feedback, or iterative, path (indicated in Fig. 8.1) must be followed. When solving real engineering design problems, it is like going back to the drawing board to seek a new alternative. Along the way back to the synthesis step we must stop, and based on our new experience, decide whether it still seems possible to satisfy the specifications established in the problem definition. For example, suppose that the problem definition for the grass trimmer specified 10 hours of cordless operation and an electric motor powered by a rechargeable storage battery is to be used. During the analysis phase, we might determine that a battery of allowable weight could not store enough energy for even an hour of typical trimming operation. It would thus not be possible to meet the original problem definition. We would be forced to consider different power sources (a gas engine possibly) or we would be faced with changing our original goals (perhaps reducing the operating time, increasing the allowable weight, or permitting a power cord).

Once a solution, or alternative solutions, to a design problem has been found which seems to be technically acceptable, the next step is to construct a working model, or *prototype*. More than

There are special types of design problems that require modification of the design process presented in Figure 8.1. One extreme case is the one-of-a-kind design. In this situation the cost of the design, prototype, and test must be completely absorbed by the one item to be constructed. Unless special arrangements have been made, the cost of the design must be held to a minimum, since it can represent a substantial fraction of the cost of the item. In fact, in some cases, *the design* of large special items such as electric power stations or bridges may be *the product* of a consulting engineering company. An electric utility or unit of government might hire the engineering firm to perform the design and then arrange at a later date for actual construction by another company supervised by the original engineers.

In the one-of-a-kind design, the cost and size of an item normally prohibits constructing a prototype and testing it completely before commitment to a particular design concept. Fortunately, design engineers in these cases are normally experienced with similar design problems, so their experience can be utilized. Another aid in these types of design problems is the use of scale models. For example, although you would not usually build a

The arch commemorating St. Louis as the "Gateway to the West," near completion in 1965. It is an example of a one-of-a-kind design—similar in nature, perhaps, to design problems previously solved by the engineering team but with its own unique difficulties. Note the special construction equipment that had to be designed and built for this project. (*Artega Photos Ltd.*)

8.1 THE DESIGN PROCESS

prototype bridge, a scale model could be constructed. In large chemical processes such as a fertilizer plant, it would also be impractical to build a prototype. However, a "pilot plant" that demonstrates the critical features of the process on a small scale would be a common way of evaluating a prospective design. One-of-a-kind designs may also be found in special mechanical equipment and machinery as used, for example, in large-scale manufacturing operations. In such a situation, the one unit being produced may be treated somewhat like a prototype, by being carefully designed but modified and redesigned until satisfactory performance is obtained.

Design problems in which only limited prototype development and testing are possible place a heavy burden on design engineers. They cannot try it out except on paper, so special ingenuity and analytical ability are important. Often analysis and computer methods are used to obtain the best possible predictions of design performance without construction of the actual device. At other times, the engineer anticipates potential problem areas in a design and builds in back-up considerations so that changes can be made rather easily if problems do arise during actual construction.

Most real design problems are somewhere in between the extremes of very large production and the one-of-a-kind situation just discussed. Most real design problems are also quite complex; they rarely involve the design of some single isolated item but are normally an interconnection of components. Each component must be designed, so it becomes a design subproblem within the original design problem. Consider the design of a jet engine, an airplane, a plant for producing nylon, or a system of flood-control dams. These are more complex design problems than the grass trimmer we have mentioned and consist of many interconnected components. Of course, components are different things to different people. To the airplane designer, a jet engine is a component; to the designer of an intercity transportation system, an airplane is a component. This also works in the other direction: a jet-engine designer considers the compressor to be a component; the compressor designer thinks of the blades as being components; etc.

The overall design of a large system of components like an airplane becomes extremely complex. The concepts of *system design*, however, can be applied to make the problem manageable. In this approach, specific components of the system are identified. The design problem for each component is then carefully defined, with the effects of that component on the overall system kept in mind. The designers of each component go to work on their design. If necessary, they subdivide the problem further, identifying the "components" of their component. A supervisory design group, often called a *systems group*, keeps track of the design problems being solved by the various groups of engineers designing each component in the larger design. This supervisory group monitors the progress of each group on meeting its design specifications and evaluates the effect of particular components on the overall performance of the design.

It is through management tools such as system design that large design projects have become possible. For intermediate-sized projects that have reduced complexity but still have many components, the systems group is replaced by one individual called a *project engineer*, who then monitors and coordinates progress of the various component designers and has overall responsibility for the design.

Testing

In the design process, testing plays an important role. In the early stages of a design, testing may be necessary to gather basic information on the operating principles of the design. Data, for example, may be obtained from models or early prototypes and then used to predict the performance of the final design. This is the purpose of wind-tunnel studies on scale models of airplanes, for example. Other types of tests may be useful simply to establish if a suggested idea has any merit whatsoever, such as the fish line and drill motor mentioned earlier for our grass trimmer. Tests performed early in the design process are simply thought of as experimental and are often informal and run at the direction of the design engineer. However, when accurate results are necessary, careful planning of the testing is required and careful attention must be given to instrumentation. The instrumentation must be sufficiently accurate to obtain satisfactory measurements of the desired voltages, forces, pressures, gas concentrations, velocities, temperatures, etc., which may be involved in the test; and suitable recording equipment may be necessary to make a permanent record of the results. Tests and instrumentation have become increasingly sophisticated, leading to an engineering speciality in itself. (Ref. 1 deals with the principles and design of instruments and Ref. 2 deals with the design of experiments.)

As an engineering design nears completion by reaching the prototype stage, the testing becomes more formal. Called *performance testing* it is intended specifically to see that the proposed design satisfies the specifications of the problem definition. In the automotive field, for example, performance tests of prototype vehicles are made to measure fuel consumption and exhaust emissions to see that these performance measures meet company and government specifications. In larger organizations, performance tests are not run by the design engineers. Special testing

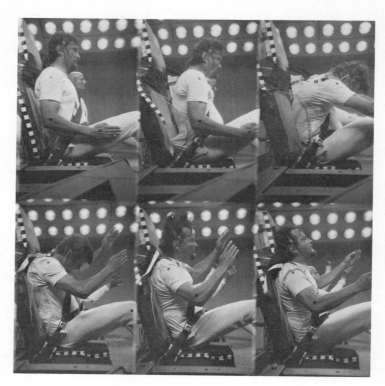

Human-factors testing applied to the study of automobile accidents. This sequence of photos is being used to compare the behavior of a person and a dummy during a low-speed impact. Validating the dummy behavior at low speed, aids in evaluating automotive designs for high-speed accidents that are too dangerous for human subjects. (*Arvin/Calspan Advanced Technology Center.*)

groups take over as the tests become more complex and more formalized. *Human-factors* testing is also often performed at this time. This type of testing is required to be certain that the device being constructed is compatible with the human body and its capabilities. Machines intended for operation by human operators must not require physical dexterity that is unreasonable or an extraordinary speed of response on the part of the operator. Operators should be comfortable during operation of the machine, with, in the case of mobile vehicles, their visibility unobscured. Human-factors testing is intended to ensure smooth interaction between human beings and machine and to produce designs that are safe and comfortable for human use.

After initial performance and human-factors testing of a design, it is common for some design changes to be necessary; but as the design nears its final stages, the prototype should be able to meet the basic performance specifications. At this point more extensive prototype testing begins, as *life tests, endurance tests,* and *environmental tests* may be necessary. These tests are normally very expensive and may be of long duration or require a special geographic location or a test chamber. Such tests are not normally performed until a design is nearly final. In this way, repetition will not be needed because of a design change resulting from a more easily run performance or human-factors test. Life tests are intended to study the performance of a device during its intended life. For example, if the life of an automobile is intended to be 100,000 mi, such is the duration of the required life test. For operation of the automobile at a speed of 50 mi/h a period of 2000 hours of continuous operation is required, which is nearly three months running 24 h/day. You can no doubt see the expense that can be involved in this type of testing. Of course, particular components of an automobile can be life-tested more conveniently in special test rigs to focus on that component. For example, an alternator used in producing electric power for the automobile can be run by an electric motor to perform its life test in a laboratory. This does not simulate all the conditions of actual use of the alternator, which is normally run by the automobile engine, but it is far less costly than testing a complete automobile. By simulating the electric loads and by providing an environmental chamber in the laboratory to duplicate extreme conditions of high and low temperatures, dust, dirt and moisture, very meaningful tests of the alternator can be conveniently performed.

A life test records only the behavior of the device through its required life. An endurance test is a test of a design which operates the device until failure occurs. This type of testing can determine how much extra life the device has. You may recall the endurance tests of light bulbs, which we presented earlier in our discussion of statistics (Example 2.4). In that case, by testing many light bulbs to failure, we were able to determine the probability of a production light bulb failing before meeting a certain lifetime requirement. We applied statistics in that example and statistics are often applied in the analysis of endurance data.

After a design is in production, two other types of testing become important: acceptance testing and quality control. *Acceptance testing* is the process of accepting delivery of a product. The product must be tested to see that it meets its performance specifications. There are two separate views about this test. In one, it is felt that materials and components of a product being manufactured must be tested as they are received to see that they will function properly when used in the product. Of course, this must be done to some extent before completing the design based on these materials and components; however, in production, a program of acceptance testing must operate continually to ensure that each item received meets the original specifications. Another important viewpoint is that an engineering staff must be aware that acceptance testing of its product will be performed by its customers. If its product is sold to another company as a component in a larger system, it may be subject to carefully planned accep-

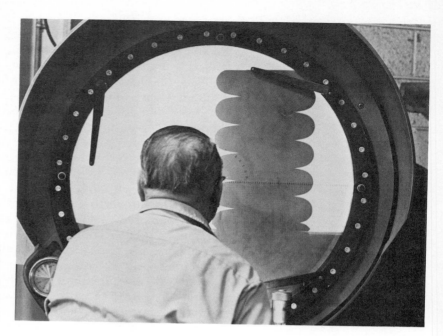

An optical comparator being used for quality-control inspection of a coil spring. By magnifying the spring to ten times life size, it may be easily checked for its dimensions and flaws. (*Associated Spring, Barnes Group Inc.*)

tance testing designed by engineers. If the product is sold to consumers, the acceptance testing is not formal, but there will be consumer complaints if the product does not meet the expected standard of quality.

Because each unit shipped by a manufacturer is subject to evaluation by its customers, the procedures of quality control testing are used by the manufacturer to monitor production. For critical items and in cases of limited production, this can mean a rather

A computer-based testing machine for electronic circuits. It is operated from a computer terminal and its sequences are programmed like a computer. When a circuit board is plugged into the test bed, it can completely evaluate the circuit and component performance in only a few seconds. (*Fairchild.*)

detailed performance test of each unit as assembly is completed. For mass-production items, this normally involves a quick check of each item after assembly; but carefully planned statistical tests are also made on the various components of the product and at different stages of production. Samples of the completed product are taken randomly for thorough performance tests and possibly for endurance and environmental tests. Quality-control testing is becoming very sophisticated as a result of efforts to perform fast, thorough, and automated tests of mass-produced but complex items. Consider, as an example, the evaluation of calculators coming from an assembly line. Quality-control tests should be performed to be certain that each functions correctly in its many mathematical operations. Computer-controlled testing has become an invaluable tool for rapidly checking the quality of complex products such as this, especially in mass production.

PROBLEMS

8.1 Grass trimmers using nylon filament as the cutting element have become very successful commercial products. They are, however, only one alternative to the solution of the grass trimming problem. List specifically some other approaches to trimming grass, including one concept that you have proposed yourself. Make a comparison of the approaches, listing advantages and disadvantages for each. Identify the one concept that seems most attractive for general-purpose uses. Discuss your conclusion. Discuss another concept that may be attractive for some special situation and should be given further consideration. See Fig. P8.1 on next page.

8.2 The use of the battery-powered electric motor for trimming grass has some unique advantages and disadvantages. List them. Identify some alternatives to this power source and list the advantages and disadvantages of these. Can a hand-powered device compete with these alternatives? Discuss some design concepts that might enhance the attractiveness of hand-powered grass trimmers.

8.3 There are variables in any design that can be adjusted or changed to improve the performance of the design. For the grass trimmer example, two obvious things that might be changed are the length of the nylon cutting filament and the speed of rotation. For this grass trimmer configuration, list all the variables that might influence trimmer performance.

8.4 Outline a test program that could be followed to study the influence of the variables in Prob. 8.3. Consider first the way in which performance of the trimmer would be evaluated. Then propose a testing procedure that would consider each of the variables mentioned and allow the best values for

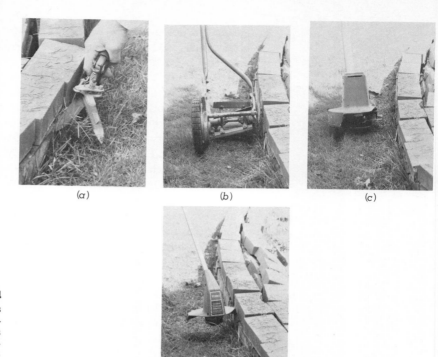

FIGURE P8·1
Alternative approaches to grass trimming that have been marketed commercially. (a) grass shears; (b) "half" mower; (c) electric rotary blade trimmer; (d) the "fish-line" trimmer.

each of the variables to be determined. What do you mean by "best"?

8.5 Assume that you are beginning a large-scale design project for a commercial electronics firm. This firm has sold at retail over 100,000 complete computer systems for use in homes. These systems consist of an input keyboard, a central processor, and a substantial memory, as well as a video- or paper-output device. You know that these computer systems are capable of taking over many monitoring, control, and routine computing tasks in the home. Despite this capability, the product's low price, and increasing sales, market surveys show that a large untapped market still exists. The number of homeowners who have computers is small (compared with the potential market), and those who do own computers use them mostly for business purposes or as a hobby.

In order to increase sales, your firm has the "need" to identify significant tasks that its computer systems can perform in the home. Your job is to list and evaluate such tasks.

Make a comprehensive and imaginative list of tasks that your firm's computer system could perform. Of course, for many of the tasks you list there are alternative solutions. For example, a simple calculator can be used for balancing a checkbook, for which a complete computer system is not necessary. For each of the tasks you have listed, identify alternatives to a computer system and determine what advantages or disadvantages the computer system might have.

8.6 Consider Prob. 8.5 with respect to uses of a computer system in the automobile.

8.7 The search for energy alternatives has become *the* American design problem of the 1980s as the *need* to become less dependent on traditional energy sources has become more and more obvious. Solar heating designs are commonly considered as alternatives for heating homes. However, it has been suggested that solar heating is not energy-efficient. That is, the energy needed to produce the solar heating equipment may be more than the energy produced by the equipment during its lifetime. If this is true, then the large-scale use of solar heating equipment may actually not reduce the United States need for electricity generated by coal and nuclear power. In fact, solar heating may not be an acceptable energy alternative. Consider this difficulty and develop your own conclusion about the acceptability of solar heating from this viewpoint. This is not a problem that should be attempted in a short period of time. It is a research problem that requires library work and outside reading. [You might begin your research by studying the article by C. Whipple, "The Energy Impacts of Solar Heating" which appeared in *Science*, Volume 208, Number 4441 (18 April, 1980).]

8.8 Alternatives to traditional engine and drive systems are being sought for automobiles to enhance their energy efficiency. Develop a list of alternatives for powering an automobile and identify the advantages and disadvantages of each. Which of these alternatives leads to the most acceptable design and in your opinion has the most potential. Base your conclusion on your own analysis and reasoning and also on information about models and prototype designs that you may locate in your engineering library.

8.9 Walking or bicycling are, of course, energy-efficient alternatives to the typical modes of American transportation. However, there are often "barriers" within a community that discourage people from walking or bicycling a distance that could otherwise be easily covered. An especially busy intersection or an expressway might be in the route, for example. Or perhaps a 3-mi roundabout drive is the quickest way from point A to B because a woods or marsh separates the two points, and there is no convenient sidewalk or bicycle path.

Identify a location in your community where you think walking or bicycling could be made more convenient. Conduct a survey of people in your community to determine if your intuition is correct and to explore possible solutions to the problem. (For example, ascertain whether improved signals and pedestrian crossings could help a traffic problem or whether a pedestrian bridge is required.) Perform a cost analysis of your suggested solution. Have you been able to identify an actual need in your community and an acceptable solution to the resulting design problem?

8.10 The acceptance of the automobile is owed, in part, to its nearly all-weather operation. List the difficulties in all-weather operation of a bicycle and suggest design strategies to overcome them. The bicycle, the rider, and the bicycle path may be considered as components of a system.

8·2 EVOLUTION OF AN ENGINEERING DESIGN

For any company and any industry, design is a continuing process. The needs of society as well as those of a customer are continually observed and anticipated. Products are redesigned and new products are developed as needs change or as new technology makes it possible to meet existing needs more effectively. Through this continuing process, the technology of our present society and most of the devices we use in day-to-day living have evolved over extended periods of time. Modern products of engineering design are nearly always the culmination of a large number of good ideas accumulated over periods of several years and even a truly revolutionary idea must be supplemented by many just plain good ideas before reaching its full potential.

In this section we will discuss an example of the engineering design process as it occurs over an extended period of time. This example should illustrate the way most complex products tend to evolve naturally through meeting changing needs and adopting available technology. The product we will consider is an actual line of automatic riveting machines manufactured by the General Electromechanical Corporation. "Gemcor" is a relatively small[1] company located in Buffalo, New York, that is dedicated to the specialized process of automatically forming high-quality riveted joints. This process is especially important in the construction of modern airplanes, which require thousands of rivets, and aircraft manufacturers from all over the world use Gemcor riveting equip-

---

[1] Gemcor employs about 325 people and has an annual sales of about $18 million. This might be compared with the Eastman Kodak Company, which employs 125,000 people and has an annual sales of $8 billion, or to the Exxon Corporation, which employs 137,000 people with sales of $80 billion (all 1980 figures).

ment in airplane production. The story of the company and the technology it has developed is a study of the design process and the natural evolution that is a part of engineering design.

A DEVELOPMENT OF AUTOMATIC RIVETING MACHINES

Gemcor's story begins about 1940 when the large-scale production of aluminum aircraft was beginning and thousands of these aircraft were to be built for use in World War II. Rivets were (and still are) the primary means of fastening together aluminum sheets, channels, and beams to form an aircraft structure. A standard rivet is made with a head formed on one end. At that time the normal procedure for installing rivets was to drill a hole through the parts to be joined, to manually push the shaft of the rivet through the hole up to the rivet head, and then to form a head on the other end of the rivet by using a pneumatic hammer and a "bucking bar." The process was time-consuming and extremely noisy. It also required substantial skill from workers doing the riveting and produced riveted joints of varying quality. The first need in improving the riveting process was to design a riveting machine that produced more uniform rivets and operated more quietly.

The pneumatic hammer operates by using compressed air to accelerate a mass which is stopped in an impact and produces a force sufficient to deform a rivet a small amount. Repeated impacts of the hammer are necessary to form the rivet head completely and to hold the structural members securely together. An alternative approach to forming the rivet head was developed which met the need for more uniform and quiet operation. This process utilized a piston and cylinder driven by high-pressure hydraulic oil. The pressure on the piston created sufficient force to squeeze the rivet into its final form with a single stroke. No impact or repeated squeezing was necessary and the process could produce nearly identical rivets time after time. This hydraulic squeezing process continues today to be the basis for the installation of rivets in modern aircraft production.

Throughout the war years (1940 to 1945) Gemcor designed and produced squeeze riveting machines in many variations for particular applications. A new need developed to go beyond quiet, uniform riveting to more productive operation with a lower cost for each installed rivet. Initial efforts to improve productivity included the multiple rivet machine, where five or more rivets could be squeezed simultaneously. However, by the late 1940s, it became apparent that the key to increased productivity was to make hole drilling, automatic feeding, and insertion of rivets a part of the riveting operation. Gemcor was able to design a machine which automatically carries out each of the steps in the riveting

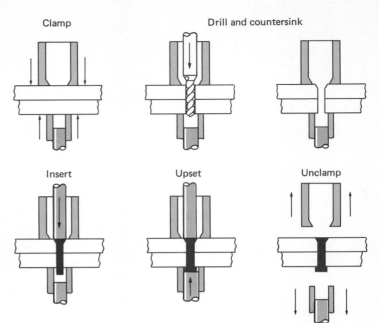

A schematic of the Drivmatic process. It begins by clamping the pieces to be riveted, drilling and countersinking the hole, and then inserting the rivet. A hydraulic squeeze then "upsets" the rivet to form the new head and the pieces are unclamped. (General Electromechanical Corp.)

process. When a rivet is to be used to join two metal sheets, the two sheets are positioned in the jaws of the riveting machine. The machine firmly clamps the two sheets together and a drill unit moves in place, drilling a precise hole through both sheets. Since the sheets are clamped tightly, the hole is especially smooth and need not be deburred after drilling. With the clamping force still applied, the drill unit is moved aside and the riveting unit is moved into place. A rivet is automatically fed into position and inserted into the drilled hole where it is hydraulically squeezed into final shape. The clamp then releases and the sheets can be removed or repositioned so that additional rivets can be installed. Gemcor refers to this riveting procedure as the Drivmatic process and it is known by that name throughout the aircraft industry.

The Drivmatic process sounds easy enough in concept, but the details of the design are complex. The riveting machine consists of many subsystems involving hydraulic, mechanical, and electric operation. Each subsystem required extensive design effort and many blind alleys were considered which required new design strategies. A particular design goal which could not be met was the goal of building a universal riveting machine to handle the wide range of riveting problems found on a total aircraft struc-

A Gemcor riveter being used for small aircraft components. Note the flat-top tables being used to support the parts during riveting. (*General Electromechanical Corp.*)

ture. The possible variations in the structure are simply too numerous to use a universal machine. Parts requiring rivets may be large, small, flat, simply curved, or compound curved. The structure may be open or have a variety of obstructions. The rivets may be large or small and have special head configurations or require two-piece fasteners. In short, the design goal of a single universal riveter had to be abandoned. Instead a family of riveting machines was developed to cover the wide range of riveting geometries and applications.

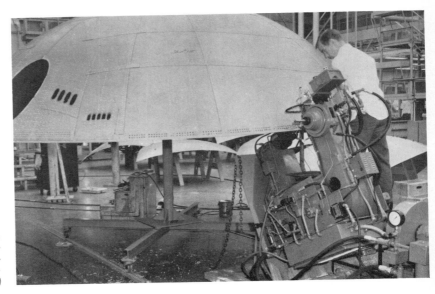

A riveter mounted on a special tilting platform to make rivet installations in dome-shaped bulkheads. (*General Electromechanical Corp.*)

A New Riveting Concept

During the mid-1950s, aircraft were being designed for increased performance and the first jet aircraft for commercial use were in production. New emphasis was being placed on reduced weight and increased strength in aircraft design. As a result of this new emphasis, aircraft designers developed the concept of the integral fuel tank. The aircraft wings became *the* fuel tank—without a liner and without an additional container inside the wing. The concept of the integral fuel tank has led to reduced weight and complexity for the wing structure and has become a standard feature of modern aircraft. Leak-proof rivets were necessary to make the integral fuel tank successful, since aircraft fuel came directly in contact with the riveted joints. In response to this need, Gemcor developed the headless rivet, or *slug*, as it has become known in the aircraft industry. The slug is inserted into a drilled hole and is hydraulically squeezed so that a rivet head is formed simultaneously on both ends. The shank of the rivet, between the heads, expands during this squeezing process. It fills the hole completely and, in fact, even expands the hole to produce an extremely tight fit, called an *interference fit*, which is completely leak-proof.

The interference fit produced with the headless rivet not only has been found desirable for eliminating leaks but has become a vital structural part of modern aircraft. Extensive testing of riveted joints made with the headless rivets has shown that they resist loosening and failure under vibratory stress and that they remain sound for the remarkably long life of the aircraft structure. This superior performance has led to the wide acceptance of the headless rivet and the Gemcor installation procedure. The automatic riveting process has, in fact, become more than a labor-saving system, because the strength of the interference-fit joint now plays an important role in the structural integrity of modern aircraft.

Incorporating Controls and Computers

Thousands of rivets must be installed in the structure of a large aircraft. The Gemcor Drivmatic process became particularly good at drilling, feeding, and squeezing rivets automatically. However, another need was recognized during the process development to further speed installation of the many rivets required. Not only was it desirable to perform the riveting automatically but it seemed natural that the aircraft parts being riveted should also be moved and positioned automatically.

A basic strategy commonly used for automatic positioning is shown in Fig. 8.2. An electric motor drives the object to be positioned. An electric sensor, possibly a potentiometer as discussed

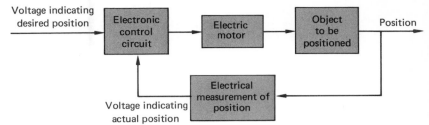

FIGURE 8·2
Schematic of an automatic positioning system.

in Chap. 5, measures the actual position of the object by producing an electric voltage proportional to the position. In an electronic control circuit, this voltage is then compared with a separate voltage which indicates the desired position of the object. The output of the electronic circuit drives the motor, which moves the object until the actual position voltage equals the desired position voltage. With this automatic control system, movements of the object may be made by simply changing the desired position voltage. After each new setting for the desired position voltage, the motor moves the object until the actual position corresponds to the desired position.

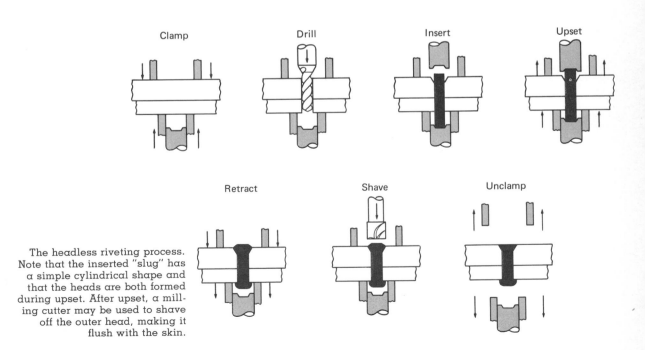

The headless riveting process. Note that the inserted "slug" has a simple cylindrical shape and that the heads are both formed during upset. After upset, a milling cutter may be used to shave off the outer head, making it flush with the skin.

8.2 EVOLUTION OF AN ENGINEERING DESIGN

Cut section of riveted joints. Left, an ordinary rivet; right, one formed from a headless rivet. Note that the ordinary rivet has not completely filled the region below the head, so two voids are present. The complete expansion seen for the headless rivet accounts for its improved characteristics. (*General Electromechanical Corp.*)

In the simplest automatic positioning systems, a worker turns a knob to change the voltage for desired position and the positioning system moves the object to the appropriate position. In more sophisticated positioners a computer produces the voltage for desired position and may play a role in the actual control circuit. A sequence of position changes may be produced by properly programming the computer.

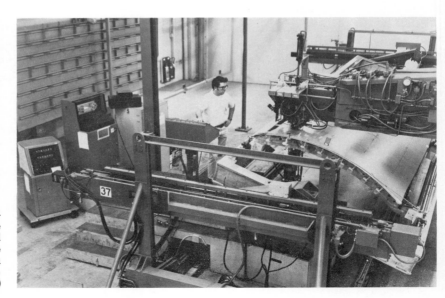

A riveting machine with a work-positioning system that moves the fuselage panel being riveted within the jaws of the riveter. The digital positioning equipment can be seen on the left. (*General Electromechanical Corp.*)

"Rosie" the riveter rolls along the tracks on the lower right, installing rivets in the wing of a jumbo jet. The computer control units and video displays can be seen onboard. (*General Electromechanical Corp.*)

Between the mid-1950s and the present day, Gemcor's automatic positioning systems have gone from the simplest arrangements to sophisticated computer control. During the 1960s numerical control methods were introduced with a sequence of riveting positions stored on a tape recording. The tape is played back during the riveting process and computer circuitry reads the desired rivet locations. As each location is read from the tape, the aircraft wing or other structure is moved to that location and a rivet is installed using the clamping, drilling, and inserting procedure of the Drivmatic process. After this, the next rivet location is then read from the tape and another automatic installation takes place.

Automation of the riveting process continues to progress toward increasing complexity. The more sophisticated riveting machines of the 1970s and 1980s stand over 20 ft high and weigh nearly 100 tons. They move along railroad tracks under the position control of an on-board computer to an accuracy of a few thousandths of an inch. The same computer may be simultaneously positioning and rotating the wing of a jumbo-jet aircraft within the jaws of the machine while installing several rivets per minute. The operator of the machine rides on the riveter and monitors the installations of rivets through instrumentation and closed-circuit television cameras focused at the riveting point.

The Actual Design Process

The 40-year history of Gemcor riveting machines is, in many ways, one extended application of the design process outlined in

Fig. 8.1. This is the case because the riveter designs have evolved over that long period of time. However, most of the riveting machines produced by Gemcor during the 40-year period were each designed for a specific customer to meet a specific need. Very few of the riveting machines have been off-the-shelf models and most have required special engineering. A design process similar to Fig. 8.1 has been required for nearly each machine produced. There are modifications to the process, however, which are typical of design work not intended for a mass-produced product. The short-term pressures of the actual marketplace and the market evaluation are not part of the design process. Instead, the design engineer is primarily concerned with the performance specifications, cost, and schedule contained in the contract for the design at hand.

The engineers' efforts as they consider and evaluate alternative solutions during the design process are dominated by the necessity of staying within the agreed-upon cost and meeting a prescribed delivery date. A prototype is not normally used as a testing ground and engineers must work directly on design of the actual machine. This does not mean that performance evaluation of the design is slighted in any way. A complete evaluation is performed directly on the machine as it is being constructed. If a problem occurs, construction stops if necessary and modifications are made in the design to eliminate the problem. A final evaluation of the finished machine is performed after assembly to be certain that it meets the specifications defined in the contract. The customer also tests the finished machine and accepts its delivery only when satisfied with the performance.

The contract for purchase of a large item such as a riveting machine is an important document for both the customer and the manufacturer. At Gemcor, people from the technical sales office, the engineering design office, and the production staff work closely together to be sure that the performance specifications, cost, and schedule are acceptable and obtainable. Often a customer must be persuaded that an investment in a costly automatic riveting machine will be profitable. The investment is justified by considering the annual rate of return (interest rate) earned by the money invested in the machine. This rate of return is earned because the machine reduces labor costs during its lifetime. More employees would be needed for rivet installations without the automated equipment. Generally an annual return of 20 percent on the money invested in a riveting machine has been sufficient to justify its purchase, especially when rivet quality is improved.

It should be clear that monitoring and managing the design process is an important task at Gemcor as it is in many other industries involving special products. Costs and schedules must be monitored carefully. As an aid to this monitoring process, Gemcor divides larger projects into work packages, each with its own cost and schedule goal. These work packages are each small enough to monitor readily, and when one package shows signs of overruning either cost or schedule, it can be given special attention. Often it can be brought back into line through some innovative redesign. However, not all projects can be rescued and Gemcor, like any other business, has had projects which have not met performance or profit goals. Postmortems have often revealed that carelessness in the design and evaluation steps of Fig. 8.1 was a contributing factor, perhaps coupled with overly optimistic decisions in the estimates leading to the original contract.

8·3 ALUMINUM AS AN ELECTRIC CONDUCTOR: A PROBLEM IN ENGINEERING DESIGN

Our discussion of automatic riveting machines in Sec. 8.2 described the evolutionary process of design. In that case the accumulation of moderate advances over a period of many years has led to a modern and sophisticated product. This is the most common process of technological advancement. However, there are occasional revolutions in technology where the significance and widespread influence of the change are very great compared with the time period in which the change takes place. Transistor electronics and Xerographic photocopy equipment are two examples of technological developments that have occurred rapidly enough and on such a large scale that they are clearly revolutionary.

When a single engineering design will have a relatively rapid and large-scale impact on the economics of an industry and the well being of its customers, it must be called a "revolutionary" design. It is important that such revolutionary design problems be identified and handled with special care to avoid consequences which can be disastrous from the viewpoints of personal injury and economic loss. In this section we will discuss the development of aluminum conductors for long-distance transmission and for residential and commercial wiring (such as houses, stores, and restaurants). This may seem to be an innocent problem in engineering design. However, because of the breadth of its impact and its rapid introduction, it has proved to be a revolutionary design. It has been charged that testing in many aspects of aluminum conductor design was inadequate, especially as applied to residential and commercial wiring. As a result, the rapid and large-scale introduction of aluminum as an electric conductor has not been completely successful.

BACKGROUND

At the present time in the United States over 95 percent of new electric transmission cables are either aluminum or steel-reinforced aluminum cable (ACSR). The performance record of these cables has been entirely satisfactory, and the cost is typically 30 to 40 percent of competing copper cables. Electric transmission cables are used to carry power over long distances (for example, from Niagara Falls to Rochester—100 km, or from Boulder Dam to Los Angeles—400 km). Such cables operate at electric potentials in the range of 115,000 to 750,000 V and carry currents of thousands of amperes.

In contrast to the eminently satisfactory performance of aluminum cable in transmission service, the performance of aluminum wire in residential branch wiring—house wiring—has been disappointing. As of 1980, 26 manufacturers of aluminum house wiring were being sued in a United States district court under Section 12 of the Consumer Product Safety Act. In this legal action, the Consumer Products Safety Commission (CPSC) alleges that aluminum branch circuit wiring systems in approximately 2 million residences present an imminent hazard of fire and of injury or death resulting from the fire. The CPSC seeks to compel each of the 26 manufacturers to notify the affected homeowners of the hazard and to repair or replace the aluminum wiring.

A tower typical of many that support the bare overhead conductors used for long-distance transmission of electric power. In this application, over 95 percent of the new installations use aluminum conductors.

Because of the adversary nature of the legal system, somebody has to lose this case, and

1. If the manufacturers lose they will have to assume the repair or replacement cost, which may aggregate to hundreds of millions of dollars.
2. If the CPSC loses, some 2 million consumers are either left with the hazard or must bear the cost of replacement or repair.

Our purpose in the remainder of this section is to show how the design process illustrated in Fig. 8.1 was applied to the design of aluminum electric transmission cables and aluminum house wiring. For house wiring, we will illustrate how skimping on the vital step of performance evaluation hid potential problems inherent in aluminum house wiring, so that the problems showed up seven design steps later in the field evaluation and have persisted over a period of 13 years.

We mention in passing that the very profound problems with aluminum house wiring arise from a seemingly trivial item: the electric resistance of screwed terminals. However, this resistance depends on a complex combination of material properties, mechanics of screwed joints, and human workmanship. These ideas will be explored further in Chapter 10 where we look at the details of engineering design for electric connections.

DESIGN SUMMARY FOR TRANSMISSION CABLE

Step 1. Identification of the Need

During the late 1950s, the price of copper began to rise sharply relative to its historical levels; and spot shortages of copper developed, so that occasionally no copper was available at any price. At the time, almost all electric wire was made of copper. A need clearly existed for a substitute material.

Step 2. Problem Definition

In this case the problem definition was simple: design a range of electric transmission cables having performance equal to or better than that of copper and using a material other than copper.

Step 3. Synthesis of One or More Alternate Solutions

Two alternate solutions were eventually considered. (1) An all-aluminum cable that imitated existing copper cable but had increased dimensions to allow for aluminum's lower electric conductivity. (2) An aluminum cable with a reinforcing core of steel.

This so-called ACSR cable was proposed for long spans to counteract unacceptable sag owing to the smaller elastic modulus (stiffness) of aluminum relative to copper.

Step 4. Analysis, Evaluation, Optimization, and Comparison
This will be illustrated below for the simple case of an all-aluminum cable.

Resistance and cost of copper cable Consider a copper cable having a cross-sectional area A of 7.60 cm². (This corresponds to a so-called 1,500,000 circular mil cable—a cable roughly $1\frac{1}{4}$ in in diameter—and a size frequently used in electric transmission.) The salient properties required for preliminary analysis are: resistance (ohms per kilometer), cost (dollars per kilometer), and mass (kilograms per kilometer).

The resistance per kilometer is computed using Eq. (5.1): $R = \rho l/A$, where for copper, $\rho = 1.70$ $\mu\Omega\cdot$cm (Table 5.1), and $l = 1$ km $= 10^5$ cm. Substituting,

$$R = 1.70 \times 10^{-6} \frac{10^5}{7.60} = 0.0224 \ \Omega/\text{km}$$

The mass is density times volume, or

$$M = dlA$$

and for copper, the mass density d is 8.99×10^{-3} kg/cm³. Substituting,

$$M = (8.89 \times 10^{-3})(10^5)(7.60) = 6760 \ \text{kg/km}$$

A kilometer of this copper cable weighs nearly 7000 kg. The present price of copper is about $0.87 per pound (or $1.92 per kilogram) so the material cost is

$$P = (6760)(1.92) \doteq \$13,000 \text{ per kilometer}$$

Comparable aluminum cable An aluminum cable having the same resistance as the copper cable is designed using

$$A = \frac{\rho}{R} l$$

where A is area in square centimeters.

Using the resitivity of aluminum (2.80 $\mu\Omega\cdot$cm) and the resistance of the example copper cable, we find

$$A = \frac{(2.80 \times 10^{-6})10^5}{0.0224} = 12.5 \text{ cm}^2$$

The aluminum cable will have a diameter of about 1.6 in, about 25 percent larger than the corresponding copper cable. Since the density of aluminum is only 2.7×10^{-3} kg/cm³, the mass of a kilometer length is

$$M = (2.70 \times 10^{-3})(10^5)(12.5) = 3380 \text{ kg/km}$$

The weight per kilometer of the aluminum cable will be just about half that of the example copper cable. Since aluminum at this time costs about $0.59 per pound (or $1.30 per kilogram), the material cost is

$$P = (3380)(1.30) = \$4390 \text{ per kilometer}$$

The material cost for copper ($13,000) is nearly three times the material cost of an aluminum cable of equal electric resistance.

Thus, the aluminum cable appears to be technically acceptable (step 5). Its dimensions are not much different from the prototype copper cable, and for comparable performance the weight is one-half and the material cost one-third of the copper cable it was intended to replace. We now continue with the remainder of the design process.

Step 6. Prototype Development—Performance Evaluation

A major problem here was to determine if aluminum cable could be manufactured in available facilities. It proved easy to manufacture aluminum wire using minor modifications of already-existing wire-drawing and cable-making machinery. When the feasibility of manufacture had been demonstrated by the production of a few dozen reels of cable (a reel typically holds $\frac{1}{2}$ mi or about 0.8 km of cable), the performance was evaluated using extensive outdoor test facilities. The outdoor tests demonstrated satisfactory performance in the face of such environmental factors as icing, temperature extremes, wind loading, and air pollution. An interesting result of this testing was that aluminum cable proved extremely resistant to intergranular chloride stress corrosion resulting from bird droppings, which had always been a problem with copper cable.

Also tested at this stage was the technology of connections. This is no small problem: a 100-km-long line with three parallel conductors would use more than 300 reels of cable, with a corresponding number of end-to-end splices. Special bolted connectors were developed to implement these splices. Line installers were

At this site, dozens of copper-to-aluminum connectors can be tested by an electric utility. This test facility is located north of San Francisco, where the ocean salt spray makes the environment unusually severe. A large number of samples is used for statistical analysis. Note how the test design emphasizes essentials and maximizes environmental stress. (*Pacific Gas and Electric Corp.*)

trained in their use, and splices were tested out-of-doors under realistic conditions of environmental, mechanical and electric stress. Since the material cost for a 100-km transmission line is measured in the millions of dollars, it made sense to spend comparable amounts on research and development for connectors and

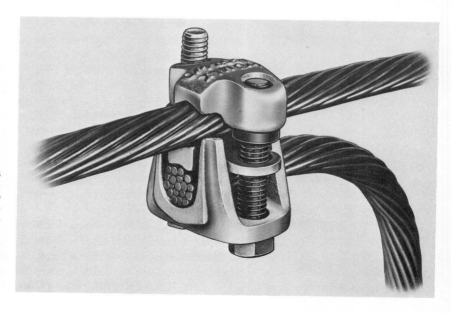

An environmentally proven aluminum-to-copper connector. Aluminum wire is usually at the top, with copper wire below. The machine screws and the connector body are aluminum alloy. This design allows the installer to control tightness with a socket wrench and to examine the finished connection. (*Burndy Corporation.*)

on training costs for connector installers. (As we shall see, the failure to give sufficient emphasis to the connector problem was the major flaw in the design process for aluminum house wiring.)

As a consequence of the careful work that had been done under step 6, aluminum transmission cable was found operationally acceptable (step 7) and was quickly adopted by utility companies, so step 8, market evaluation, was just a formality. It was also economically acceptable, (step 9). Step 10, select best alternative, was also simple: use all-aluminum cable for short spans and ACSR for longer spans. Step 11, finalize design details, was largely a matter of selecting cable diameters to suit customer needs. Production and marketing facilities (step 12) were quickly established and field evaluation (step 13) is still going on. Results from the field have been so favorable, however, that essentially all (>95 percent) of new transmission lines are now built of aluminum.

ALUMINUM HOUSE WIRING: THE PENALTY FOR SHORTCUTS

In contrast to the success which aluminum cable has had in the electric transmission industry, its record in house wiring has been a dismal failure. Over the 8-year period from 1967 to 1975 aluminum house wiring was involved in over 500 accidents: of those 39 were major house fires, resulting in at least $250,000 of property damage and 12 deaths. What accounts for this history of failure? Basically the problem is that all wiring is a *system* in which *devices* are connected together. In house wiring, for example, the devices are such items as wall receptacles, wall switches, outlet boxes, and fuses or circuit breakers. In the system, wire (whether aluminum or copper) is merely a *component*. All the reported problems with aluminum house wiring have one factor in common: *in a system all components must be compatible.* The heart of the "aluminum house wiring problem" is a set of basic incompatibilities between aluminum wire and the brass- or zinc-plated steel screws used to connect the wire to the other system components. These incompatibilities will be explored further in Chap. 10, but it bears mentioning that the areas of incompatibility are metallurgical (thermal expansion and creep), chemical (oxidation and corrosion), and mechanical (problems of torque and pressure in bolted joints). The performance of workers under actual field installation conditions also plays a role. Hence, what appears at first glance to be an electric problem actually has metallurgical, chemical engineering, mechanical engineering, and industrial engineering aspects. Even civil engineering plays a role, since the original motivation for the use of aluminum house wiring was to reduce the cost of housing construction.

Our discussion on the penalty for shortcuts is keyed to Fig. 8.1.

Where the procedures specified in the figure were successfully followed, we will be brief in order to concentrate on the steps where omissions and shortcuts led to later failures in the field.

Step 1. Need
Do something about high cost and uncertain supply of copper wire.

Step 2. Problem Definition
Find a substitute for copper.

Step 3. Alternate Solution
Use aluminum in place of copper, but with possible increased dimensions to allow for aluminum's lower electric conductivity.

Step 4. Analysis, Evaluation, Optimization, and Comparison
One aspect of design which we have not considered to this point is the following: when material is produced in standard shapes and sizes, it is good design practice to use them.

In the United States, wire is produced in a series of standard diameters called the American Wire Gauge (AWG) Standard. Some typical AWG dimensions are listed in Table 8.1 below. Hence, in carrying out step 3, the wire industry followed good design practice by selecting wire sizes corresponding to the AWG standard.

As can be seen from the table, a *decrease* of two AWG numbers *increases* the cross-sectional area by about 58 percent. Since the conductivity of copper is about 65 percent greater than that of aluminum, the wire industry recommended a simple decrease of two AWG sizes when substituting aluminum for copper wiring: the increased area would approximately compensate for the decreased conductivity.

As an example, consider a 100-ft run of #14 AWG copper wire; a

TABLE 8·1
AWG STANDARD WIRE SIZES

Wire size, AWG	Diameter, mm	Cross-sectional Area, cm^2
8	3.26	0.0835
10	2.59	0.0527
12	2.05	0.0330
14	1.63	0.0209

size usually specified for 15-A service. The resistance of 100 ft (3048 cm) of #14 copper is

$$R = \rho \frac{l}{A} = 1.7 \times 10^{-6} \frac{3048}{2.09 \times 10^{-2}}$$
$$= 0.248 \ \Omega$$

The corresponding resistance of 100 ft of #12 aluminum is

$$R = \rho \frac{l}{A} = 2.8 \times 10^{-6} \frac{3048}{3.30 \times 10^{-2}}$$
$$= 0.258 \ \Omega$$

The voltage drop across 100 ft of wire at 15 A, from Ohm's law [Eq. (5.2)] is $v = Ri$, or $(0.248)(15) = 3.72$ V for #14 copper and $(15)(0.258) = 3.87$ V for #12 aluminum. Rounded to one significant figure, these are both of order 4 V and are, in practice, indistinguishable.

Step 5. Technical Acceptability

Using resistance and voltage-drop calculations similar to the example just given, substitution of aluminum for copper (with a decrease of two AWG sizes) appeared technically acceptable and was approved by Underwriters Laboratories[2] in 1966.

Step 6. Prototype Development—Performance Evaluation

This vital step was unfortunately not completely carried out in the United States (but it was in Britain, where the results clearly indicated the problems of interfacing aluminum wire with copper components). Hence, the question of whether it was operationally acceptable was too quickly answered yes in the United States and on the basis of too little evidence.[3]

Steps 8 to 12. Cost Evaluation Clearly Favored Aluminum

For example, the Sears, Roebuck Catalog of 1967 lists #12 AWG aluminum wire at a price which is one-third the price of #14 AWG

---

[2] Underwriters Laboratories, Inc., is an independent, not-for profit testing organization. It tests and lists electric and other products which meet certain standards of the industries involved. See Chap. 1, Sec. 1.4, for a further discussion of industrial standards. Many insurance underwriters, government agencies, building owners, and electricians recognize the UL Listing Mark as evidence of acceptability in installations involving possible hazards to life or property.

[3] The British tests took 7 years in all; negative results were obtained in 14 months.

copper. (Recall that #12 is two sizes *larger* than #14 but has nearly the same resistance per 100 ft.) Aluminum, then, was so economically acceptable that in 1969 some 70 percent of the new house wiring installed in the United States was aluminum. Some 800,000 mobile homes were wired with aluminum.

The steps "select best alternative," "finalize design details," and "establish production and marketing facilities" were all combined into one. Since production of aluminum wire in a standard AWG size appeared to be (and, in fact, was) the best alternative, aluminum house wiring was produced using designs and production and marketing facilities which already existed for copper wire.

Production and marketing facilities were so quickly established that in 1971 (the year of peak production) over half a billion feet of aluminum wire was produced. Aluminum wiring appeared to be driving copper off the market.

Step 13. Field Evaluation

All was not well with aluminum house wiring, however. As early as 1966 Underwriters Laboratories began to receive reports of aluminum wire failures in the field. In general, these reports concerned failure of aluminum-to-copper connections, usually involving overheating. In order of increasing severity, the reports described the problem as

Warm or hot component
Smell of burning; component charred, melted, or burned
Smoking, arcing
Burning, possibly with structural damage

The answer to the question, "Is everything OK" (see Fig. 8.1) was clearly no. The problems were major; and although the synthesis of alternate solutions was attempted, after several trips around the loop involving steps 3, 4, and 5 the answer to the question, "Does this still seem possible?" turned out to be no.[4] This decision appears to have been reached around 1974.

The goals (safe house wiring at a reasonable cost) could not be changed. Meanwhile, costs of parts of houses other than wiring had increased so greatly that the high cost of copper seemed less important in the 1970s than it had in the 1960s. Consequently, the quit decision was made; aluminum house wiring fell from its 70 percent penetration of the home market and is no longer being offered for home use. The only remaining activity involves the ques-

---

[4] "Possible" in this sense means politically, economically, and psychologically, as well as technically.

In tests in a simulated building wall at the U.S. National Bureau of Standards, a receptacle and its enclosure are nearly destroyed by a glowing connection when a steel screw is used to connect aluminum wire. Approximately one-third of the nonmetal parts of the receptacle were consumed, the plastic outlet box charred through and deformed, and the polyvinyl chloride insulation on the wire destroyed to a distance of 10 to 13 cm from the box. Temperature inside the box reached 398°C (748 °F). (*U.S. National Bureau of Standards, 1977.*)

tion, "How do we repair the 2 million existing installations and who is to pay for these repairs?"

Thus the action has moved from the engineering arena to the legal arena.

8·4 SAFETY AS A VITAL PART OF PERFORMANCE EVALUATION

The example of aluminum house wiring should have impressed you with the importance of product and system safety as a performance criterion. A complete evaluation of product performance—including safety—is especially necessary when a "revolutionary" design is being considered, when human lives are involved, or when large-scale production is expected.

In the recent book *Products Liability and the Reasonably Safe Product* [3], two professors of law and two professors of engineering outline how safety should be evaluated during the design process. They emphasize that the designer must consider not only use but also possible misuse of the product. Factors which the design engineer must often consider are the choice of materials, safety-mandated design alterations, safety warnings, and instructions for the safe installation and use of the product.

The design engineer should first ascertain that the design meets the requirements of applicable industry and government standards. In addition, the design team should:

1. Examine forseeable use and misuse of the product.
2. Ask: Where will the product be used? Who will use it?
3. Identify possible hazards. Subjectively estimate their likelihood of occurrence and the seriousness of the consequences.
4. Identify measures which will eliminate or reduce the likelihood of hazardous events. These measures may be design changes or warnings or instructions.
5. Ask: Will design changes introduce other hazards? Will they reduce the usefulness of the product? Will they unreasonably increase its cost?

When these actions have been taken, the designer may be able to decide which safety features to include in the final product. Comparison with similar products is often helpful.

The designer of a grass trimmer might reasonably expect the tool to be used for other garden work. For example, it may be used to chop leaves for a compost heap. The designer may expect the consumer to use an electric drill to mix paint or to drive a sanding disk or an improvised reamer. Professional chefs sometimes use propane brazing torches to brown the tops of lemon meringue pies. It is the designer's responsibility to consider safety not only with respect to the product's principal use but also with respect to reasonably forseeable alternate uses.

With the grass trimmer, for example, which is used on wet grass, on slopes, and near gravel driveways and walks and by men, women, and children, the users are *not likely* to wear safety shoes and safety glasses. Therefore, likely hazards center about the trimmer's rotating parts. They may pick up hard objects and hurl them through the air. Moving parts may entangle or cut the user's hands or feet. If the trimmer is electric-line-driven, there may be shock hazards; if it is battery-driven, there may be acid hazards; if it is gasoline-driven, there may be explosion or burn hazards. Having identified these potential hazards, the designer must use a combination of experience and ordinary common sense to assess their likelihood and the potential consequences. The combination of likelihood and serious consequences motivates the experienced design engineer to devise countermeasures. Countermeasures will preferably be design changes that eliminate or greatly reduce the hazards. Warnings are usually a second-best choice. For grass trimmers, design changes might include guards or automatic shutoffs. Battery hazards can be reduced by sealed, explosion-proof batteries. In some cases, however, design countermeasures are not possible. For example, gasoline is inherently explosive and although alcohol might be considered as a substitute, its use could impair the cold-weather performance of the engine. The designer of a gasoline-powered product would thus have to warn the user of the explosion hazards and explain how to avoid them.

A warning sign carefully designed to meet the three requirements of an effective warning: identify the hazard (electrocution); describe how to avoid it (keep clear); and indicate the consequences of not heeding the warning (injury, death, etc.). The symbol on the right is the international symbol for electricity. (*Farm and Industrial Equipment Institute.*)

Warnings themselves must be properly designed. They must be prominent, legible, and intelligible to all possible users and must be durably constructed to endure a possibly hostile environment (hot, cold, wet, dirty). The life expectancy of the warning must equal or exceed the life of the product. Some industrial engineers specialize in the design of warning notices. It is a complex task.

If the design team is convinced that the design changes have not introduced other hazards and that the product is not only safe but also useful and competitive in cost,[5] they will consider that it has passed the safety test of its performance evaluation. Having put safety first, they will have satisfied the very first of the fundamental canons of the Code of Ethics: "Engineers shall hold paramount the safety, health and welfare of the public in the performance of their professional duties."

REFERENCES

1. Doebelin, E. O., *Measurement Systems: Application and Design*, McGraw-Hill, New York, 1983 (in press).
2. Lipson, C., and Sheth, N. J., *Statistical Design and Analysis of Engineering Experiments*, McGraw-Hill, New York, 1973.
3. Weinstein, A. S., Twerski, A. D., Piehler, H. R., and Donaher, W. A., *Products Liability and the Reasonably Safe Product*, Wiley, New York, 1978.

---

[5] "Safe" and "cost" have special meanings here. Safety is always relative: motorcycles must be safe compared with other motorcycles, automobiles safe compared with other automobiles. Similarly, cost must include total cost, including insurance costs, the cost of lawsuits, the cost of product recalls, and so forth.

9 MATERIALS, NUTS, AND BOLTS*

The preceding chapter presented a broad view of engineering design and the general nature of problems encountered by design engineers. As stated in the chapter, details are an important consideration in design and engineers have a special ability to handle the technical details of design. In this chapter, and through the remaining chapters, we will be focusing on some of those details.

This chapter is devoted to two important realities an engineer must face. First, engineering designs can be built only with real materials; second, even the simplest devices are based on theoretical principles. Section 9.1 introduces the study and measurement of material properties, which are topics of interest to all branches of engineering. Section 9.2 deals with the operation of nuts, bolts, and screws and presents some of the considerations necessary in designing a bolted connection.

9·1 MATERIALS

In nearly every field of engineering there are people who specialize in materials. They study the structure, properties, and production of materials which are most related to their field of engineering. For example, civil engineers study soils, concrete, and asphalt, since they are directly concerned with the design of roads and buildings. They may also deal with some types of steels as used in large structures. Mechanical engineers, on the other hand,

---

* Before proceeding with this chapter, readers should understand the basic ideas of statics contained in Sec. 4.1.

deal with a larger variety of steels and metals since these materials are common in the different types of machinery which they design. Chemical engineers tend to work with plastics, thermal insulation, and materials in which chemical composition and chemical processes have a dominant effect. Electrical engineers are very much involved in the study of materials for their electric properties, including electric insulators and semiconductor materials for use in electronic components.

Materials engineers of all types are playing an expanding role in modern technology because many important technological developments have hinged on materials research. Synthetic fibers used in clothing and other textiles is just one example. The development of microelectronics and inexpensive computers is another, since the discovery and production of materials with special electrical characteristics was required. Material breakthroughs can also have substantial impact in somewhat surprising areas. In energy, for example, improved thermal insulation can help reduce energy consumption for heating and air conditioning. Increasing material strength at high temperatures is an important step in improving the efficiency of combustion processes. A 10°C rise in the operating temperature of a jet engine could be a result of improved materials, which, in turn, would reduce engine fuel consumption by about 1 percent. The study of materials is clearly a very important part of present-day engineering, so that ignorance about material behavior and characteristics could substantially reduce an engineer's effectiveness.

In previous chapters we have already touched on a few of the important properties of materials. We mentioned electric resistivity (Table 5.1), fluid viscosity (Table 6.1), thermal conductivity (Table 6.3), and others. In this section we will discuss a few additional characteristics of materials and mention some of the procedures for testing them. We will focus on their mechanical properties. For a more exhaustive description of the very wide range of material properties than we have been able to present here, you should browse through Ref. 1. You will find as you progress through your engineering career that you will become quite familiar with such references and with material properties important for your special area of interest. You will learn why engineering materials behave as they do and may even become involved in creating new engineering materials.

MECHANICAL PROPERTIES

The most basic test of the mechanical properties of a solid is the tensile test—a test that is run on most types of materials used in the construction of structures and machines. The purpose of the test is to study the strength and behavior of the material under a

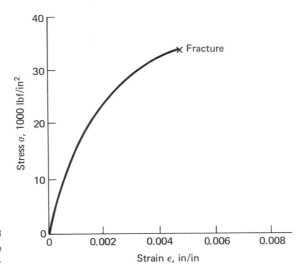

FIGURE 9·3
Stress-strain graph for a brittle cast iron.

their high strength, despite the possibility of fracture with little warning. Sometimes such materials are "backed up" by a ductile material (as in a shatterproof windshield), so that fracture of the more brittle material will not result in a complete failure. At other times, usually when little danger is involved, the disadvantages of a brittle material may be simply accepted. This is the case with respect to glassware and china dishes, which can be easily broken but maintain a scratch-resistant surface and a durable, polished finish because of the compressive strength of their material.

Compression

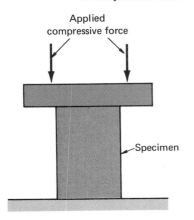

FIGURE 9·4
A compression test.

Our discussion of mechanical properties has focused on the tensile test and the failure of a material while being pulled apart. Materials can also fail by being squeezed or compressed. The same type of testing machines used for tensile testing may also be used to test materials in compression. No special devices for holding the specimen are necessary; it may be simply squeezed between two flat surfaces, as we see in Fig. 9.4. However, the determination of accurate stress-strain data in compression becomes complicated by the possibility of buckling, which prevents the use of a specimen with a long test section, as shown in the tensile specimen of Fig. 9.1. The long test section reduces the influence of end effects and permits good stress-strain data to be obtained in tension. Used in compression, a specimen with a long test section could buckle at modest load, as shown in Fig. 9.5. This tendency to buckle is not solely a material property. It depends on the length and cross section of the specimen, as we will discuss

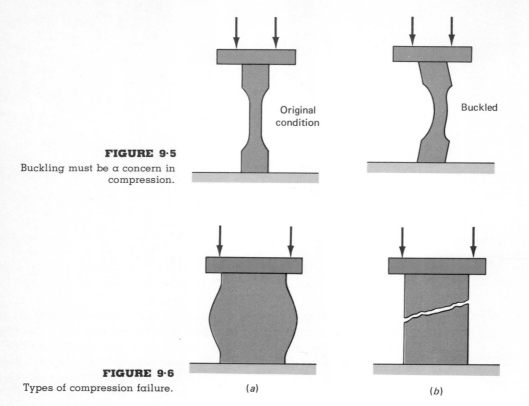

FIGURE 9·5
Buckling must be a concern in compression.

FIGURE 9·6
Types of compression failure.

further when we consider structures in Chap. 11.

Carefully selecting the dimensions of a shorter specimen of the type in Fig. 9.4 allows some reasonable stress-strain data to be obtained in compression. The influence of end effects is minimized and simple compressive failure of materials can be observed without buckling. The same linear behavior and elastic region found in tension occurs for compression and Hooke's law $\sigma = E\epsilon$ can be applied. Ductile specimens fail by yielding, with essentially the same value for S_y as is found in tension. The ductile specimen becomes deformed by bulging in the center, as shown in Fig. 9.6a. It tends to flatten rather than actually fracture. Brittle specimens show little yielding and ultimately fail along a plane at an angle to the applied load, as shown in Fig. 9.6b. Brittle materials often have a considerably larger value of compressive strength (S_c) than tensile strength, as can be seen in Table 9.2.

TABLE 9·2
COMPRESSIVE AND TENSILE STRENGTHS OF SOME BRITTLE MATERIALS

Material	Compressive Strength S_c, lbf/in^2	Tensile Strength S_t, lbf/in^2
Cast iron	110,000	30,000
Concrete	5,000	400
Glass	130,000	10,000
Cast aluminum alloy*	40,000	45,000

* The compressive value shown for cast aluminum corresponds to yield strength.
Note: 1000 lbf/in^2 = 6.895 MPa.

PROBLEMS

Material properties given in Tables 9.1 and 9.2 may be used in the problems below.

9.1 A steel rod with a diameter d = 25 mm is pulled in tension. At what force will the rod begin to yield and at what force should actual fracture be expected? The rod is made of low-carbon steel.

9.2 A cable is to be used to support a maximum load of 2000 lbf. It is desired to have a cable diameter of $\frac{1}{4}$ in. Will a low-carbon-steel cable be able to support this load without yielding? Will medium-carbon steel be satisfactory for supporting the load?

9.3 A wire made of copper is being pulled in tension with a force of 50 N. What will be its change in length under this load for each length and diameter below?

(a) L = 50 m, d = 2 mm   (c) L = 100 ft, d = 0.05 in
(b) L = 10 m, d = 4 mm   (d) L = 10 ft, d = 0.10 in

9.4 Will unalloyed aluminum wire be able to support a load of 50 N without deformation if it has a diameter d = 2 mm? A diameter d = 0.05 in?

9.5 A concrete slab has a rectangular cross section 8 by 20 in. What is the largest force this slab could support if it is loaded in tension? If it is loaded in compression and buckling does not occur?

9.6 A 10-m cable is needed to support a load of 50,000 N. What cable mass would be required to just support this load without permanent deformation if the cable were made of (a) low-carbon steel, ρ = 7800 kg/m^3; and (b) wrought aluminum alloy, ρ = 2800 kg/m^3?

9.7 As in Prob. 9.6, a 10-m cable is needed to support a load of 50,000 N. However, the deflection under this load must now be less than 5 mm and, of course, it should not yield. What cable mass is necessary in this situation for (a) low-carbon steel and (b) wrought aluminum alloy?

9·2 OPERATION OF BOLTS AND SCREWS

Section 8.2 describes a process of installing rivets in aircraft structures and the development of automatic riveting machines. A rivet is a type of fastener which forms a nearly permanent joint and requires rather special equipment for installation. More familiar to most students are probably bolts and screws, which are called *threaded fasteners*. They are considerably easier to install than are rivets. In this section we describe some of the technical aspects of bolts and screws. Mechanical and civil engineers probably specify the majority of threaded fasteners. However, you will see by an electrical application in Chap. 10 that these fasteners are of interest in every branch of engineering.

The need for a bolted or screwed joint can arise from the apparently simple question of how to fasten two parts together. If a bolted joint is to be used, it must be designed. Its design is a common subproblem which occurs regularly within larger design problems. The size and number of bolts must be selected for the joint along with the tightening specifications and the bolt material. This can be a complex process with important implications. (For example, the DC-10 crash at Chicago on May 25, 1979, focused nation-wide attention on the three bolts used in supporting engines on that aircraft.)

THEORY OF THREADED FASTENERS

Machine screws are normally made of steel or other metal and are threaded into a matching hole in a structure or machine part. This matching hole must be "tapped" to form the proper threads. A tapped hole is readily made in a machine shop and is often used in fastening machine parts together when substantial strength is required. "Self-tapping" types of screws that can cut their own threads are normally used only in thin materials for relatively light loads.

A bolt is simply a screw intended to be threaded into a nut. A nut is not an inherent part of a machine; it is a separate item made only to mate with the bolt. Normally two or more parts to be fastened are clamped together by tightening them between the nut and the head of the bolt. In cases where strength is a primary consideration, the bolt and nut are made of high-quality steel which can have a yield strength S_y of 150,000 lbf/in^2.

Threads

Figure 9.7 shows a simple screw, threaded into a solid fixed plate. This screw is right-handed—that is, when it is turned clockwise, as indicated, it will move upward and into the plate. The distance

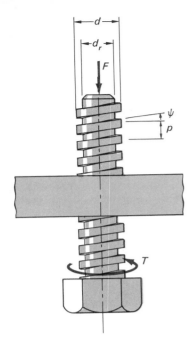

FIGURE 9·7
Characteristics of screw threads.

p between adjacent threads is called the *thread pitch* and one full rotation of the screw will move it into the plate a distance equal to the pitch.

The size of a screw (or bolt) is normally defined by specifying the outer diameter d and the pitch of the thread. Other details, such as the root diameter d_r, the cross-sectional area, etc., can then be obtained from widely available tables which describe the dimensions of standard-sized screws. In the metric system, standard screws are defined directly by the diameter and pitch expressed in millimeters as, for example,

$$M10 \times 1.5$$

The M identifies the metric designation and the numbers refer to an outer screw diameter of 10 mm and a pitch of 1.5 mm, respectively. In the unified screw thread system commonly used in the United States, a screw is specified by its outer diameter in inches and the number of threads per inch, which is equal to the inverse of the pitch in inches. The screw designated

$$\tfrac{5}{8}\text{–18 UNF}$$

has an outer diameter of $\tfrac{5}{8}$ in and 18 threads per inch ($p = \tfrac{1}{18}$ in). This is a fine thread screw, as indicated by the UNF in the designation. A similar coarse thread screw is designated as $\tfrac{5}{8}$–11 UNC, which has 11 threads per inch ($p = \tfrac{1}{11}$ in). Many of the standard-size threads are shown in Table 9.3.

We can begin to understand the nature of screw threads by inspecting Fig. 9.8, where one revolution of the thread is shown

TABLE 9·3
SOME OF THE STANDARD THREAD SIZES

Metric Threads			American National Threads–UNC			American National Threads–UNF		
Diameter d, mm	Pitch p, mm	Effective area, mm^2	Diameter d, in	Threads per inch	Effective area, in^2	Diameter d, in	Threads per inch	Effective area, in^2
4	0.7	8.83	0.164	32	0.0140	0.164	36	0.0147
6	1	20.2	$\tfrac{1}{4}$	20	0.0318	$\tfrac{1}{4}$	28	0.0364
8	1.25	36.8	$\tfrac{5}{16}$	18	0.0524	$\tfrac{5}{16}$	24	0.0580
10	1.5	58.3	$\tfrac{3}{8}$	16	0.0775	$\tfrac{3}{8}$	24	0.0878
12	1.75	84.7	$\tfrac{7}{16}$	14	0.106	$\tfrac{7}{16}$	20	0.119
16	2	157	$\tfrac{1}{2}$	13	0.142	$\tfrac{1}{2}$	20	0.160
20	2.5	246	$\tfrac{9}{16}$	12	0.182	$\tfrac{9}{16}$	18	0.203
24	3	354	$\tfrac{5}{8}$	11	0.226	$\tfrac{5}{8}$	18	0.256
30	3.5	563	$\tfrac{3}{4}$	10	0.334	$\tfrac{3}{4}$	16	0.373
36	4	820	1	8	0.606	1	12	0.663

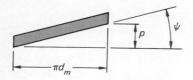

FIGURE 9·8
An unwrapped thread.

unwrapped. The vertical rise of this unwrapped thread equals the pitch p and the horizontal length equals the mean circumference πd_m, where d_m is the average of the outer thread diameter d and the root diameter d_r shown in Fig. 9.7, so that $d_m = (d + d_r)/2$. The thread is inclined at an angle ψ where

$$\psi = \tan^{-1} \frac{p}{\pi d_m} \qquad (9.4)$$

The effort needed to turn a screw is equivalent to pushing a load up or down a surface inclined at the angle ψ.

Required Torque

In Fig. 9.7, the torque T is being applied to a screw to advance it upward while the force F is pushing downward at the tip of the screw. In the discussion below we will be applying the principles of statics to determine the torque required to turn the screw in terms of the force which must be overcome. The analysis leads to an important equation which is a basis for the calculation of tightening torques for screws and bolts.

Figure 9.9a shows our unwrapped screw thread in contact with its mating thread fixed in the plate. The force F at the tip of the screw acts vertically on the screw thread, pushing it into contact with the mating thread. The horizontal force F_u results from the applied torque and tends to push the screw thread uphill. A free-body diagram of the thread is shown in Fig. 9.9b. In addition to F and F_u, the normal force F_n acting between the threads is shown as well as the friction force which tends to prevent sliding, $F_f = \mu F_n$, where μ is the coefficient of friction. To calculate the value of F_u required for movement, we sum forces in the x and y directions shown in Fig. 9.9b:

FIGURE 9·9
Forces acting on a thread.

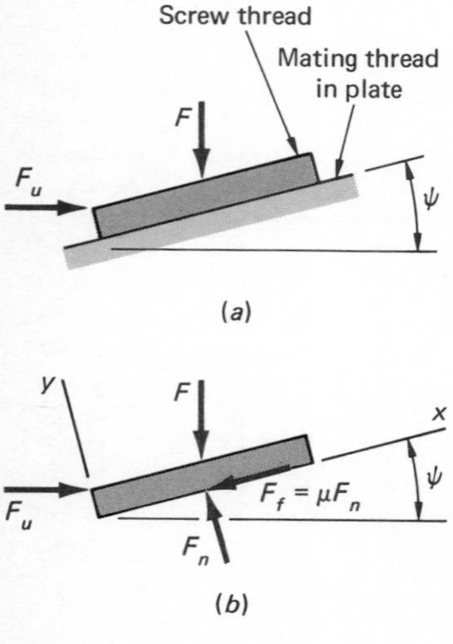

$$\Sigma F_x = 0$$
$$F_u \cos \psi - \mu F_n - F \sin \psi = 0 \qquad (9.5)$$
$$\Sigma F_y = 0$$
$$F_n - F_u \sin \psi - F \cos \psi = 0 \qquad (9.6)$$

Equation (9.6) may be solved to find $F_n = F_u \sin \psi + F \cos \psi$. Substituting this into Eq. (9.5) and rearranging we find

$$F_u = \frac{\mu \cos \psi + \sin \psi}{\cos \psi - \mu \sin \psi} F \qquad (9.7)$$

This last equation now defines the force F_u which must be applied at the thread of the screw to slide it upward against the applied force F. This force occurs at a distance $d_m/2$ from the center of the screw and would be provided by the torque applied with a wrench to the head of the screw. The required torque to advance the screw is $T_u = F_u (\tfrac{1}{2} d_m)$ or, finally,

$$T_u = \frac{1}{2} \frac{\mu \cos \psi + \sin \psi}{\cos \psi - \mu \sin \psi} F \, d_m \tag{9.8}$$

In the numerator of Eqs. (9.7) and (9.8), $\mu \cos \psi$ reflects the effort needed to overcome friction; $\sin \psi$ indicates additional effort needed to overcome the component of the force F along the threads, which opposes sliding.

Analysis can also be performed for the downward motion of the screw in Fig. 9.7, obtaining the torque necessary to lower the screw in the same direction as the applied force. The result, obtained in a similar way to Eq. (9.8), is

$$T_d = \frac{1}{2} \frac{\mu \cos \psi - \sin \psi}{\cos \psi + \mu \sin \psi} F \, d_m \tag{9.9}$$

The threads of a machine screw have a nearly triangular shape.

An important difference between Eqs. (9.8) and (9.9) is seen in the second term of the numerator, where $-\sin \psi$ occurs in the equation for T_d. This is the case because the applied force F is acting downward. It helps to overcome friction between the threads and reduces the required turning torque. In the case where $\mu \cos \psi$ is less than $\sin \psi$ (equivalent to $\mu < \tan \psi$), the torque T_d will be negative. This means that the load can overcome the friction and wants to "slide downhill" by itself. No applied torque would be necessary to lower the load in Fig. 9.7, but a torque *would* be needed to hold the screw in position.

When a screw is used for a fastener, it should be self-locking. That is, after it is tightened, it should not back out of the hole. This is done by keeping the angle ψ small, ensuring $\mu > \tan \psi$, so that self-locking action occurs for real bolts and screws. Note that bolts and screws used as fasteners do not use square threads as we have been discussing for our simplified analysis. Threads which are nearly triangular are used because they are more convenient to manufacture. Our analysis is no longer completely accurate, but the concepts still apply. The thread angle in the triangular thread tends to increase frictional effects, making it easier to produce a self-locking screw. However, the increased frictional effect makes the screw harder to turn and reduces the screw efficiency. For this reason square threads (despite being more difficult to manufacture) are often found in power screws designed to lift or move loads in machines. An automobile "scissors" jack is a simple example of a power-screw application.

This automobile jack is raised and lowered by turning the power screw, which has nearly square threads.

A THREADED CONNECTION

In Fig. 9.10a a large mass is being supported by a lifting "eye" installed at the top of the mass. The force F_w equals the weight of the mass and is provided by a hook or cable attached to the eye. We are going to analyze the threaded connection holding the eye to the mass to determine some of the details which should be considered in the design of a threaded connection. The cross-sectional view of Fig. 9.10b shows the connection more clearly. A hole has been drilled and tapped in the mass. The lifting eye and its threaded shaft are a single part, acting like a screw threaded into the hole so that the base of the eye is pressed against the mating surface of the mass. We are now concerned with how the force F_w is transmitted through the connection to support the mass and what screw size is required to support a given load.

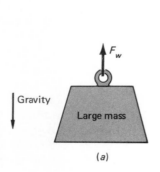
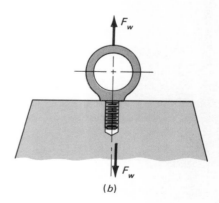

FIGURE 9·10
Lifting a load.

(a) (b)

380 MATERIALS, NUTS, AND BOLTS

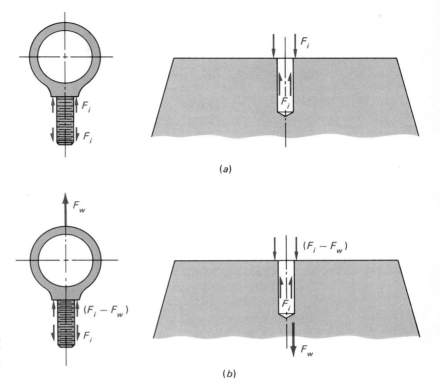

FIGURE 9·11
Free-body diagram of the "eye" and tapped hole. (a) Unloaded; (b) loaded.

As discussed in Sec. 4.1, the interaction of forces is best observed in a free-body diagram. Free-body diagrams for the eye and upper portion of the mass are shown in Fig. 9.11. The unloaded case in Fig. 9.11a indicates the condition after tightening the connection and before applying the lifting force F_w. The tightening torque has produced an initial force F_i between the threads of the screw and the threads of the hole. This force pulls on the shaft of the screw and squeezes the base of the eye against the upper surface of the mass so that F_i acts upward on the base of the eye and downward on the mating surface of the mass. We have shown two arrows to represent F_i at the various locations in this figure to emphasize that the forces are distributed around the contact areas and do not occur only at a single point. Note that the forces are equal and opposite in both portions of Fig. 9.11a, so that the assembly is in equilibrium. The shaft of the screw is in tension, while the base of the eye and the upper surface of the mass are in compression.

Force F_w has been applied to the eye in Fig. 9.11b and is being transmitted through the connection to lift the load. The free-body diagrams show how the transmission of force can occur without any change in the tension of the screw shaft. The force F_w may instead produce a reduction in the compressive force between the base of the eye and its mating surface. The initial compressive force F_i becomes $F_i - F_w$. The connection will remain solid as long as F_w does not exceed F_i. For many practical purposes, the design of a screwed joint to support a given load amounts to providing a sufficient force in the screw when it is initially tightened. This means that: (1) the screw must be large enough to provide the initial force without failing, (2) compressive forces should not be large enough to crush any mating surfaces, and (3) a proper tightening torque must be specified to produce the desired initial force.[2]

The maximum force permitted in the shaft of a screw may be calculated by knowing the yield strength of the material and the cross-sectional area of the shaft. It is common to tighten a screw to an initial force corresponding to about 80 percent of the yield strength. Possible compression failure or crushing where the mating surfaces of a screwed joint are in contact may be predicted from the yield strengths of the materials and the size of the contact area (the base of the eye in this case). When compressive stresses are above the yield strength, washers of stronger material may be used to spread out the load and prevent surface damage.

Equation (9.8), which we derived previously, shows a relationship between the force in a screw and the torque needed to turn it. The torque was related to the product of the force and the screw diameter by a complicated constant of proportionality. The derivation of Eq. (9.8) was for a square thread and neglected friction between the head of the screw and the clamped material. More detailed analysis can account for these additional factors. When this analysis is completed and numerical dimensions for actual screws are included, the results show that the proportionality constant is approximately 0.2. The tightening torque for a screw or bolt may thus be calculated from the simple expression

$$T = 0.2 \, F_i \, d \qquad (9.10)$$

for an initial force F_i and a diameter d. Experimental studies have confirmed this equation showing a standard deviation of 10 to 15 percent [2]. Example 9.2 shows how Eq. (9.10) and the guidelines above can be applied to design a simple threaded connection.

PROBLEMS

9.8 Determine the thread pitch and effective cross-sectional area for each of the following screw sizes

---

[2] Other problems, such as "stripping" of the threads, may require attention, and in some special cases, changes in the screw force may be important especially if the load is to be applied and removed many, many times. We will not attempt to discuss these topics here.

Example 9·2

An electric equipment cabinet is to have a lifting eye installed so that it may be moved easily by an overhead crane. The cabinet contains heavy transformers and weighs 1000 lb. A lifting eye which fits the crane hook has a base diameter of 1 in. The lifting eye, its threaded shaft, and the mounting plate for the tapped hole are all made of low-carbon steel. Specify the size of the screw threads and the tightening torque and check the compressive contact stresses. A factor of safety of 5 is desired.

Solution

Factor of safety refers to how much extra strength should be furnished. For a factor of safety of 5 we multiply the 1000-lbf load by 5 and design for 5000 lbf.

Since a screw is normally tightened to about 80 percent of its yield strength, an initial force of

$$F_i = 0.8 \, S_y A$$

can be obtained from a tightened screw, where A is the cross-sectional area. For low-carbon steel (Table 9.1), $S_y = 30,000$ lbf/in² and thus

$$A = \frac{F_i}{0.8 \, S_y} = \frac{5000}{(0.8)(30000)} = 0.208 \text{ in}^2$$

From Table 9.3, a standard screw thread of $\frac{5}{8}$–11 UNC has an effective area of 0.226 in² and will be adequate. The tightening torque necessary [Eq. (9.10)] will be

$$T = 0.2 \, F_i \, d = 0.2 \, (5000) \, (0.625)$$
$$= 625 \text{ in·lbf}$$

A torque wrench should be used to measure and apply this torque during installation.

We check the compression stress between the eye and mounting plate by using the contact area. Subtracting the hole area from the area of the base of the eye, we find the contact region to have an area

$$A_C = A_{\text{base}} - A_{\text{hole}}$$
$$= \frac{\pi \, 1^2}{4} - \frac{\pi \, 0.625^2}{4}$$
$$= 0.479 \text{ in}^2$$

The compressive contact stress is then

$$\sigma = \frac{F_i}{A_C} = \frac{5000}{0.479} = 10{,}400 \text{ lbf/in}^2$$

This is less than the $S_y = 30{,}000$ lbf/in² for low-carbon steel, so damage to the surface should not occur. A washer will not be needed to protect the plate.

    (a) M4 × 0.7   (d) $\frac{1}{2}$–13 UNC
    (b) M16 × 2   (e) $\frac{1}{4}$–28 UNF
    (c) $\frac{1}{4}$–20 UNC (f) $\frac{3}{4}$–16 UNF

9.9 A screw with thread size M6 × 1 is being rotated in a hole with matching threads. What linear distance will the screw advance for a rotation of
    (a) 1 revolution   (c) 180°
    (b) 2 revolutions (d) 45°

9.10 Repeat Prob. 9.9 for a screw thread of $\frac{1}{4}$–28 UNF.

9.11 Figure P9.11 shows a simplified tensile testing machine driven by two square-thread power screws. When a torque T is applied to these screws, as shown, the crosshead tends to move upward. This creates a tensile force in the specimen whose ends are firmly held by "chucks." The screws have four threads per inch with outer diameter $d = 1.50$ in and root diameter $d_r = 1.25$ in. What torque T is necessary to produce a tensile force of 5000 lbf in the specimen when the friction $\mu = 0.1$? Remember that there are two screws sharing the load.

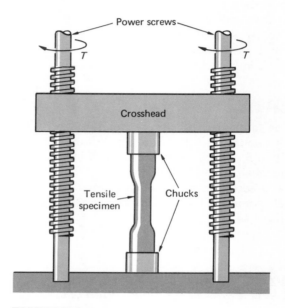

FIGURE P9·11

9.12 On a tensile testing machine, after a tensile load has been produced in a specimen, it is desirable to maintain it with the drive motor turned off so that torque $T = 0$. This will be possible if the screws are self-locking. Are the screws self-locking in Prob. 9.11? What torque must be applied to the screws to *remove* an applied tensile force of 5000 lbf?

9.13 Repeat Example 9.2 for a cabinet mass of 500 kg and a hook base with a diameter of 30 mm. Other conditions are the same. Specify a standard metric thread and appropriate tightening torque.

9.14 Six bolts seal the cylinder head on the pressure vessel of Fig. P9.14. The inside diameter of the vessel is 10 in and the pressure inside the vessel may be as high as 100 lbf/in² gage. If a safety factor of 5 is desired, select a standard thread size which will provide a satisfactory design. What tightening torque should be used? Assume a yield strength of $S_y = 80{,}000$ lbf/in² for the bolt material.

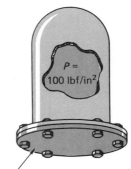

FIGURE P9·14   Cylinder head

REFERENCES

1. Lynch, C. T. (ed.), *Handbook of Materials Science*, CRC Press, Cleveland, 1974.
2. Shigley, J. E., *Mechanical Engineering Design*, McGraw-Hill, New York, 1977.

10 ELECTRIC CONNECTIONS WITH WIRE-BINDING SCREWS*

As mentioned in the introduction to Sec. 8.3, the profound problems encountered with aluminum house wiring arose from a seemingly trivial item: the electric resistance of terminals using wire-binding screws.[1] As pointed out there, the resistance depends on the interaction of the mechanics of screwed connections with a complex of other factors. Since you have just been introduced to the mechanics of screwed connections—a subject that is central to our discussion—we are now in a position to explore the design of this important type of electric connection.

10·1 THE NEED FOR RELIABILITY IN ELECTRIC CONNECTIONS

Our technological society depends on the functioning of mind-boggling numbers of electric connections. For example, the Bell Telephone System makes an estimated one billion (10^9) new electric connections every year. Every telephone conversation you make depends on hundreds of these connections working properly. The number of connections used in house wiring is smaller, but still impressive: one estimate is 20 million (20×10^6) new connections each year. Steady passage of current from the fuse box through several intermediate junction boxes and the switch to

---

* Background material for this chapter may be found in Sec. 2.3, where log-log graphs are discussed; Sec. 4.1, where statics is covered; and in Secs. 5.1 and 5.2, where simple electric circuits are described. Section 8.3 introduces the electric connection problem. Sections 9.1 and 9.2 discuss material properties and screws.

[1] A wire-binding screw is simply a machine screw used to hold (bind) a wire against a contact plate.

your reading lamp is dependent on a dozen or so electric connections working properly; many of these connections are pressure connections using wires held in place by wire-binding screws.

The reliability required is impressive. For example, if only one out of a thousand house-wiring connections is defective, the potential exists for the creation of 20,000 [$(20 \times 10^6)/(1000)$] new consumer problems per year. Some of these problems may be serious; house fires, for example. Hence, with human life at stake, even 99.9 percent reliability may not be good enough.

In this section, we will first describe some of the electrical phenomena at a metal-to-metal contact. We will show that these phenomena are very closely tied into certain mechanical phenomena, such as plastic deformation (Sec. 9.1) and stress relaxation. We will demonstrate that although good connections can be made using either copper or aluminum wire, the design requirements for an aluminum connection are more stringent. We will conclude: reliable connections *can* be made to aluminum wire, but only if the connections are carefully designed (by engineers) and made (by craftspersons). If due care is not exercised, however, the possibility of failure *does* exist; and as we have seen, a relatively small percentage of failures may have profoundly serious social and economic consequences. We hope that you keep this in mind as you read the discussion which follows.

10·2 ELECTRIC CONTACT BETWEEN METAL SURFACES

The pioneering work on analysis of electric contacts was done by the British physical chemists Bowden and Tabor, who presented their results in the *Proceedings of the Royal Society of London* in 1939 [1]. Bowden and Tabor pointed out that even when very carefully polished and made as flat as possible, metal surfaces are still rough on a microscopic scale. Thus, when two metal surfaces are pressed together, the area of *actual* contact is very much less than the apparent area. As shown schematically in Fig. 10.1a, a wire (which was round before tightening) is flattened on its top and bottom surfaces by the plastic deformation which occurs as a result of the axial tightening force. On a macroscopic scale, as shown in Fig. 10.1b, the top of the wire and the bottom of the screwhead then appear flat. However, on a microscopic scale, as shown in Fig. 10.1c, hills and valleys are present on both surfaces. Contact occurs only at the tops of the hills, and a large fraction of the metal surface is separated by a distance which is small compared with the size of the screwhead but large compared with the dimensions of atoms and molecules.

Consequently, the electric resistance of a contact depends not

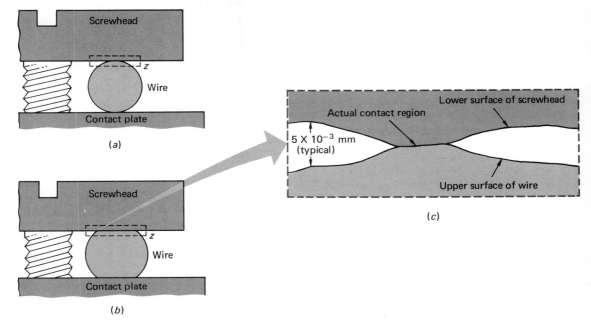

FIGURE 10·1
Cross sections of contact regions between a wire-binding screw and a wire.

at all on its apparent size as visible to the unaided eye but rather on the actual, but invisible, contact areas where matching "hills" touch.

If only part of the metal surface is available for contact, the streamlines of electric current must constrict, as shown schematically in Fig. 10.2. The effect of this constriction extends from the surface into the interior of the metal over a distance—the effective length—which is comparable to a diameter of the contact area. The effective length is typically large compared with the invisible space between the two contacting surfaces. Figures 10.1 and 10.2 suggest that the electric resistance of a contact must arise mostly from the constriction in the current streamlines and that an equation similar to Eq. (5.1) could be used to relate contact resistance to the material resistivity and the actual contact area if an effective length for the contact region were known. That is,

$$R = \rho \frac{l_{\text{eff}}}{A_{\text{eff}}} \tag{10.1}$$

where R = contact resistance
A_{eff} = area of actual contact
l_{eff} = effective length

Under certain ideal conditions (with a circular, isolated contact area, where the contact area is very small compared with the

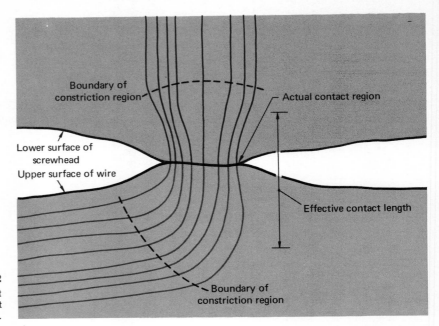

FIGURE 10-2
Constriction of electric current flow near a small area of contact between two metal surfaces.

apparent contact area) l_{eff} can be calculated, and is about 0.8 times the diameter of the actual contact area.[2] In practice, however, the contact areas are not circular and not isolated (there are typically 3 to 30 effective areas per contact) and consequently l_{eff} must be determined from experimental results.

The engineer Ragnar Holm [2] has provided a simple method for estimating the actual contact area. Suppose that the two contacts are being pressed together by a force F. The average pressure at the points of actual contact is the force divided by the actual contact area, so that

$$\bar{p} = \frac{F}{A_{\text{eff}}} \tag{10.2}$$

where $\bar{p}$ = average pressure at actual contact surfaces
F = total force acting on upper contact
A_{eff} = actual contact area (not the apparent contact area)

In practice, relatively large forces are used. From Eq. (10.2), $\bar{p}$ tends to be large, but it cannot be very much larger than the yield strength, for if it were, the hills in Fig. 10.1 would be crushed,

---

[2] This result was obtained in 1873 by the Scottish mathematical physicist James Clerk Maxwell [3]. Maxwell is more famous for his work on the foundations of electromagnetic theory (which led in time to radio, radar, and television), but his calculation of l_{eff} was part of his work on the practical development of an international standard for the ohm.

increasing the effective area until the pressure fell to a value at or below the yield strength. Using this reasoning, Holm suggests replacing $\bar{p}$ in Eq. (10.2) by the yield strength and solving for the effective area:

$$A_{\text{eff}} = \frac{F}{S_y} \tag{10.3}$$

For l_{eff}, we can use dimensional analysis or Maxwell's result[3] as a guide and postulate

$$\begin{aligned} l_{\text{eff}} &= c\sqrt{A_{\text{eff}}} \\ &= c\sqrt{\frac{F}{S_y}} \end{aligned} \tag{10.4}$$

where c is a dimensionless (but not necessarily constant) multiplier. Substituting Eqs. (10.3) and (10.4) into Eq. (10.1),

$$R = \frac{c\rho\sqrt{S_y}}{\sqrt{F}} \tag{10.5}$$

Equation (10.5) predicts that good contacts (small R) will result from using:

Good conductors (small ρ)
Soft conductors (small S_y)
Large forces (large F)

Note that the apparent contact area does not appear at all in Eq. (10.5). The resistance depends only on the material constants ρ and S_y and the total force F. In practice, the multiplier c must be determined experimentally.

EXPERIMENTAL RESULTS

Figure 10.3 shows some experimental data on the resistance of copper and aluminum contacts. Resistance is given in microohms (1 $\mu\Omega$ = 10^{-6} Ω) and force in newtons (1 N = 0.225 lbf). The 1939 data of Bowden and Tabor [1] are compared with a least-squares-fit straight line. Note that over a two-decade range [0.5 to 50 N (roughly 2 oz to 11 lbf)], the data behave like the function[4]

---

[3] Maxwell showed that for a circular area, l_{eff} is proportional to the diameter. For a circle, however, the diameter (and therefore l_{eff}) is proportional to the square root of the area.
[4] Recall that on log-log coordinates, R = (constant)($F^{-1/2}$) will be a straight line with negative slope. This can be seen in Sec. 2.3.

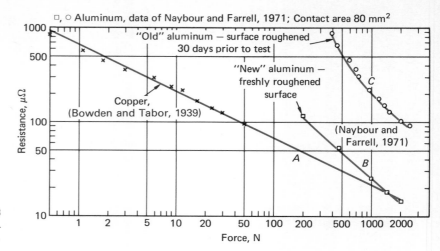

FIGURE 10·3
Resistance of copper and aluminum contacts.

$$R = \frac{\text{constant}}{\sqrt{F}} = (\text{constant})(F^{-1/2}) \tag{10.6}$$

suggested by Eq. (10.5).

In electric house wiring, "good" contacts have resistances in the range of 200 to 500 $\mu\Omega$, corresponding to the resistance of a few inches of #14 copper or #12 aluminum wire. For copper, Fig. 10.3 shows that

> Good contacts can be obtained with modest contact forces [2 to 10 N (0.5 to 2.25 lbf)].
>
> Copper contacts obey the simple theory presented in the previous section: there is a steady and predictable decrease in resistance as contact force is increased.

For aluminum, the situation is more complicated. Figure 10.3 also shows some data on aluminum published in 1973 by Naybour and Farrell [4]. Note the effect of surface preparation. At 500-N force, the resistance of "old" surfaces (roughened 30 days before the test) is over 10 times higher than the resistance of "new" (freshly roughened) surfaces. Thus for aluminum,

> Good contacts require relatively large forces, especially if the surfaces are old [500 to 1000 N (112 to 225 lbf)]. The forces required for good contacts are approximately 100 times larger for aluminum than for copper.
>
> Aluminum contacts can have as low a resistance as copper contacts, but only for new surfaces and large loads (> 2000 N).
>
> Aluminum contacts do not obey a simple $F^{-1/2}$ law; although resistance does decrease with increasing load, it does so in a complicated way. It also depends on surface preparation.

In order to understand why the aluminum contact behaves as it does, we need to look more closely at the aluminum surface, and in particular at its oxide coating.

ALUMINUM AND ITS OXIDE, ALUMINA

Aluminum does not rust, because its surface is always protected by a thin tenacious film of alumina, Al_2O_3. If bare metal is exposed, as in machining operations, the oxide film reforms in seconds. The oxide layer[5] grows to a thickness of about 10 nm (1 nm $= 10^{-9}$ m). Oxide growth then stops, because the tenacious film of oxide isolates the metal from the oxygen in the air. The tenacious oxide film is a help in protecting the surface from corrosion, but it is a hindrance in making good connections. Alumina is, unfortunately, a very good insulator. Even a 10-nm layer of alumina can prevent the flow of electric current. Hence, as Mittleman [5] put it in 1969,

> Before a good electrical joint can be made on an aluminum conductor, the oxide films must be removed or penetrated, so that bare metal surfaces, essentially freed of films, are in intimate contact with one another. Once the contact is initiated, air must be excluded from the joint to prevent reoxidation.

One simple way to penetrate the oxide layer is by plastically deforming the aluminum conductor. The mechanics of the process were described by Bowden and Tabor [6]. Aluminum is very soft and ductile. You can cut it with a knife. The oxide, alumina, is very hard—harder than quartz—and brittle. The behavior of the oxide is thus "analogous to that of a thin film of ice on mud." A load sufficient to deform the ductile aluminum substrata will break through the brittle oxide coating. To see how an increasing load causes plastic deformation and breakthrough of the aluminum oxide, refer again to Fig. 10.3.

First, consider curve C, which shows data for old aluminum surfaces, that is, surfaces cleaned with an abrasive and then allowed to stand in air for 30 days prior to making contact. Resistances for contacts between old aluminum surfaces are an order of magnitude higher than for copper, at the same contact forces. For example, when force is 500 N (112 lbf), a copper contact has approximately 30 $\mu\Omega$ resistance, compared with over 500 $\mu\Omega$ resistance for a contact between old aluminum surfaces. This is clearly the effect of a mature oxide film built up on the aluminum surfaces by 30 days exposure to air. As the force is increased, the actual contact area increases and the resistance consequently de-

---

[5] The oxide layer is invisible; its thickness is about one-fiftieth the wavelength of visible light.

creases; but since much of the actual contact area remains covered by an insulating layer of oxide, the contact resistance remains high even with relatively large force.[6]

Now consider curve B, which shows data for new aluminum surfaces, that is, surfaces cleaned with an abrasive immediately before contact was made. Tests were run over the force range of 200 to 2000 N. Even in the few seconds which elapsed between cleaning and contact, some thickness of Al_2O_3 has formed. Consequently, at the lower forces, resistance is high relative to copper. From the data in the 1000-to-2000-N range, however, we see that new aluminum surfaces can form contacts which are as good as copper if the contact force is large enough.

What happens to the metal when the force exceeds 2000 N? For the aluminum data in Fig. 10.3, the pressure based on the total, or apparent, contact area is $(F/80) \times 10^6$ Pa. When F exceeds 2000, the pressure exceeds 25 MPa. Referring to Table 9.1, we find that this is close to the yield strength of annealed aluminum. Obviously, as the force increases above 2000 N, yielding will occur over a larger and larger fraction of the contact surface. As mentioned previously, the oxide layer is hard and brittle and cannot follow the deformation of the ductile aluminum substrata. Thus, as shown in Fig. 10.4, the oxide layer is broken and metal-to-metal contact occurs.

In our discussion thus far, we have assumed that the contact force is constant. In practice, contact force may vary as a result of dynamic effects such as shock and vibration. Temperature changes may cause expansion and contraction of the contact structure and hence lead to temperature-dependent forces.

Extensive experimentation by Naybour and Farrell [4] and by Holm [2] has shown, however, that good aluminum-to-aluminum contacts can be made and maintained (even with variable contact forces) just so long as the forces remain above 1000 N. Based on what we have learned so far, we may set down some preliminary basic design rules.

In designing aluminum contacts,

> Specify that contact surfaces should be cleaned immediately before contact is made.
> Design the contact to maintain a clamping force in excess of 1000 N (1 kN) for the design life of the contact.

The requirement for *maintaining* a 1-kN clamping force brings us to the subject of stress relaxation.

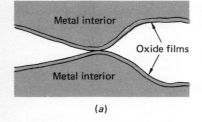
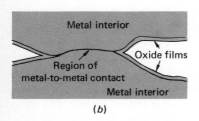

FIGURE 10·4
Effect of oxide films on aluminum contacts.

---

[6] The data of curve C are for 80-mm² contacts. That is roughly the area of your thumbnail. A force of 1000 N corresponds to 225 lbf. To get an idea of the contact pressures used to obtain curve C, imagine a 225-lb (100-kg) person standing on your thumbnail!

TABLE 10·1
LIFE TESTS ON HIGH-FORCE, LOW-STRESS ALL-ALUMINUM CONNECTORS

Sample	Joint Resistance, $\mu\Omega$					Final Joint Temperatures
	Number of Temperature Cycles					
	0	166	1072	2006	7300	
1	79	83	88	81	78	48 °C
2	120	122	130	122	117	50 °C
3	92	98	117	95	93	52 °C

contact area. In aluminum house wiring, on the other hand, the wires and connectors have been relatively small, and the apparent contact areas are correspondingly small. This puts great demands on the craftsperson in the field. If he or she does not get the wire wrapped correctly, or doesn't get the torque just right, the aluminum connection may fail. Copper is much more forgiving: as shown in Fig. 10.3, forces over the very wide range of 1 to 2000 N will allow good (less than 500-$\mu\Omega$) contacts. Thus, with copper, the installer has a margin for error which is enormous compared with the margin allowed by aluminum.

Nevertheless, properly trained workers working with properly designed connectors *can* make reliable connections to aluminum conductors. For example, the data in Table 10.1 were taken by Naybour and Farrell [4] using tests in which the current was "cycled," that is, repeatedly turned on and off to duplicate real-life conditions. When on, the current was large enough to produce 130°C (266°F) temperatures in the conductor. When the current was off, the conductor was allowed to cool to 20°C (68°F).

Note that initial resistances are on the order of 100 $\mu\Omega$ and remain at this level through the testing period. On the other hand, if the design principles we have described are violated, some (but not all) samples of a particular design may fail, as shown in

TABLE 10·2
LIFE TESTS ON LOW-FORCE, HIGH-STRESS, ALL-ALUMINUM CONNECTORS

Sample	Joint Resistance, $\mu\Omega$					Final Joint Temperature
	Number of Temperature Cycles					
	0	166	1072	2006	7300	
4	87	103	120	117	125	71°C
5	206	310	680	Gross overheating: connector removed		
6	126	167	190	237	275	85°C

10.3 AN ACTUAL DETAILED DESIGN

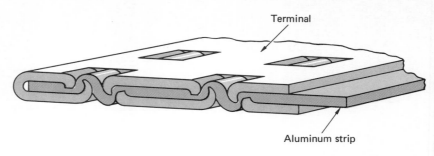

FIGURE 10·8
Design for large area of contact between aluminum conductor and terminal. Sharp lances grip the aluminum strip material and are embedded deeply into it. The lances pierce the metal, eliminating the need for conductor cleaning. Moreover, the oxide that forms on the metal's surface does not affect the operation, because the teeth are embedded. (*AMP, Inc.*)

Table 10.2. Note that the connector with the highest initial resistance was the one that failed. As the test progressed, the resistance increased steadily until after 2006 temperature cycles the connector was removed from the test rack, presumably for reasons of laboratory safety.

Aluminum is a good conductor, is in plentiful supply, and is relatively cheap. These factors will someday motivate a manufacturer to see a need to reintroduce it for house wiring. Perhaps you will be the engineer who designs a line of reliable connectors for this important application. Perhaps your design will resemble Fig. 10.8.

PROBLEMS

Electrical and mechanical properties given in Tables 5.1, 8.1, and 9.1 may be used in the problems below.

10.1 Assume that 7,000,000 automobiles are sold per year in the United States. Estimate the total number of new electric connections thus created. Remember that there are two or three connections for each light bulb, one for each sparkplug, and so forth. You can use any typical car as the basis for your estimate. In order to simplify this problem, assume that the car has no radio, cassette player, or other similar electronic equipment.

10.2 If the probability of failure of an electric connection is only 1/100 per connection per year, estimate the number of electric connection failures in the 5-year lifetime of one typical automobile.

10.3 Assuming that contact resistance (ohms) depends only on material resistivity (ohm-centimeters), yield strength (pascal), and force (newtons), derive Eq. (10.5) directly from dimensional analysis.

10.4 Using data from Fig. 10.3, at a force of 50 N, determine the constant c in Eq. (10.5) for copper.

10.5 Using the "new" aluminum, freshly roughened surface data of Fig. 10.3 at a force of 500 N, determine the constant c for aluminum. Compare with the value found for copper in Prob. 10.4.

10.6 In house wiring, contacts are designed to have resistance equal to or less than a few inches of wire. For #14 AWG copper and #12 AWG aluminum wire, verify that these resistances are less than 10^{-3} Ω. What is the logic behind this design principle?

10.7 What is the stress imposed on the shaft of the wire-binding screw in Fig. 10.7? Assume a cross-sectional area of 8.83 mm^2 and a force of 1000 N. How does this stress compare with the yield stress of materials such as carbon steel, wrought aluminum alloy, and yellow brass, which are commonly used to make such screws?

10.8 Show that a torque of 17 in·lbf applied to the screwhead shown in Fig. 10.7 will cause a compressive stress sufficient to exceed the compressive strength of pure aluminum wire, thus crushing it.

10.9 Using Figs. 10.5 and 10.6, determine the time required for the contact force to decrease below 1000 N when a torque of 8 in·lbf is applied to the screwhead shown in Fig. 10.7. Assume that stress relaxation curves for initial stresses below 60 MPa are parallel to the 60-MPa curve.

10.10 It has been suggested that there are few connection problems with respect to aluminum wire of size 8 AWG or larger. Repeat the design process of Sec. 10.3 and come to an independent opinion. In particular, estimate the range of torques which yield satisfactory connections. Compare this range with that found in Sec. 10.3 for #12 AWG wire. Assume that the wire-binding screw is metric size M8 × 1.25 and has a 16-mm head diameter. Dimensions of metric screws are found in Table 9.3 and standard AWG sizes are given in Table 8.1.

REFERENCES

1. Bowden, F. P., and D. Tabor, "The Area of Contract between Stationary and Moving Surfaces," *Proceedings of the Royal Society*, vol. 169, series A, no. 939, pp. 391–413, February, 1939.
2. Holm, Ragnar, *Electric Contacts: Theory and Application*, New York, Springer-Verlag, 1967.
3. Maxwell, James Clerk, *Treatise on Electricity and Magnetism*, Clarendon Press, Oxford, 1873.
4. Naybour, R. D., and T. Farrell, "Degradation Mechanisms of Mechanical Connectors on Aluminum Conductors," *Proceedings of the Institution of Electrical Engineers*, vol. 120, no. 2, pp. 273–280, February, 1973.

5. Mittleman, Joseph, "Connecting Aluminum Wire Reliably," *Electronics*, vol. 42, pp. 94–98, December 8, 1969.
6. Bowden, F. P., and D. Tabor, *The Friction and Lubrication of Solids,* Clarendon Press, Oxford, 1950.
7. Underwriters Laboratories, Inc., Ad Hoc Committee's Statement on the Use of Aluminum Conductors with Wiring Devices in Electrical Wiring Systems, March, 1973.

11 STRUCTURES AND BRIDGES*

Engineering structures have played an important role in the history of civilization. The pyramids of Egypt and the Parthenon of Greece are remarkable remants of the ancient world and more recent structural marvels include the Eiffel Tower and the Golden Gate Bridge. Engineering structures are not only buildings and bridges, however. Machines, vehicles, and even furniture are also structures, and design methods for them may be based on structural analysis techniques. The construction of jumbo-jet aircraft and that of the modern light-weight automobile are two recent examples of the important role played by the methods of structural analysis.

In this chapter we cannot, of course, make you an expert in structural analysis. But we will demonstrate some of the ideas and concepts which are applied in the design of structures. In particular, we will try to illustrate how engineers combine the quantitive methods of statics and material science to design safe and economical structures.

11·1 MODES OF FAILURE

In designing any type of structure, the designer must begin by considering the various ways, or "modes," in which the elements of the structure might fail. The next step is to be certain that the structural elements are strong enough so that the modes of failure cannot actually occur during use of the structure. In this section

---

* For this chapter, you should be familiar with statics as introduced in Sec. 4.1 and the basic properties of materials as discussed in Sec. 9.1.

we will discuss four of the most common failure modes and some of the simple calculations that can be used to predict them.

Calculations are important because the most economical way to check the strength of a structure is with a pencil and paper. It is extremely expensive to build a real structure and test it. For this reason, structural analysis methods and computer methods for applying structural analysis are among the most sophisticated analytical tools used by engineers. These mathematical methods make it easy to test a design without actually building it. The methods pay off, because many designs can be conveniently tried before a safe and economical design is selected for actual construction.

We will now investigate some of the ways in which a long bar of low-carbon steel could fail. This type of steel, often referred to as *structural steel*, is representative of the material used in many real structures. We will consider a piece of steel with a length of 5 m and a rectangular cross section of 50 mm × 100 mm (about the size of a two by four). Recall from Table 9.1 that the typical yield strength for this type of steel is

$$S_y = 30{,}000 \text{ lbf/in}^2 = 207 \text{ MPa}$$

and that its modulus of elasticity is

$$E = 30 \times 10^6 \text{ lbf/in}^2 = 2.07 \times 10^5 \text{ MPa}$$

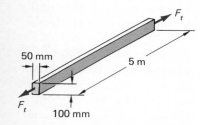

FIGURE 11·1
A steel bar loaded in tension.

Let's now look at some of the ways this piece of steel could fail if used in a structure.

YIELDING IN TENSION AND COMPRESSION

Figure 11.1 shows the 5-m steel bar loaded in tension by the force F_t. If this bar fails in the tension mode of failure, the stress produced by the force F_t will exceed the yield strength and the steel will be permanently stretched. The force at which this failure begins is

$$F_t = S_y A \qquad (11.1)$$
$$= (207 \times 10^6)(0.050 \times 0.100) = 1.04 \times 10^6 \text{ N}$$

so the tension force F_t must be less than 1040 kN to prevent failure.

If the bar were exposed to a compression force F_c as shown in Fig. 11.2, it could be deformed if F_c produced a compressive stress greater than the yield strength. This would be a failure by compressive yielding, and just as for tension yielding of the bar would begin at

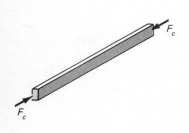

FIGURE 11·2
A steel bar loaded in compression.

$$F_c = S_y A = 1040 \text{ kN} \qquad (11.2)$$

However, buckling is a more likely form of compression failure for long structural members and normally occurs before compressive yielding.

BUCKLING

The failure of a long, slender member as a result of buckling is familiar to anyone who has attempted to lean on a yardstick (or meter-stick). As the load slowly increases, the member remains straight until a critical load is reached, whereupon it suddenly buckles to take on the bent shape shown in Fig. 11.3. After that point a modest increase in load will then cause a total failure of the member. The development of an equation to calculate the force at which buckling occurs is a rather complicated problem in elastic stability found in most strength of materials textbooks [1,2]. The resulting equation, however, is quite simple and is widely used by design engineers for many types of buckling problems. Buckling of the bar in Fig. 11.3 is expected for a compressive force greater than

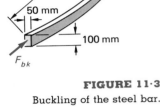

FIGURE 11·3
Buckling of the steel bar.

$$F_{bk} = \pi^2 \frac{EI}{L^2} \qquad (11.3)$$

where F_{bk} = buckling force, N (or lbf)
E = the modulus of elasticity, Pa (or lbf/in²)
L = length of the bar, m (or in)
I = cross section moment of inertia, m⁴ (or in⁴)

For a bar of solid rectangular cross section, the moment of inertia

$$I = \frac{bh^3}{12} \qquad (11.4)$$

where b and h are the dimensions of the rectangle, with h being the height of the cross section in the direction of possible buckling.

Consider our steel bar and the possibility of buckling in the direction $h = 0.100$ m with $b = 0.050$ m. The moment of inertia is

$$I_{100} = \frac{bh^3}{12} = \frac{(0.050)(0.100^3)}{12} = 4.17 \times 10^{-6} \text{ m}^4$$

and the force at which buckling occurs would be

$$F_{bk100} = \pi^2 \frac{EI}{L^2} = \pi^2 \frac{(2.07 \times 10^{11})(4.17 \times 10^{-6})}{5^2}$$
$$= 341 \text{ kN}$$

This is the force at which the bar would buckle in the 100-mm direction. However, in the direction of $h = 0.050$ m and $b = 0.100$ m, we find

$$I_{50} = \frac{bh^3}{12} = \frac{(0.100)(0.050^3)}{12} = 1.04 \times 10^{-6} \text{ m}^4$$

and the corresponding buckling force is

$$F_{bk50} = \pi^2 \frac{EI}{L^2} = \pi^2 \frac{(2.07 \times 10^{11})(1.04 \times 10^{-6})}{5^2}$$
$$= 85.0 \text{ kN}$$

This is the force at which buckling of the steel bar would occur in the direction of the 50-mm dimension, as shown in Fig. 11.3.

To find out how the steel bar would fail if a compressive force were actually applied at its ends, we examine the three possibilities calculated above: (1) it yields in compression at a force $F_c = 1040$ kN; (2) it buckles in the 100-mm direction with $F_{bk100} = 341$ kN; or (3) it buckles in the 50-mm direction with $F_{bk50} = 85.0$ kN. Clearly the bar will fail like a yardstick, buckling in the thin direction, as indicated in Fig. 11.3. The force level for buckling in the thick direction (341 kN) or for yielding (1040 kN) cannot be reached, because the bar will buckle first in the 50-mm direction when the compressive force reaches 85.0 kN. Buckling is thus expected in the thinnest direction of a member where the value of I in Eq. (11.3) is the smallest. Normally only this buckling force and the force for yielding [Eq. (11.2)] need be computed to find the compressive load at which a structural member will fail.

BENDING

Another way that our steel bar may be deformed is by bending due to a force applied perpendicular to the bar, as shown by F_{bn} in Fig. 11.4. This force causes the bar to act as a beam. It is easily deflected, with the bottom surface of the bar going into tension and tending to stretch while the upper surface is in compression. If the force is applied at the midpoint of the solid rectangular bar, the outer surfaces of the beam have the following maximum stress, as may be shown by strength of materials methods [1,2].

$$\sigma = \frac{hL}{8I} F_{bn} \tag{11.5}$$

where σ = stress, Pa (or lbf/in²)
L = length, m (or in)
h = height of cross section, m (or in)

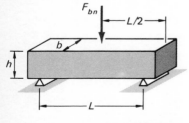

FIGURE 11·4
A beam loaded at its midpoint.

F_{bn} = force at the midpoint, N (or lbf)
I = cross-section moment of inertia, m⁴ (or in⁴)

For a rectangular cross section, $I = bh^3/12$ just as in buckling; but now b and h must be as shown in Fig. 11.4. The width across the beam is b and the beam height is h.

With Eq. (11.5) we can compute the force which will fail the steel bar when it is loaded at its center as a beam. In this equation, corresponding to Fig. 11.4, consider $L = 5$ m, $h = 0.050$ m, and $b = 0.100$ m, so that $I = 1.04 \times 10^{-6}$ m⁴. Failure will occur by deforming the beam if the stress from Eq. (11.5) exceeds the yield strength. For low-carbon steel where $S_y = 207$ MPa, this corresponds to a bending force

$$F_{bn} = \frac{8I}{hL}\sigma = \frac{(8)(1.04 \times 10^{-6})}{(0.050)(5)}(207 \times 10^6)$$
$$= 6.89 \text{ kN}$$

The modes of failure we have considered for our steel bar are summarized in Table 11.1, wherein it can be seen that the largest load can be supported if the bar is in tension, while the calculated failure in bending shows the smallest of the loads at failure. This is typical, since tension is an efficient form of loading and it is difficult to support a bending load. When large loads must be supported by beams, the beam height is made large to increase the moment of inertia I while the beam length L is kept as small as possible. The effects can be easily seen in Eq. (11.5), where decreasing L and increasing I both reduce the stress σ. In real structures, I beams, channels, and hollow members are often used as beams because their shapes have a large moment of inertia but require only a relatively small amount of material.

TABLE 11·1
MODES OF FAILURE FOR A LOW-CARBON STEEL BAR
(50 mm × 100 mm × 5 m)

Description of Load	Type of Failure	Failure Force, kN
Tension	Yields, permanently stretched	$F_t = 1,040$
Compressive	Buckles	$F_{bk} = 85.0$
Bending, center load, $h = 50$ mm	Yields, permanently bent	$F_{bn} = 6.9$

PROBLEMS

11.1 The solid members of rectangular cross section listed below are made of wrought aluminum alloy with $S_y = 207$ MPa and $E = 0.690 \times 10^5$ MPa. Determine the force at which failure is expected in each case for a compression load and indicate the expected mode of failure, either buckling or yielding.
(a) $b = h = 25$ mm, $L = 1$ m
(b) $b = h = 25$ mm, $L = 0.1$ m
(c) $b = 4$ cm, $h = 2$ cm, $L = 5$ m
(d) $b = 8$ cm, $h = 2$ cm, $L = 5$ m

11.2 A solid beam of rectangular cross section is shown loaded at its center in Fig. P11.2. Determine the maximum stress expected in the beam for
(a) $x = 1$ in, $y = 2$ in, $L = 4$ ft
(b) $x = 2$ in, $y = 1$ in, $L = 4$ ft
(c) $x = 2$ in, $y = 2$ in, $L = 4$ ft
(d) $x = 2$ in, $y = 1$ in, $L = 40$ ft

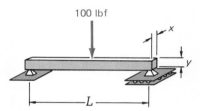

FIGURE P11·2

11.3 Both cables shown in Fig. P11.3 have diameters of 10 mm and a yield strength of $S_y = 207$ MPa. Can these cables support the 1000-kg mass shown in the figure without permanent deformation?

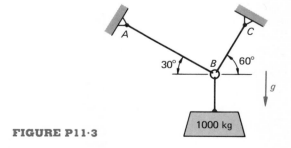

FIGURE P11·3

11.4 Consider the cable arrangement discussed in Prob. 11.3. Determine the maximum mass that the cables can support without yielding.

11.5 In Fig. P11.5, the mass produces a compressive force in the aluminum rod and a tensile force in the aluminum cable. The aluminum alloy used has a yield strength of $S_y = 30,000$

lbf/in² and an elastic modulus $E = 10 \times 10^6$ lbf/in². Determine the cable diameter and the size of a square rod which will *just* support the load.

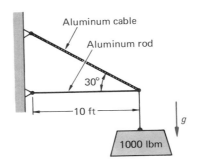

FIGURE P11·5

11.6 In Prob. 11.5, size the cable and rod for a factor of safety of 4. This may be done by assuming the supported mass to be four times its actual value, or 4000 lbm.

11·2 TRUSS STRUCTURES

Table 11.1 summarizes our failure mode analysis for a typical piece of structural steel. One must conclude from this table that bending long members should be avoided or that especially careful design is necessary when it does occur. Truss structures are based on the pinned triangular element discussed in Sec. 4.1. A major advantage of truss structures is that each member of the truss is normally loaded only in tension or compression and not in bending.

The usefulness of this concept can be seen in Fig. 11.5a, where a truss structure is acting as a bridge. The load F at the center is supported through tension and compression in the members and none of the members must support a bending load. Of course, the force F could be applied between pins, as in Fig. 11.5b. This situation is normally avoided in design because the member where the force is applied is loaded in bending. But even in this case, the bending length is short compared with the total span, and none of the other truss members is loaded in bending.

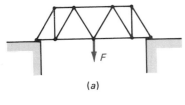

(a)

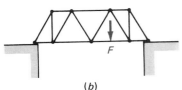

(b)

FIGURE 11·5
A truss bridge.

DESIGN CALCULATIONS

The design of a truss structure consists of many subproblems. Some of these are difficult mathematical problems beyond the scope of this book; others are very special problems of construc-

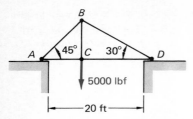

FIGURE 11·6
A truss with a 5000-lbf load.

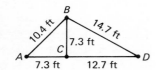

FIGURE 11·7
Truss dimensions.

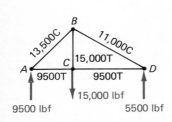

FIGURE 11·8
Load on each member.

tion and fastening which have been solved mostly by the ingenuity of generations of structural engineers. We cannot discuss all the details of structural design here, but we can expose you to some of the basic concepts.

Consider the simple structure shown in Fig. 11.6 with a 5000-lbf load. This is the maximum load expected to be applied to the structure. We must size the members of this structure to support the load with a factor of safety of 3. A factor of safety is used to consider the unexpected. For example, the maximum load may turn out to be larger than our expectations, the material may not be as strong as expected, and our calculations may not account for every detail of the actual structure. Hence, the calculations are carried out below as if the load were 15,000 lbf. In this design, all the members are to be made from the same size square bar of low-carbon steel, which comes only in $\frac{1}{4}$-in-size increments.

There are three steps in solving this problem: (1) the length of each member must be calculated; (2) the force loading each member must be determined; and (3) the member requiring the largest cross section must be identified so that the proper bar size may be specified. The length of the members can be calculated by trigonometry. For example, an equation for the length of member BC in Fig. 11.6 is

$$\frac{BC}{\tan 45°} + \frac{BC}{\tan 30°} = 20$$

$$BC = 7.3 \text{ ft}$$

This dimension is indicated in Fig. 11.7 along with the lengths of all the other members which can be similarly calculated. The force analysis for this truss was performed previously using the methods of statics in Example 4.3. We now multiply the expected load of 5000 lbf by 3 to achieve our desired factor of safety and design for a load of 15,000 lbf. The force on each member with the 15,000-lbf load is as shown in Fig. 11.8 based on the results of Fig. 4.8c. If we now size the members to support these loads, it will be adequately strong.

The structure will fail if one of the bars in the structure yields or if buckling occurs. If the truss members have equal area and a yield failure takes place, it will be in member BC, which carries the largest load in either tension or compression. The yield strength of low-carbon steel is $S_y = 30,000$ lbf/in², so that an area of

$$A = \frac{F}{S_y} = \frac{15,000}{30,000} = 0.50 \text{ in}^2$$

is required for the cross section of that member. This can be furnished by a $\frac{3}{4}$-in square bar ($A = 0.56$ in²).

If a buckling failure occurs it can occur *only* in a compressively loaded member. Members *AB* and *BD* are the only compression members; hence we check only these two for buckling. Based on Eq. (11.3), $F_{bk} = \pi^2 EI/L^2$. This can be used to calculate the required I to support a given force:

$$I = \frac{L^2}{\pi^2 E} F_{bk} \tag{11.6}$$

For member *BD* we substitute $L = 14.7$ ft from Fig. 11.7 and $F_{bk} = 11{,}000$ lbf from Fig. 11.8 along with $E = 30 \times 10^6$ lbf/in² for steel. We obtain from Eq. (11.6)

$$I_{BD} = \frac{(12 \times 14.7)^2}{\pi^2 (30 \times 10^6)} 11{,}000$$
$$= 1.2 \text{ in}^4$$

as the required moment of inertia. A similar calculation for member *AB* yields $I_{AB} = 0.71$ in⁴. Member *BD* thus requires the largest size to prevent buckling. From Eq. (11.1), $I = bh^3/12$, and with a square bar $b = h$ so that a value of $I = 1.2$ in⁴ requires

$$b = (12\,I)^{1/4}$$
$$= (12 \times 1.2)^{1/4}$$
$$= 1.95 \text{ in}$$

which means that to the nearest quarter inch, a square bar with 2-in sides must be selected to prevent buckling of member *BD*. If all members of equal size are desired, as stated earlier, the members must be made of the 2-in steel bar.

After evaluating the truss design based on 2-in steel bars, it may be decided that the truss is too heavy and may be wasteful of material. One way to save weight would be to use the calculated $\frac{3}{4}$-in steel bars for all tension members. Since these members are not subject to buckling, the smaller size would be adequate. Another approach to saving weight would be to custom-design the size of each member of the structure for its own load if this were acceptable for construction and aesthetic purposes. Buckling members could utilize I beams or hollow cross sections to improve the moment of inertia with less material. It is also possible to investigate different truss configurations and not accept Fig. 11.6 without careful study. In actual practice many truss configurations might be considered, computing the weight necessary for a safe design in each case and selecting the lightest configuration for actual construction. Of course, economy of construction and appearance are factors that must also be considered.

PROBLEMS

11.7 The simple structure of Fig. P11.7 is to be designed to have aluminum alloy members of square cross section with $S_y = 207$ MPa and $E = 0.690 \times 10^5$ MPa. The members should all have the same cross-sectional size. Determine that size if the structure is to just support the load indicated.

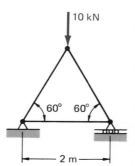

FIGURE P11·7

11.8 The truss in Fig. P11.8 is made of square steel bars 60 mm on a side. The steel has $S_y = 207$ MPa and $E = 2.07 \times 10^5$ MPa. Determine the force F that will just cause failure of the structure due to buckling or yielding of a member. Which member is expected to fail first and how?

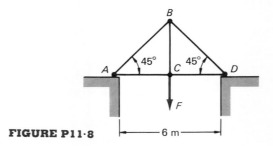

FIGURE P11·8

11.9 Square steel bars are available in multiples of $\frac{1}{4}$ in only. Select a single cross-sectional size for the structure shown in Fig. P11.9 to obtain a factor of safety of 3. The steel has $S_y = 30{,}000$ lbf/in^2 and $E = 30 \times 10^6$ lbf/in^2. Discuss your result compared with the design for Fig. 11.6 in this section.

11.10 A certain lot of balsa wood has a tensile strength $S_t = 3000$ lbf/in^2, a compressive strength $S_c = 1900$ lbf/in^2, and an elastic modulus $E = 700{,}000$ lbf/in^2. The small structure of Fig. P11.10 is to be built from this balsa wood using members with square cross sections of equal size. What size members will allow the structure to just support the load shown?

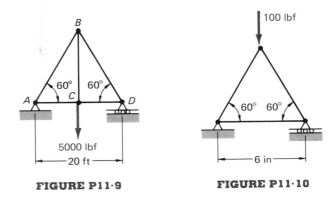

FIGURE P11·9   **FIGURE P11·10**

11.11 The properties of balsa wood depend on its density, which varies within individual trees. Table P11.11 summarizes the properties shown in Refs. 7 and 8. From this data, a graph may be plotted to obtain values for S_t, S_c, and E at densities different from those indicated in the table.

Obtain a supply of balsa wood and measure its density. Repeat Prob. 11.10 for the properties of your available wood. Based on these results, build the structure in Fig. P11.11a using triangles with 6-in sides so that it will support a load of at least 200 lbf (100 lbf per side). In building your structure, pay particular attention to the joints. Gusset plates should be made by gluing thin balsa sheets together. Joints are then formed with liberal amounts of wood glue, sandwiching the members between gusset plates, as shown in Fig. P11.11b.

Your structure should be able to support your weight. Test it to failure, if possible, using laboratory equipment to measure its maximum load at failure. Compare this load and the failure mode to your calculations.

TABLE P11·11
BALSA WOOD PROPERTIES

Balsa Density lbm/ft³	S_t, lbf/in² (Parallel to Grain)	S_c, lbf/in² (Parallel to Grain)	E, lbf/in² (Elastic Modulus)
6	1400	750	300,000
11	3000	1900	700,000
15.5	4500	3000	1,000,000

11.12 Design and build a balsa wood truss bridge to span a 15-in gap. Your bridge should be no taller than 6 in and able to support a load of 100 lbf placed at the top center of the bridge. Design the bridge to have minimum weight while still sup-

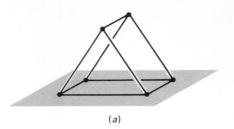

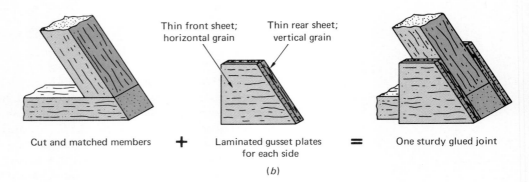

FIGURE P11·11

porting the required load. Use Table P11.11 and the suggestions of Fig. P11.11b in your design and construction. A bridge weight less than 50 g is excellent.

Test your model bridge to failure using laboratory equipment. Compare its maximum load and mode of failure to your calculations. A typical bridge and failure are shown in Fig. P11.12.

FIGURE P11·12

11·3 SOME ACTUAL BRIDGES

In our previous discussion, we presented a highly simplified model of a truss bridge so that we could perform design calculations similar to those applied by practicing structural engineers. We will now look at a few real bridges to discuss the types of structures used in actual applications and some of the reasons why they are used. For further information on this topic there are many interesting books (References 3 to 6 are good examples) on the history and theory of bridges which you might consult.

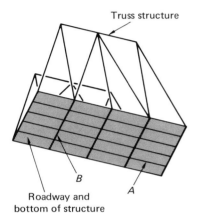

FIGURE 11·9
Bottom view of a typical truss bridge.

TRUSS BRIDGES

A typical truss bridge does not support a simple single load as we considered in our discussion of Fig. 11.6. It supports a roadbed on which traffic travels across the bridge. Figure 11.9 shows a fairly realistic sketch of a truss bridge. In this figure the roadbed is supported directly on several longitudinal beams like the one labeled A in the figure. The longitudinal beams are in turn supported on heavier lateral beams like the one labeled B. The lateral beams run from the truss structure on one side of the roadway to the truss on the opposite side. Each of the lateral beams is connected to the truss structure at a lower pinned joint, so the load of the roadbed is applied to the truss at the pins and does not tend to bend any members of the truss.

Truss bridges are used in practice for bridge spans of intermediate length. A bridge supported only by beams would be used for shorter spans. Arches, suspension bridges, and other types are used for longer spans. When a river or gorge is too wide for a single span of a particular type of bridge, intermediate supports,

An actual truss bridge, which actually does not have pinned joints. Solid "gusset plates" are riveted to the members, joining them together where the pins would ideally be located. Even in this situation, however, the pinned force analysis works reasonably well.

A view of the underside of a truss bridge, which should be compared with Fig. 11.9. Here you can see one of the larger lateral beams connecting to a "pinned" joint of the truss at the upper right of the photograph. The smaller beams, running the length of the bridge, are supported by the lateral ones. Additional structural members that may be seen form an X-shaped pattern and control some of the more complicated failure modes.

or "piers," may be used to form a multiple-span bridge. In those locations where intermediate supports can be installed at low cost, each span of a bridge will be relatively short. When intermediate supports can be provided only with considerable difficulty and expense, fewer supports can be used and each span must be longer.

SUSPENSION BRIDGES

The modern suspension bridge is one of the most spectacular of engineering achievements. The characteristic soaring towers and webs of suspension cables have become landmarks at locations throughout the world. The suspension bridge is an efficient structure based on perhaps the most basic way of bridging a gap—stretching a rope across it. The "rope" has now become a very heavy steel cable and is held not only at each side of the gap but also at intermediate supports. Figure 11.10 shows the components of a suspension bridge. The main supporting cables follow the

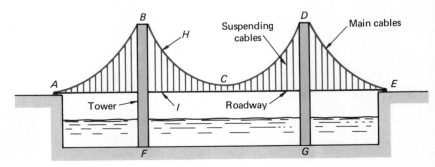

FIGURE 11·10
Characteristics of a suspension bridge.

416 STRUCTURES AND BRIDGES

A view of the Golden Gate Bridge, which spans the entrance to the San Francisco Bay. It is a suspension bridge with a longest span of 4200 ft. Its supporting towers rise 700 ft above sea level. (Karen Rogers.)

path *ABCDE*. Suspending cables, such as *HI* in Fig. 11.10, are attached to the main cables and to the large steel beams which directly support the roadway. The main cables and the suspending cables are loaded only in tension, which as we have seen is generally the most effective way of supporting a load. The towers (*BF* and *DG* in Fig. 11.10) are loaded in compression due to the cable forces pulling downward at *B* and *D*. They must be heavy, sturdy structures which can support the compressive load and also resist the side forces produced by winds and currents.

Suspension designs are used for the longest of bridge spans. No other type of bridge has been built with a span longer than 600 m (2000 ft). The Verrazano Narrows suspension bridge at the entrance to the New York city harbor has a span of 1300 m (4260 ft) between its towers. It was completed in 1964 and as of 1980 has the longest unsupported span ever constructed. The second longest span is part of the Golden Gate Bridge in San Francisco, which is also a suspension bridge.

ARCH BRIDGES

The arch is based on a principal just the opposite of the suspension bridge design. The arch supports its load in compression not in tension, as in the supporting cables of a suspension bridge. Compressive loads are ideal for construction with stone or brick because these brittle materials, like concrete, are strong in compression and relatively weak in tension. Stone and brick were the building materials of our oldest civilizations. As a result, bridge designers for centuries have used the principals of the arch, which were recognized by the peoples of ancient Egypt and Babylon as early as 2000 B.C.

Arch bridges today are commonly built of concrete and steel,

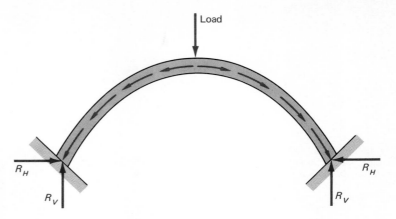

FIGURE 11·11
An arch transmits its load by compression.

often involving steel truss work. The longest spans are nearly 600 m (2000 ft) in length but still rely on the basic principal of a curved structure supporting its load in compression, as indicated in Fig. 11.11, where a load applied at the center of the arch is transmitted through compression to the foundations at its base. The foundations must provide both a vertical reaction force R_V and a horizontal reaction force R_H. The horizontal force can be very large for long spans and in some situations may be a design problem. However, the horizontal reaction force can be conve-

Nero's Arch at the entrance to the ancient olympic stadium in Olympia, Greece, showing clearly how a basic arch can be formed by assembling blocks of stone in a very simple way. The arch was built in 67 A.D. to honor Nero's attendance at the original olympic games.

Arch bridges crossing the River Arno in Florence, Italy. The center bridge, supporting the small buildings, is the Ponte Vecchio, literally, "old bridge." Constructed during the Middle Ages, such distinguished visitors as Leonardo da Vinci, Michelangelo, and Dante crossed the bridge, along with countless Florentines.

A single steel arch span crossing the Niagara River just below Niagara Falls. Use of a gorge wall to provide a foundation can be seen on the left of the photograph.

niently provided by the walls of a canyon or gorge. For this reason many of the longest arch bridges are constructed in such locations.

Bridges are often beautiful structures. Engineers' pride in their profession may often be renewed when they see bridges such as those in the illustrations included here and can understand their design principles.

11.3 SOME ACTUAL BRIDGES **419**

A pair of steel arch bridges formed from truss structures. These central spans are "through" arches in which the roadway is suspended from the supporting arch.

REFERENCES

1. Shames, I. H., *Introduction to Solid Mechanics,* Prentice-Hall, Englewood Cliffs, N.J., 1975.
2. Beer, F. P., and E. R. Johnston, Jr., *Mechanics of Materials,* McGraw-Hill, New York, 1981.
3. McCullough, C. B., and E. L. Thayer, *Elastic Arch Bridges,* Wiley, New York, 1931.
4. Strauss, J. B., *The Golden Gate Bridge,* Golden Gate Bridge and Highway District, 1938.
5. Parnell, J. P. M., *Man the Builder,* Crescent Books, New York, 1977.
6. Troitsky, M. S., *Cable Stayed Bridges*, CLS Publishing, London, 1977.
7. Draffin, J. O., and C. W. Muhlenbruch, "On Mechanical Properties of Balsa Wood," *Proceedings of the American Society for Testing Materials,* vol. 37, part 2, pp 582–587, 1937.
8. Lippay, A., "Balsa—Tough Lightweight Challenges Foams," *Materials Engineering,* vol. 65 (June), pp 70–73, 1967.

12 ELECTRIC FILTERS*

In Sec. 7.5 we presented the concept of frequency response as used in the analysis of dynamic systems. Electrical engineers often use frequency-response ideas in communications work and in the design of electric circuits. Engineers who work with electric instrumentation may also become involved with the frequency response of electric circuits and electronic devices. In this section we present two examples of these types of design problems. One example involves the design of a circuit intended to isolate an electric signal of a particular frequency from other signals at nearby frequencies. The second example is a somewhat common instrumentation problem, where an intruding or unwanted electric signal must be eliminated. Both examples are typical of engineering practice, the second of which is actually from the experience of one of the authors.

It may seem that these examples are rather complicated because the situations in which electric filters are required are very involved. However, to an engineer who normally works with these concepts, the problems described below are fairly routine. The discussion leading up to the actual filter design examples is quite lengthy, but we must present the necessary background to help you understand why electric filtering is needed.

---

* The material in this chapter is based on concepts presented earlier in Chaps. 5 and 7: Secs. 5.1 and 5.2 introduce the basics of electric circuits; most of Chap. 7 furnishes information on dynamic systems and Secs. 7.4 and 7.5, in particular, on electric systems and frequency response.

12·1 A TUNED CIRCUIT

Speech and hearing provide the most common form of human communication. The human voice produces sounds, or audio signals, ranging in frequency from about 70 to 5000 Hz. The human ear, sensitive to audio signals over the frequency range of 20 to 20,000 Hz, not only has been used for communication but has developed during the evolution of humans as an aid in hunting and defense. Human beings have come to depend on audio signals for convenient communication and for entertainment in civilizations throughout the world.

Unfortunately the distance over which a human voice or a stringed instrument can be heard is limited, normally to a hundred meters or so, mainly because of the nature of sound transmission in air and the need to actually move molecules of air to produce a sound. A great deal of energy is required to create sounds which can be heard for miles. Electromagnetic waves are a much more useful way of transmitting communication signals over long distances and are, in fact, the basis of radio transmission. However, electromagnetic waves must be produced electrically and, owing to a variety of factors including antenna size, are practical only at frequencies much higher than typical audio frequencies. A great deal of sophisticated but now everyday technology has been used to combine audio communication and entertainment with electromagnetic signal transmission.

In the following section we will discuss the basic concepts of AM (amplitude modulation) radio transmission as an example of radio communication.[1] We will then focus on the tuning circuit of a receiver as one of the many radio circuits requiring filter design.

PRINCIPLES OF AM RADIO

The first step in coupling audio signals to electromagnetic waves is to convert the sound of a voice or instrument into an electric voltage, or signal. This is done by a microphone, which may be a variable resistance device as mentioned in Sec. 5.3. As shown in Fig. 12.1, the output of the microphone V_A is a relatively low frequency electric voltage which is proportional to the sound pressure incident on the microphone. Electric circuits are used to process V_A and to combine it with V_C, another electric voltage of higher frequency produced by the oscillator circuit in the radio transmitter. This voltage V_C, called the carrier signal, oscillates at the radio station's transmission frequency. For commercial AM

---

[1] AM is the first method used to impress audio information on the electromagnetic wave and is still used on many broadcast bands including commercial radio and citizens' band (CB) radio.

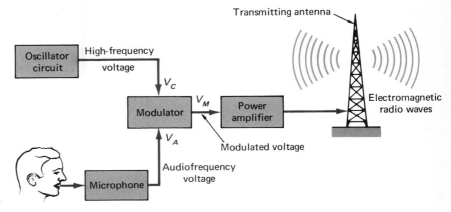

FIGURE 12·1
Typical radio transmission.

radio stations in the United States, carrier frequencies are between 535 and 1605 kHz. The low-frequency audio signal is combined with the carrier signal by amplitude modulation. A schematic version of the modulator is shown in Fig. 12.1 with inputs V_A and V_C. The modulator is an electronic circuit which carries out a type of multiplication and addition so that the output voltage from the modulator is

$$V_M = V_C + V_C V_A \qquad (12.1)$$

The effect of this modulation process on the signals can be seen in Fig. 12.2, which presents a graph of (a) a carrier signal sine wave V_C versus t, (b) the lower frequency signal V_A versus t, and (c) the modulator output V_M versus t. In the modulated signal we can see the amplitude of the carrier signal changing in proportion to the low-frequency signal, so we say that the carrier has been amplitude-modulated. In actual applications, the carrier signal frequency is nearly a thousand times higher than the audio frequencies of interest.

The modulation process produces an electric signal at a useful transmission frequency which contains the original audio information. The next step shown in Fig. 12.1 is amplification of the modulated signal to a power level sufficient to drive the transmitting antenna. This is where a station's power output comes into effect. The maximum power rating for AM stations in the United States is 50 kW. Stations with this transmission power have the largest transmission range and can often be received at distances of over 1000 mi. The purpose of the amplifier is to produce the proper electric current in the antenna oscillating at the frequency of the carrier and modulated by the audio signal. Recall that a current in a conductor produces a magnetic field, so that an oscillating current within the antenna produces an oscillating mag-

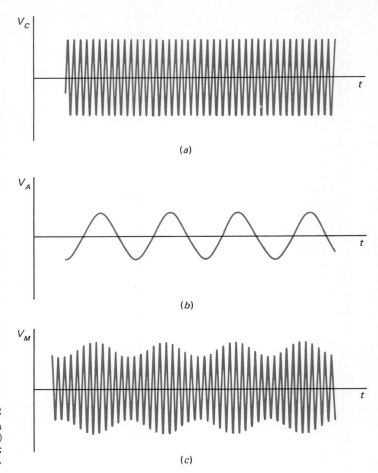

FIGURE 12·2
Development of an amplitude-modulated signal. (a) The carrier; (b) The audio signal; (c) The modulated signal.

netic field. This results in the electromagnetic wave which actually transmits the AM signal.

The radio receiver shown in Fig. 12.3 is on the other end of the radio system. The receiving antenna is a conductor immersed in an atmosphere filled with electromagnetic waves. Each wave tends to produce an electric current in the antenna and the tuning circuit of the receiver. The tuning circuit has been designed to select the one signal of interest to the listener and to reject the others. It is an electric filter based on the principle of resonance described in Sec. 7.5. Listeners tuning a radio adjust the resonant frequency of the tuning circuit so that it matches the frequency of the radio signal they want to receive. The output of the tuning circuit is the modulated signal V_M, indicated in Fig. 12.3, which is directly related to the transmitted signal of the selected station. A demodulating process then extracts the audio-frequency signal

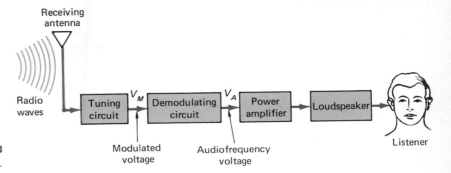

FIGURE 12·3
A typical radio receiver.

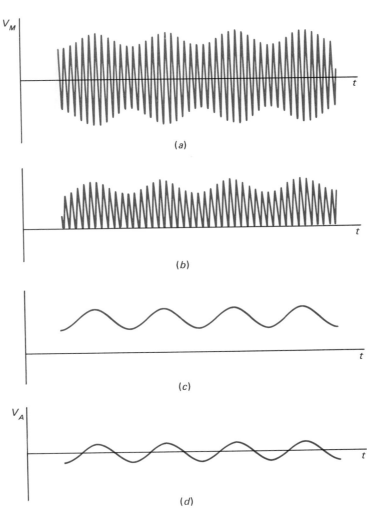

FIGURE 12·4
The demodulation process. (a) Modulated signal; (b) rectified; (c) filtered; (d) recovered audio signal.

12.1 A TUNED CIRCUIT **425**

from the modulated signal. Some of the details of a typical demodulation are indicated in Fig. 12.4. The first step is to electronically *rectify the signal,* that is, remove the negative half from each oscillation of the modulated signal. The result (which may be accomplished with a *diode*) is shown in Fig. 12.4b. The next step is to filter and remove the high-frequency carrier. An RC filter (discussed later in Sec. 12.2) can perform this, yielding the results shown in Fig. 12.4c. Finally, the remaining "offset" voltage is removed (sometimes in the normal amplification process) and the original audio signal is recovered, as in Fig. 12.4d. When the electromagnetic coil of a loudspeaker is driven by this voltage, motions of its paper cone reproduce the original sound for the listener.

DESIGN OF A TUNING CIRCUIT

We will now focus on the tuning circuit of a radio receiver—one that is simple and similar to that suggested as a home electronics project in Ref. 1. The receiver schematic shown in Fig. 12.5 although electronically simple may seem very complicated to you. We will mention a few of the components before moving on to the tuning circuit design.

The heart of this receiver is a general-purpose audio amplifier. It is powered by a 9-V battery and amplifies its input voltage v, as shown, to drive a small speaker. The diode at the amplifier input is used in the demodulation process. The adjustable capacitance C_A is an important part of the tuning circuit. It is a special electronic device, controlled by the voltage V, which has the effect of a

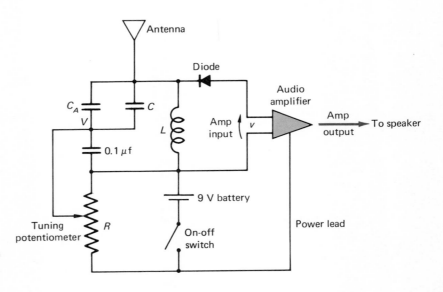

FIGURE 12·5
A receiver circuit.

variable capacitance. This receiver is tuned by adjusting the potentiometer R to change V and thus change the capacitance C_A. The particular adjustable capacitor used in this circuit may be adjusted from a lower value of 5 pF (5×10^{-12} F) to a higher value of 15 pF. Our design problem here is to choose the remaining components of the tuning circuit (the inductor L and the fixed capacitor C) so that the receiver can be tuned across the frequency range of citizens' band (CB) radio. To include all the CB channels, we need a frequency range of from 26.96 MHz (26.96×10^6 Hz) to 27.23 MHz.

A model of the receiver tuning circuit is shown in Fig. 12.6. This simplified model strips away the more complicated aspects of the receiver and concentrates on the portion of interest. The antenna is thought of as a current source and develops the current i_{in}, which we think of as the input to the tuning circuit. This current is produced by the action of the electromagnetic waves on the antenna. However, every local radio station produces electromagnetic waves and signals of many frequencies are included in i_{in}. The tuning circuit must be set to resonate at the frequency of the station desired by the listener so that the tuning circuit output voltage v_{out} in Fig. 12.6 contains only the signal of that station and signals of other frequencies are rejected. In our case we must be able to adjust the natural frequency of our tuning circuit from 26.96 to 27.23 MHz to provide an adequate tuning range.

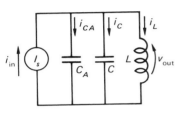

FIGURE 12·6
Model of the tuning circuit.

ANALYZING THE CIRCUITRY

We now write the differential equation relating v_{out} to i_{in} for Fig. 12.6. We begin by making the summation of currents

$$i_{in} - i_{CA} - i_C - i_L = 0$$

and rearranging to get

$$i_{CA} + i_C + i_L = i_{in} \tag{12.2}$$

For the capacitors $i_{CA} = (C_A)(dv_{out}/dt)$ and $i_C = (C)(dv_{out}/dt)$, so we substitute into Eq. (12.2) to find

$$C_A \frac{dv_{out}}{dt} + C \frac{dv_{out}}{dt} + i_L = i_{in} \tag{12.3}$$

For the inductor $v_{out} = L(di_L/dt)$. If Eq. (12.3) is differentiated we obtain

$$(C_A + C)\frac{d^2 v_{out}}{dt^2} + \frac{di_L}{dt} = \frac{di_{in}}{dt}$$

In this equation we can replace di_L/dt by $(1/L)(v_{out})$ and rearrange to get

$$L(C_A + C)\frac{d^2v_{out}}{dt^2} + v_{out} = L\frac{di_{in}}{dt} \quad (12.4)$$

Note that this is a second-order differential equation relating the output voltage to the input current. Any resistance present in the circuit of Fig. 12.6 has been neglected in developing Eq. (12.4), so unlike Eq. (7.48) or (7.25), no first derivative term shows up on the left side and no damping is present. Of course, resistance naturally occurs in the wires of the circuit and in the inductor; some small amount of damping will thus be present but much less than the $\zeta = 0.1$ curve shown in Fig. 7.33. A strong resonance will be found at the natural frequency of the circuit. This will reject undesired frequencies and allow only the desired station's signal to appear in v_{out}. As in our work with Eq. (7.48), the coefficient of the second derivative term defines the natural frequency. Here

$$\frac{1}{4\pi^2 f_n^2} = L(C_A + C)$$

which yields a natural frequency of

$$f_n = \frac{1}{2\pi\sqrt{L(C_A + C)}} \quad (12.5)$$

SELECTING L AND C

We now design the tuning circuit so that the natural frequency computed from Eq. (12.5) equals 26.9 MHz at the highest value of the adjustable capacitance C_A (15 pF) and 27.3 MHz at the lowest value (5 pF). As C_A is adjusted, the tuned circuit will then cover the CB range with a little room to spare. We solve for L and C by applying Eq. (12.5) for the two extremes

$$26.9 \times 10^6 = \frac{1}{2\pi\sqrt{L[(15 \times 10^{-12}) + C]}}$$

and

$$27.3 \times 10^6 = \frac{1}{2\pi\sqrt{L[(5 \times 10^{-12}) + C]}}$$

We then obtain two simultaneous equations to solve for L and C:

$$LC + (15 \times 10^{-12})(L) = 3.50 \times 10^{-17}$$

and

$$LC + (5 \times 10^{-12})(L) = 3.40 \times 10^{-17}$$

The solution of these equations, obtained by solving for L and the product LC simultaneously, yields $L = 0.100$ µH and $C = 335$ pF.

These will be appropriate values for L and C to use in the first construction of the circuit. It should be noted, however, that capacitance values involved in this design are quite small and the influence of the amplifier circuit and even the wiring arrangements may cause variations in the circuit's behavior. After these calculations, construction of a "breadboard," or prototype, receiver would be the next step in the design process. Minor adjustments to L and/or C would be made until proper operation of the receiver was obtained.

PROBLEMS

12.1 Prepare a graph showing an amplitude-modulated voltage V_m as a function of time. The carrier signal has a frequency of 500 kHz with an amplitude of 5 V. The audio signal has a frequency of 10 kHz and an amplitude of 1 V. The graph should show at least a full cycle of the audio signal with the actual modulated signal sketched in. Detailed point-by-point plotting is not necessary.

12.2 Frequency modulation is another common technique for transmitting audio signals at radio frequencies. Visit the library and find out how FM works. Draw a set of curves similar to those in Fig. 12.2 which show a carrier signal, an audio signal, and an FM-modulated signal.

12.3 Figure P12.3 is a simple model of a tuned circuit. For each case below, determine the value of L or C necessary for the circuit's natural frequency to match the specified value of f_n. [Equations (12.4) and (12.5) can be applied here by setting $C_A = 0$.]
(a) $f_n = 100$ MHz, $L = 0.2$ µH, $C = ?$
(b) $f_n = 100$ MHz, $C = 50$ pF, $L = ?$
(c) $f_n = 750$ kHz, $L = 200$ µH, $C = ?$
(d) $f_n = 1600$ kHz, $C = 50$ pF, $L = ?$

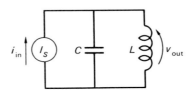

FIGURE P12.3

12.4 The standard AM broadcast band covers the range from 535 to 1605 kHz. An adjustable capacitor to be used in tuning the radio across this band has a lowest capacitance value of 30 pF. For the tuning circuit of Fig. P12.3, what should be the highest value of the adjustable capacitor and what fixed value of the inductance is necessary to cover the band?

12.5 The circuit of Fig. P12.5 is similar to the one shown in Fig. 7.18 and described by Eq. (7.48). This tuned circuit is to be used to amplify any signal contained in the voltage v which matches the circuit natural frequency of approximately 5 kHz. Use Fig. 7.33 to estimate the amplification at this frequency if (a) $R = 6.3\ \Omega$ and (b) $R = 32\ \Omega$.

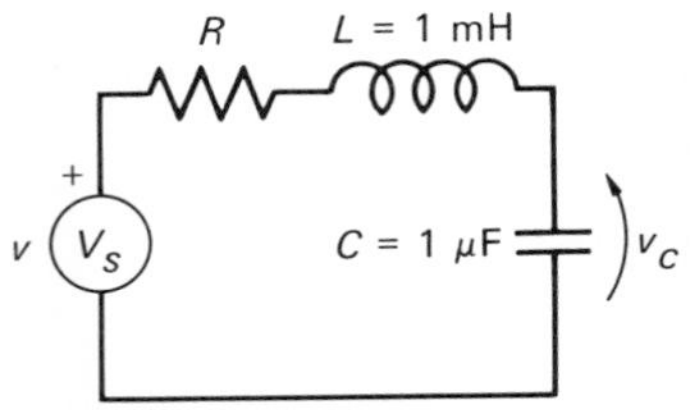

FIGURE P12·5

12.6 Use a computer program similar to FREQ2 in Fig. 7.31 to verify your results for Prob. 12.5. Let $v = \sin 2\pi f_n t$ and obtain the output v_c at various values of time. Compare the amplitude of the oscillations in v_c to the oscillation amplitude in v. Consider both $R = 6.3\ \Omega$ and $R = 32\ \Omega$. Note that the step size Δt must be very small compared with $1/f_n$ to obtain proper results.

12.7 Radio transmission in the frequency range of 21.00 to 21.75 MHz is restricted to amateur use and international broadcasting according to regulations adopted in Geneva, Switzerland, in 1959. This is called the "fifteen meter" shortwave band to indicate the approximate wavelength of the carrier signal. Based on Eq. (12.5), select values for the inductor L and capacitor C which will allow the simple receiver circuit in Fig. 12.5 to be tuned over the frequency range of the 15-m band as given above.

12·2 REMOVING A HIGH-FREQUENCY SIGNAL

This filter-design problem was part of the development of a laboratory simulator for studying automobile controls. Figure 12.7 is a schematic drawing of the simulator. A test subject, or "driver," sits in the simulator and operates a steering wheel and pedals for brake and accelerator control. In many cases, hand controls for

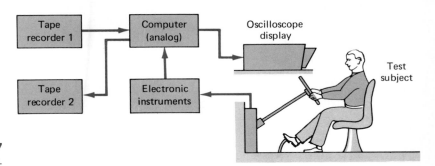

FIGURE 12·7
An automobile simulator.

handicapped people have been installed in the simulator for evaluation. In fact, the purpose of the simulator was to allow a controlled study of automotive hand controls used by handicapped people who must drive without the use of their legs [2, 3].

The simulator consists of the front portion of an automobile with electric instrumentation installed to measure steering wheel and accelerator pedal position (using potentiometers) as well as brake-line hydraulic pressure (using strain gages which are also a variable-resistance device). This instrumentation converts the movements of the automobile controls into electric voltages. As shown in Fig. 12.7, the voltages from the electronic instruments are fed to a computer. The computer contains equations which calculate the motion of the simulated automobile based on the driver's actions. These motions can be recorded on tape recorder 2 for later evaluation. Tape recorder 1 contains a prerecorded tape of "desired" vehicle motions which provide the input to the simulator. As the subject drives the simulator, he attempts to make the computed vehicle motion based on his actions duplicate the desired motion played from tape recorder 1. Inside the computer, voltages representing computed motion are subtracted from voltages representing desired motion. The difference, or "error," is presented to the subject on an oscilloscope. During a test run on the simulator, subjects try to hold the error as close to zero as possible.

This simulator has been used in various control studies involving steering, as well as braking and acceleration. We will focus on a simplified study of acceleration control to illustrate the filter-design problem of interest here. As a person attempts to control the acceleration of our simulated vehicle, the computer is continually performing its calculations. Positive acceleration is computed when the driver steps on the accelerator pedal. The computer includes a time delay in the acceleration to represent the behavior of the engine. Negative acceleration (deceleration) is computed directly from the measured brake-line pressure as the subject operates the brake pedal. The oscilloscope display

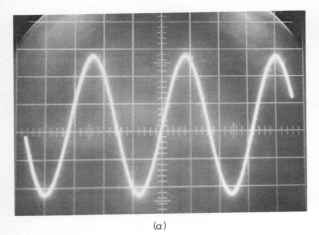

(a)

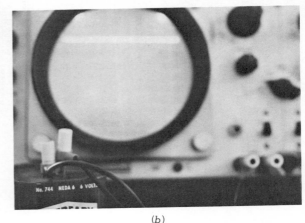

(b)

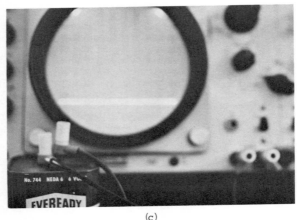

(c)

An oscilloscope continually plots a graph of an electric voltage vs. time. The plot is displayed on the face of a television-like tube as shown in (a), where a sinusoidal voltage is presented. If a constant voltage is applied to the input of the oscilloscope, a straight line results. This can be seen in (b), where a battery provides +6 V to the input, and in (c), where −6 V is the input.

watched by the subject contains a single horizontal line. Stepping on the accelerator produces upward motion of the line and stepping on the brake moves the line downward. However, the desired acceleration played from tape recorder 1 in Fig. 12.7 also tends to move the oscilloscope line. The line is centered at zero on the oscilloscope when the computed acceleration equals the desired acceleration. During a test run, the subject continually tries to hold the line at zero by applying the accelerator or brake as necessary.

The purpose of this type of simulator test is to evaluate the performance of both the subject and the vehicle controls. In this case, the evaluation can be performed by comparing the computed vehicle acceleration during the test with the desired acceleration played from recorder 1. Computer processing of tape recordings made during the course of the test can be performed afterward and can show how quickly and how accurately the subject has been able to control the simulated vehicle.

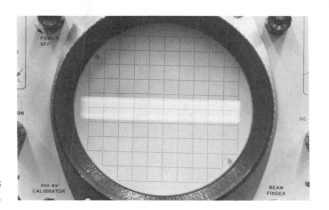

FIGURE 12·8
Inadequate oscilloscope display.

THE FILTER-DESIGN PROBLEM

During the development of the simulator, a great many design problems had to be considered and solved. The particular problem we will discuss here was not discovered until the final stages, after the simulator was assembled and the first attempts to check out its operation were being made. As shown in Fig. 12.8, a clear horizontal line did not show up on the oscilloscope. Instead the line was thick and fuzzy and not at all suitable for the simulator display. Many unsuccessful attempts were made to locate the source of this problem within the instrumentation and the computer circuitry. Finally, close inspection of the oscilloscope with a different time scale revealed more detail than was previously noticed.

As shown in Fig. 12.9, the fuzziness appeared to result from a high-frequency sine wave at about 900 kHz. However, the amplitude of the sine wave was not constant; it varied noticeably, thereby providing a clue to the source of the problem. As we have

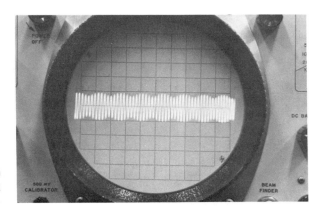

FIGURE 12·9
An amplitude-modulated signal at about 900 kHz.

discussed, the AM radio band in the United States ranges from 535 to 1605 kHz and is based on amplitude modulation of a sine-wave carrier. To check the possibility that a radio signal was the cause of the problem, an AM radio was set near the oscilloscope and tuned to various stations near 900 kHz. A local popular music station was broadcasting music which clearly corresponded to the signal on the oscilloscope. It was later determined that the station's transmission tower was within a mile of the simulator laboratory. Somewhere in the wiring and electronics of the simulator there was apparently a hidden antenna receiving the unwanted signal.

After the nature of the unwanted signal was established, it was decided to explore the possibility of removing it with a filter. It was necessary, however, to design a proper filter which met the need of removing the signal without affecting the responsiveness of the oscilloscope display. For a simple first-order filter to be used, its time constant must be large enough to do the filtering but not so large as to noticeably slow down the display.

Figure 12.10 shows a schematic of the filter installation. The resistor and capacitor of the first-order filter are installed at the input connections of the oscilloscope. This is done because the wires themselves, coming from the computer, may be picking up part of the interfering signal. We want to remove the signal by placing the filter as close to the display as possible.

The electric circuit in Fig. 12.10 is identical to the circuit we have considered previously in Fig. 7.17. The input-output relationship is described by the differential equation

$$RC \frac{dV_{out}}{dt} + V_{out} = V_{in} \tag{12.6}$$

where the input voltage V_{in} comes from the computer circuitry and the output voltage V_{out} is the input to the oscilloscope which is actually displayed on the screen. Recall that the time constant in Eq. (12.6) is $\tau = RC$. If V_{in} suddenly changes from one constant voltage to another, 63 percent of the corresponding change in V_{out} will be completed within τ seconds. Human reaction time is normally about one-tenth of a second. Certainly, if we design the circuit so that τ is equal to 0.001 s, it will respond quickly enough so

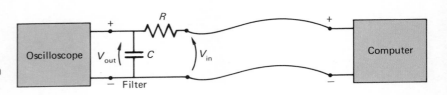

FIGURE 12·10
The filter installation.

that the test subject will be unable to notice any delay or hesitation in the display as he steps on the brake or accelerator. The question now is: will this τ remove the 900 kHz disturbance contained in the voltage V_{in}? For this circuit the ratio of the peak amplitude of an output sine wave to the input peak amplitude is given by the graph of A/A_0 which has been presented in Fig. 7.29 for a first-order system in terms of the product $f\tau$. For this case considering $f = 900$ kHz and $\tau = 0.001$ s, we find

$$f\tau = (900{,}000)(0.001) = 900$$

This number falls far to the right on the graph in Fig. 7.29 and clearly A/A_0 is a small number. We actually have obtained $A/A_0 = 0.00018$. This means that if a disturbance with a peak amplitude of 1 V is found in V_{in} at a frequency of 900 kHz [so that, $V_{\text{in}} = \sin(2\pi)(900{,}000t)$], the resulting peak amplitude of V_{out} would be only 0.00018 V.

The choice of $\tau = 0.001$ s very adequately removes the 900-kHz signal, since the original fuzziness in Fig. 12.8 had a magnitude of less than 1 V and since the very small filtered voltage of 0.00018 V could not even be noticed on the oscilloscope as used in this study. The value of $\tau = 0.001$ s is also small enough so that the test subject would not be aware of any delay while operating the simulator.

CHOOSING R AND C

Once the filter time constant τ has been selected, it is necessary to choose particular values for the resistance and capacitance. A wide range of values for R and C will work adequately in this design, but the range is limited. The limitations occur because of real, nonideal characteristics of both the computer output circuitry and the oscilloscope input circuitry.

The computer output circuitry cannot provide large amounts of power. Any attempt to draw a large current from the output reduces the output voltage. Figure 12.11a shows a model of the out-

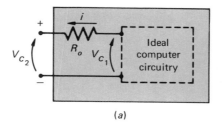
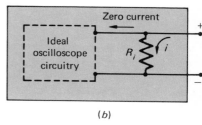

FIGURE 12·11
Models of input and output resistance.

(a)   (b)

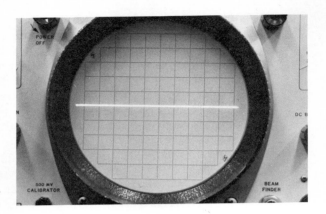

FIGURE 12·12
The oscilloscope display after filtering.

put circuit with an output resistance R_o. The results of the computer's calculations are represented by the voltage V_{C1} in the figure. The actual output voltage $V_{C2} = V_{C1} - R_0 i$; and as long as the current i is small, the output voltage V_{C2} will accurately represent the computer calculated V_{C1}. The smaller the value of the output resistance R_0, the more ideal is the behavior of the output circuit. For the circuits used in this simulator, the output resistance is approximately 100 Ω.

The input circuit of the oscilloscope also contains some important details. The oscilloscope cannot operate without drawing a small amount of current when a voltage is applied to it. This amount of current is governed by the input resistance of the oscilloscope R_i shown in Fig. 12.11b. The *ideal* oscilloscope, as shown in the figure, does not require any current input. Generally $R_i = 1 \times 10^6$ Ω = 1 MΩ for oscilloscopes. This is a large input resistance and only the smallest amount of current flow actually takes place. However, if an oscilloscope is used to measure the output of

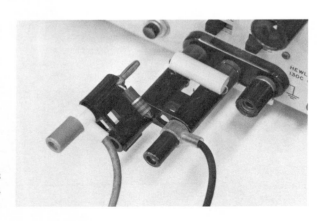

FIGURE 12·13
The RC filter at the oscilloscope input.

a circuit with a similarly large output resistance, voltage measurements can be distorted.

Selection of the filter resistance R (Fig. 12.10) may be conveniently made so that R is both large enough to avoid interaction with the output resistance R_o (Fig. 12.11a) and small enough to avoid interaction with R_i (Fig. 12.11b). A range of R greater than 1000 Ω and less than 100,000 Ω might be selected here, since $R_i = 1$ MΩ and $R_o = 100$ Ω. A value of $R = 10,000$ Ω, in the center of this range, was actually used for the filter. The corresponding value of capacitance is calculated to satisfy the chosen $\tau = 0.001$ s. In this case,

$$\tau = RC = 0.001$$
$$C = \frac{0.001}{10,000}$$
$$= 1 \times 10^{-7} \text{ F}$$
$$= 0.1 \ \mu\text{F}$$

so that a capacitance of 0.1 μF is needed. This is a readily available capacitance value. A capacitor larger than, say, 10 μF might be somewhat expensive or difficult to find. If an inconveniently large capacitance resulted from the calculation above, new values of R or τ could be considered to avoid having the large capacitor.

The final appearance of the display and the simple filter at the input to the oscilloscope are shown in Figs. 12.12 and 12.13, respectively.

PROBLEMS

12.8 In Fig. 12.10, the voltage v_{in} has been zero for a long time. It suddenly changes to $v_{in} = 2$ V. If $RC = 0.001$ s, how long will it take for v_{out} to reach
(a) 1.26 V (c) 1.90 V
(b) 1.72 V (d) 0.63 V
Figure 7.6 should be helpful in solving this problem.

12.9 In Fig. 12.10, any oscillations in v_{in} at a frequency f are to be reduced so that v_{out} shows less than 20 percent of the v_{in} oscillation amplitude. What is the smallest value of the time constant $\tau = RC$ which will be satisfactory if
(a) $f = 1$ Hz   (c) $f = 1$ kHz
(b) $f = 10$ Hz (d) $f = 1$ MHz
Figure 7.29 should be helpful in solving this problem.

12.10 In Fig. 12.10, when a sudden change occurs in v_{in} it is necessary that v_{out} complete 95 percent of its resulting response within 0.025 s. It is also desirable that any oscillation in v_{in}

at a frequency of 100 Hz be reduced in magnitude so that v_{out} shows less than 20 percent of the oscillation amplitude. What range of values for τ will allow both these requirements to be met?

12.11 Select one value for τ which satisfies both conditions defined in Prob. 12.10. Obtain computer solutions to verify that your choice does indeed fulfill both the response requirement of 0.025 s and the 20 percent oscillation requirement. You will need to write two computer programs for the solution of Eq. (12.6)—one similar to Fig. 7.4 and the other similar to Fig. 7.25. For the response requirement you may set $v_{in} = 1$ V and start with $v_{out} = 0$ in your computer program. For the oscillation requirement, let $v_{in} = \sin 200\pi t$ in your program. In both cases be certain that Δt is small enough to give you accurate results. Have your programs print results at convenient time intervals so that you obtain enough information to plot graphs of your response in both cases without an excessive amount of data. Present your results on carefully plotted graphs of v_{out} versus time.

12.12 Check the results of your solution to Probs. 12.10 and 12.11 in the laboratory. Use the circuit of Fig. P12.12 with the assistance of a laboratory instructor. Choose appropriate values for R and C. Set the function generator to "square wave" to obtain the response to a sudden change in v_{in}. Set it to "sine wave" and adjust the frequency to 100 Hz to check the requirements on output oscillation size.

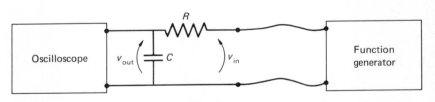

FIGURE P12·12

12.13 The circuit in Fig. P12.13 is to be used in an application where v_{in} contains signals at 100 and 1000 Hz. Select the filter components R and C to "pass" the 100-Hz signal and "filter" the 1000-Hz signal. For an input of $v_{in} = \sin 200\pi t$, the output v_{out} should show oscillations of at least 1.0 V peak to peak. For an input of $v_{in} = \sin 2000\pi t$, the output oscillations should be less than 0.3 V peak to peak. The differential equation for the system is

FIGURE P12·13

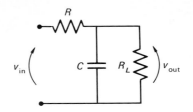

$$RC \frac{dv_{out}}{dt} + \left(1 + \frac{R}{R_L}\right) v_{out} = v_{in}$$

The load resistor $R_L = 10$ kΩ. The resistor R in your design should be no smaller than 500 Ω and your capacitance C no larger than 1 μF.

REFERENCES

1. Lewart, C. R., "Build a Legal In-Flight Airline Receiver," *Popular Electronics*, vol. 11, no. 5, 1977.
2. Anger, R. T., "An Analysis of Automotive Hand Controls for the Physically Handicapped," M. S. thesis, Mechanical Engineering Department, SUNY at Buffalo, 1976.
3. Anger, R. T., and Mayne, R. W., "The Performance of Automotive Hand Controls," presented at the 1978 Winter Annual Meeting of the American Society of Mechanical Engineers, San Francisco, ASME Paper No. 78-WA/DSC-38, 1978.

13 THE PRODUCTION OF OIL*

Oil plays a critical role in our modern society. It satisfies approximately half of our present demand for energy and we depend on it almost completely for transportation energy because gasoline is, of course, refined from oil. In addition, oil is also essential for the production of many fertilizers, plastics, artificial fibers, and other petrochemicals which have become an integral part of our daily lives.

The world supply of oil is unfortunately limited, so that we are running out of oil at an increasingly rapid rate. Our ability to prolong the life of the world's oil resources depends on controlling the consumption of oil, discovering new oil fields, and effectively removing the oil from existing oil fields. Opportunities for all types of engineers exist in each of these efforts. In this section, we concentrate on the last point only—the removal of oil from natural reservoirs within the earth's surface. We begin with a summary of the world oil situation and then discuss some of the engineering design problems which are a part of oil production. We apply in this section many of the concepts from the fluid-thermal sciences introduced in Chap. 6.

Petroleum and mining engineers play the most important role in oil production, but they work side by side with many other types of engineers—chemical engineers who understand the chemistry, fluid mechanics, and heat and mass transfer problems of oil production; mechanical engineers who may also be involved with such problems but more often work on the machinery required in

---

* This chapter presents an application of the fluid mechanics material contained in Sec. 6.1, which should be studied carefully before preceding, especially with Sec. 13.2. Some general familiarity with the various other topics of Chap. 6 is also helpful here.

oil production; civil engineers who work with soils, geological formations, and drilling processes; electrical engineers who are concerned with electric motors and control problems and also with the sophisticated instrumentation now needed in oil production. In this industry, as in most engineering practice, the engineering disciplines overlap and blur together as people tackle real problems unrestricted by arbitrary academic boundaries.

13·1 BACKGROUND

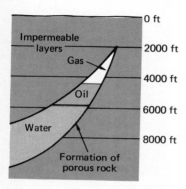

FIGURE 13·1
A natural-oil reservoir in porous rock.

People often mistakenly think of an oil well as a hole drilled into the ground to a large pool of oil where a pump may be installed to simply remove the oil from the pool. Oil is customarily found trapped in large porous rock formations just as water is trapped in a sponge. In fact, the word petroleum results from a combination of the Latin words *petra,* for "rock," and *oleum,* for "oil"—or "rock-oil" (in contrast, of course, to olive oil, cod liver oil, etc.). In most operating oil fields, the oil is trapped in sandstone, limestone, or similar sedimentary rock from which the oil can be removed with reasonable effort. On the other hand, the oil shale found in the American west contains pores which are so small that it is not currently economical for use in oil production.

The removal of oil from porous rock is certainly more complex than is pumping it from a pool. The process is complicated by the underground location of the oil. Depths of 5000 to 10,000 ft are not uncommon, making it difficult to identify the exact size and shape of oil-bearing rock formations. At these depths, the weight of the surface material above creates high pressures of perhaps several thousand pounds per square inch. The temperature also increases at a rate of nearly 2°F per 100 ft of depth and temperatures over 150°F are often found in active oil fields.

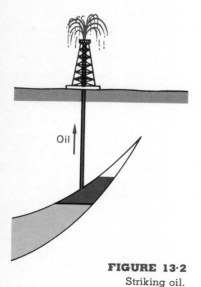

FIGURE 13·2
Striking oil.

A typical oil deposit in porous rock is shown in Fig. 13.1. The oil has natural gas above it and water below it, all of which is trapped in a porous layer surrounded by an impermeable rock formation and is normally at elevated levels of pressure and temperature. Figure 13.2 shows a successful oil strike. A well has been drilled through the impermeable layers to enter the portion of the porous layer containing oil. At this point, a "gusher" could occur, but nowadays great precautions are taken to prevent such spillage of oil. The well is brought under control and for some time may produce oil under its own pressure. However, the removal of oil from the porous layer reduces the pressure there until the oil will no longer flow to the surface. At that time a pump is installed (Fig. 13.3a) to continue production from the well by lifting the oil to the surface. Under this condition, the natural pressure remaining in the gas or water is still sufficient to force the oil through the

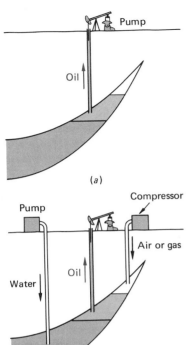

porous rock to the bottom of the well where the pumping action takes place. This natural process of producing oil from a reservoir is called "primary recovery," which is able to recover up to 25 percent of the oil originally present in the reservoir.

"Secondary recovery" methods are used when natural pressures are not adequate to force the oil to the base of the well from the surrounding porous rock. As shown in Fig. 13.3b, the pressure in the reservoir is increased by forcing water, air, or gas from the surface down to the porous layer. A pump (for the water) or a compressor (for air or gas) is used to inject the fluid and provide the increased pressure. This is costly, but the improved pressure levels will typically squeeze out another 25 percent of the oil in the original deposit. However, 50 percent of the oil is left behind that is too viscous to be forced from the porous rock economically.

"Tertiary recovery" methods have been developed to remove additional oil from the reservoir by reducing its viscosity, a process that may involve heating the oil in the reservoir by forcing steam from the surface down to the porous layer. The steam increases oil temperature and decreases its viscosity so it can flow to the base of the well. The process, shown in Fig. 13.3c, has the potential of removing most of the oil from many oil reservoirs. It is very costly, however, since large amounts of steam must be heated. As a result, the process to date has not been widely used. Correspondingly, not many oil wells have ever produced more than half the oil which they originally contained.

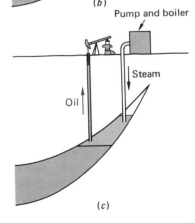

FIGURE 13-3
(a) Primary, (b) secondary, and (c) tertiary recovery.

HOW LONG WILL THE OIL LAST?

Since we are discussing the production of oil, we must also consider one of the major concerns of modern society—the depletion of the world's oil supply. Many studies have been (and continue to be) performed to determine how much time we have to develop alternative energy sources. Reference 1 reports on one important international study completed in 1978. This study, which considers only the non-Communist world, attempts to forecast the supply and demand for oil in the coming years. Several different models are considered in the study to represent oil consumption, discovery, and production. We will discuss just one case here, but all yield similar discouraging results.

If the consumption of oil increases slowly, the economists involved in the study consider it reasonable to expect a consumption increase of 2.5 percent per year until 1985. After 1985 a reduced increase of 1.8 percent per year is predicted. Based on 1975 oil consumption, this means that the consumption of oil projected for future years is given by the solid curve in Fig. 13.4.

The rate of oil discovery used in the forecast is based on the history of oil discoveries over the past 50 years, as shown in Fig. 13.5.

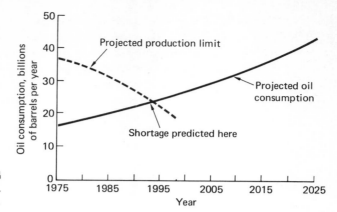

FIGURE 13·4
Projections of annual oil consumption and production limit.

Here we see that the rate at which oil reserves have been discovered has averaged about 15 billion barrels per year. The conservative assumption made in the study is that discoveries will be more difficult in the future, averaging only 10 billion barrels per year. The recoverable oil reserves in the non-Communist world were estimated at 555 billion barrels at the end of 1975. If the annual oil consumption is as projected by Fig. 13.4 and new reserves are discovered at a rate of 10 billion barrels per year, the reserves will decrease, as shown by the curve in Fig. 13.6. From this curve it seems that oil reserves will be adequate until beyond the year 2000. Recall, however, that the world's oil reserves are in porous rock layers and cannot be removed at will. A reasonable limit on the production of oil fields is one-fifteenth of the recoverable reserve per year. A higher production rate could damage some fields and reduce the ultimate yield. Based on this limitation, the maximum possible oil production in any given year can be projected by taking one-fifteenth of the reserves projected in Fig. 13.6. The result is the production limit curve shown as a dashed line in

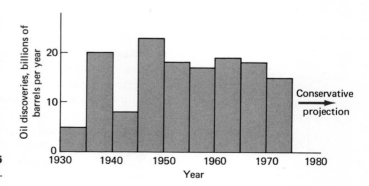

FIGURE 13·5
Rate of new oil discoveries.

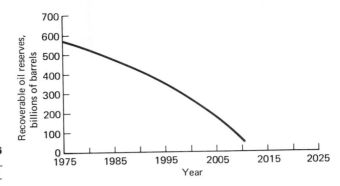

FIGURE 13·6
Projected oil reserves of the non-Communist world.

Fig. 13.4. The intersection of the two curves in Fig. 13.4 predicts that production will fall short of demand in about 1994.

Of course, all the information leading to the 1994 prediction is based on educated guesswork. If consumption grows at a slower rate, if unexpected oil reserves are discovered, or if more complete oil recovery occurs with the widespread use of secondary and tertiary methods, the 1994 date will be moved back. However, production-control policies of oil countries could lead to very severe problems at much earlier dates. It should be clear that the problem of oil dependence is immediate and continues to deserve immediate attention.

PROBLEMS

13.1 A major difficulty in projecting the supply of natural resources such as petroleum involves the assumptions made in the projections. The sensitivity of the projections to the assumptions must be evaluated. As an example of this, consider two alternative assumptions for the consumption of oil. Working with Figs. 13.4 and 13.6, consider (a) that oil consumption grows at a constant rate of 5 percent per year from 1975 and (b) that there is zero growth in oil consumption from that date. When will shortages be predicted under each of these conditions, with other assumptions unchanged?

13.2 Test the sensitivity of the oil-reserve projections to the discovery of new oil reserves. Consider the possibilities that new reserves are discovered at a rate of (a) 15 billion barrels per year and (b) 5 billion barrels per year starting from 1975. Modify Fig. 13.6 and then Fig. 13.4 to reflect these adjusted discovery rates, other assumptions being unchanged. When will oil shortages be predicted in each of these cases?

13.3 Assume that a new process for removing oil from producing wells is developed in 1990. This process effectively provides a sudden increase in recoverable oil reserves of 50 percent. Modify Figs. 13.6 and 13.4 to reflect this possible development with other conditions unchanged. What is the new prediction for oil shortages?

13.4 Visit an appropriate library and consult recent issues of *Scientific American* as well as other references. What has happened to world oil consumption, production, and discovery since the article in Ref. 1 was written? What are the most recent projections for oil shortages?

13·2 SOME DESIGN CALCULATIONS

There are obviously countless numbers of engineering design problems involved in the recovery of oil from producing wells. These range from the design of the special machinery and structures used in oil production to the design of electrical instrumentation for carrying out careful measurements of an oil reservoir's characteristics. We will focus here, however, on the fluid mechanics problems that are an inherent part of oil production in dealing with the flow of oil from the underground oil deposit to the surface.

Let's begin with the very basic problem of determining the size of the tubing needed to bring the oil to the surface and deciding whether the natural oil pressure is sufficient to produce a desired oil-flow rate. The problem is shown in Fig. 13.7. We will deal with it first in its simplest form and then describe some realistic complexities. In Fig. 13.7, the bottom hole pressure (BHP) is the oil pressure at the base of the well. The oil-flow rate toward the surface is q and the tubing has a diameter d. Before any design decisions can be made for this well, we must have a procedure for calculating the BHP required to achieve a specified flow rate knowing the depth of the well, the tubing diameter, and the properties of the crude oil in the reservoir.

ANALYSIS OF REQUIRED PRESSURE

The flow of fluid up the tubing requires an elevated pressure at the bottom for two reasons. First, the weight of the oil column needs to be supported by a pressure force even if no flow occurs. This was considered earlier in our discussion of Fig. 6.2. Second, in order to overcome fluid friction and maintain a flow rate in the tube, additional pressure is necessary, as we found in our discussion of Fig. 6.5. In order to calculate the total required pressure, we consider the two effects separately. To obtain the pressure difference ΔP_c needed to support the fluid column, we apply Eq. (6.3) so that

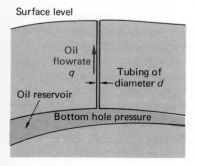

FIGURE 13·7
Flow of oil to the surface.

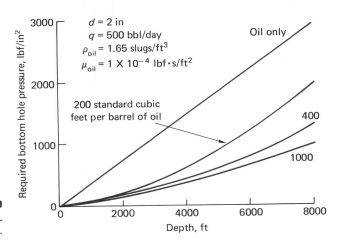

FIGURE 13·10
Required pressure for oil-gas mixtures.

13.10 is one example of the results for the case of $d = 2$ in and $q = 500$ bbl/day. The first curve on this graph, for oil alone, is precisely the curve shown in Fig. 13.8 for $d = 2$ in. However, the remaining curves in Fig. 13.10 include natural gas mixed with the oil where the proportion of gas is given in units of standard cubic feet per barrel of oil. One standard cubic foot of gas is the amount of gas which has a volume of 1 ft³ at a temperature of 70°F and standard atmospheric pressure. In Fig. 13.10, the more gas that is mixed with the oil, the lower is the required pressure to lift it, because increased gas content reduces the density of the mixture and correspondingly reduces $\Delta P_c = \rho g h$.

The effect of gas content on reducing required lifting pressure can be very important. It allows some oil wells to produce which could not if only pure oil were in the reservoir. In fact, gas or air may often be introduced purposely in some oil wells to artificially create this effect and reduce the required BHP. This process, discussed in Ref. 2, is called "gas lift." It can eliminate the need for a pump and has other beneficial effects.

Another example of the results of pressure calculations for mixtures is presented in Fig. 13.11, which shows pressure-vs.-depth curves for mixtures of oil, water, and gas. Water is more dense than is the oil considered here. As a result, the curves in Fig. 13.11 show generally higher required pressures than do the curves of Fig. 13.10 for similar conditions. However, because it is also a liquid, the influence of water on these curves is much less significant than the effect of the gas.

In actual oil-well design, careful pressure and flow calculations are performed that consider the mixture of both gas and water with the oil. These calculations could involve the extensive use of graphs such as those presented here. However, computers are now used extensively to predict the behavior of the oil both

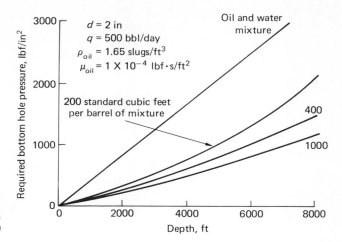

FIGURE 13·11
Required pressure for oil, water, and gas. (50 percent oil to 50 percent water.)

during its upward flow and also within the porous rock of the oil reservoir. In the discussion of Sec. 13.3 we will see one way a computer can be used to predict an oil-field's behavior.

PROBLEMS

13.5 Determine the bottom hole pressure necessary to just support the fluid column in an oil well under the following conditions:
  (a) The fluid is oil with specific gravity of 0.85 and the depth is 1000 m.
  (b) The fluid is oil and the depth is 2000 ft.
  (c) The fluid is 50 percent oil and 50 percent water and the depth is 1000 m.
  (d) The fluid is 50 percent oil and 50 percent water and the depth is 2000 ft.

13.6 Determine the bottom hole pressure that will produce the specified flow of oil to the surface under the conditions defined below. You may use a specific gravity of 0.85 and a viscosity $\mu = 1 \times 10^{-4}$ lbf·s/ft² = 0.0048 N·s/m² for a light, warm oil.
  (a) q = 1000 bbl/day, h = 4000 ft, $d = \frac{3}{4}$ in
  (b) q = 80 m³/day, h = 1000 m, d = 20 mm
  (c) q = 2500 bbl/day, h = 6000 ft, d = 1.50 in
  (d) q = 800 m³/day, h = 1500 m, d = 40 mm

13.7 A certain oil well is 1500 m in depth and a pipeline of length 1000 m is used to transport its oil horizontally along the surface after it reaches the top of the well. The vertical and surface pipelines may be considered to be one continuous pipe and the diameter d = 40 mm. What is the BHP necessary to maintain a flow rate of 800 m³/day up through the well and along the surface? A specific gravity of 0.85 and a viscosity $\mu = 0.0048$ N·s/m² may be assumed for the oil.

13.8 An oil well 5000 ft deep with a pipe diameter $d = \frac{3}{4}$ in has a bottom hole pressure of 2500 lbf/in² gage. What is the well production q in barrels per day? Consider the well to be flowing 100 percent oil with a specific gravity of 0.85 and a viscosity $\mu = 1 \times 10^{-4}$ lbf·s/ft². A convenient procedure for solving this problem is to assume a flow rate and calculate the corresponding required pressure. Make corrections in the assumed flow rate until the calculated pressure agrees with the value of 2500 lbf/in² specified above.

13.9 Use the techniques of Sec. 2.3 to develop a series of equations that reasonably represent the curves showing f versus Re in Fig. 6.7. Consider three regions: $700 < \text{Re} < 3000$, $3000 < \text{Re} < 10^5$, and $10^5 < \text{Re} < 10^7$. Use the equations for straight lines on a log-log graph.

Based on the equations developed above, write a computer program which will accept fluid properties, specified flow rate, well depth, and pipe diameter as inputs and calculate the required bottom hole pressure as the output.

You may work in SI or customary American units as desired. Use your program to solve Probs. 13.6a and c or Probs. 13.6b and d corresponding to your choice of units.

13.10 Using portions of your computer program for Prob. 13.9, develop a new computer program which will accept fluid properties, well depth, pipe diameter, and bottom hole pressure as input and calculate the well flow rate as output. Check your program by confirming various points on Figs. 13.8 and 13.9. Use it to solve Prob. 13.8.

13.11 Write a computer program which will determine required pipe diameter for an oil well from inputs of desired flow rate, bottom hole pressure, well depth, and oil properties. Use Figs. 13.8 and 13.9 to check its operation.

13.3 TERTIARY RECOVERY BY STEAM FLOODING

A major problem in oil production involves the need to remove oil from the porous layers in which it naturally resides. Tertiary recovery methods, as mentioned earlier, enhance oil production by reducing oil viscosity. The tertiary recovery method, called steam flooding, reduces the oil viscosity by increasing its temperature. Table 13.1 shows typical viscosities of a heavy crude oil at various temperatures and indicates that even modest temperature

TABLE 13·1
TYPICAL VISCOSITY OF HEAVY CRUDE OIL AT VARIOUS TEMPERATURES

Temperature, °C	Viscosity, N·s/m²
25	4.2
50	0.67
75	0.097
100	0.028
150	0.0064
200	0.0032
250	0.0019

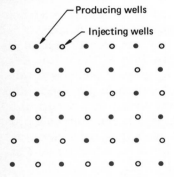

FIGURE 13·12
A field of producing and injecting wells.

increases will result in significant viscosity reductions. This is especially important since the temperature of a somewhat depleted well may typically be about 30°C, where the viscosity is quite high.

The steam-flooding process, like any engineering process, requires design. In this case, decisions must be made about how the steam should be injected into the oil, how much steam should be injected, etc. Also, since steam is actually a mixture of a liquid water with water vapor, the percentage of water vapor may be adjusted. The design decisions are complex, and experimenting on an actual oil field is extremely expensive. For this reason computer simulation is often used to guide such decisions. We will now look at an example of the way computer simulation can be used in this situation. We will discuss Ref. 3, which applies a computer simulation program to the steam-flooding problem. The program is much too complex to discuss in detail, but you should be able to understand its purpose as a design tool and the importance of its results.

A tertiary recovery installation is not really as simple as is indicated in Fig. 13.3c. An oil field actually consists of a grid of many wells drilled to the same oil reservoir and covering a large area. Viewing a typical tertiary installation from above, as indicated in Fig. 13.12, we see that each production well is surrounded by steam-injection wells. A simple "five-point" arrangement of wells is indicated in Fig. 13.12, the basic pattern of which is shown more clearly in Fig. 13.13a. Each production well is surrounded by four injection wells which move the oil toward the production well. As shown in Fig. 13.12 this five-point pattern is repeated many times over in an oil field. More complicated patterns, such as the seven-point pattern of Fig. 13.13b, are also used, but they require additional drilling, since six injection wells are used for each production well.

The first item considered by Gomaa in Ref. 3 was to see which pattern—the five- or seven-point—was more effective. He constructed computer models of the oil reservoirs based on the basic well patterns shown in Fig. 13.13. The entire oil field was represented by a well-chosen triangular slice. For example, triangle ABC in Fig. 13.13b represented the entire seven-point pattern. This could be done because line AB represents a line of symmetry between the two injection wells on either side of well B and similarly, line AC is a line of symmetry between well B and its neighboring injection well. Also line BC is a line of symmetry with one of the adjacent seven-point patterns in the oil field. The behavior of the entire field essentially results from many repetitions of the behavior within triangle ABC.

The slice of the oil reservoir was considered to be three-dimensional, as shown in Fig. 13.14 for the seven-point arrangement (a similar model was also used for the five-point configuration). To study the flow of the steam into the oil reservoir and the heating and flow of oil within the reservoir, the slice was divided into smaller compartments, as indicated by the grid lines in Fig. 13.14. Note that on the triangular slice, the reservoir depth is divided into four separate layers and that each layer is in 15 portions. The 15 portions are shown clearly for the top layer in Fig. 13.14. With each of the four layers in 15 portions, a total of $4 \times 15 = 60$ compartments was considered. The computer model used equations to determine the flow of oil and steam between each of the 60 compartments. This flow depends on the nature of the porous media (its "porosity" and "permeability") and also on the viscosities of the oil and steam which in turn depend on the temperature. Thus it was necessary to compute the temperature in each compartment of the model by considering the heat flow between compartments and the heat lost from the reservoir through its top and bottom surfaces.

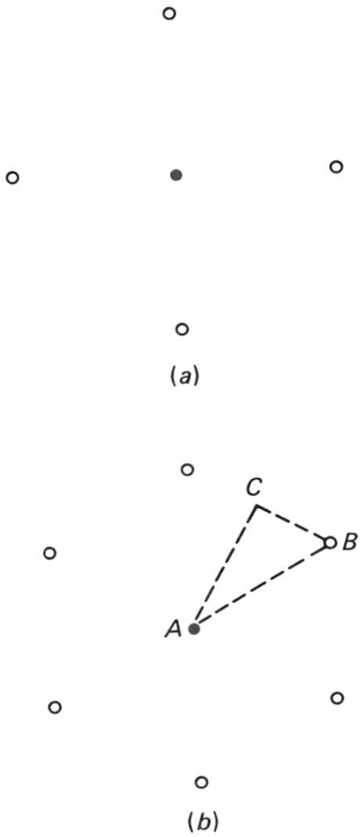

SOME RESULTS OF THE COMPUTER STUDY

One purpose of the study by Gomaa [3] was to compare the performance of the five-point and seven-point well configurations in Fig. 13.13 and also to consider the effect of well spacing and steam-flow rate. It was rapidly discovered by the computer simulations that the five-point and seven-point well arrangements were equivalent and that the most dominant factor in producing oil from the reservoir was the total steam-flow rate. This means

FIGURE 13·13
Well patterns for steam injection.

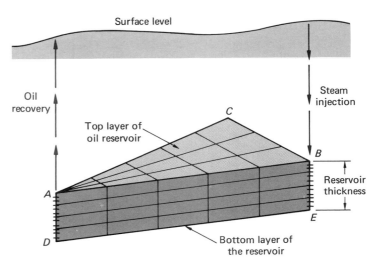

FIGURE 13·14
Model for simulation of the reservoir.

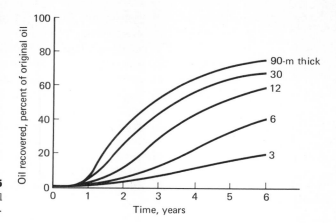

FIGURE 13·15

Effect of reservoir thickness on oil recovery.

that considerable variations in the well configuration and spacing will not show substantial changes in oil production as long as the same steam-flow rate is used.

Figure 13.15 shows the effect of reservoir thickness on oil production for fixed steam-flow rate conditions as determined by the computer model. This figure shows curves of oil production over a several-year period with different reservoir thicknesses. It can be seen that thinner reservoirs produce less effectively than do thick ones. This was found to occur because of heat transfer from the top and bottom surfaces of the reservoir. Because of this heat transfer, thinner reservoirs retain only a small percentage of the heat contained in the steam to warm the oil reservoir. Thicker reservoirs retain a larger percentage of the steam's heat. In fact, for reservoirs thicker than 50 m, the heat loss is nearly negligible. This effect can be seen in Fig. 13.15, where the curves for thicknesses of 30 and 90 m are very close together.

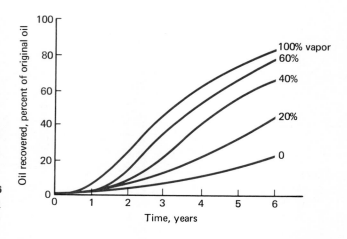

FIGURE 13·16

Effect of percent vapor on oil recovery.

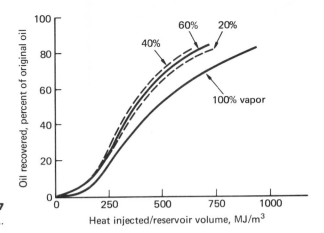

FIGURE 13·17
Oil recovered vs. heat injection.

Steam is actually a mixture of water vapor and liquid water. Oil production for a given amount of steam (by weight) depends on the percentage of vapor contained in the steam, as can be seen from the computer results in Fig. 13.16. The higher the vapor percentage, the higher the oil production. This is deceptive, however, since a higher vapor percentage means a higher heat content (because the vapor contains more heat than the liquid). Figure 13.17 shows the effect of vapor percentage on oil production with "heat injected" plotted on the horizontal axis instead of "time," as in Fig. 13.16. The curves in Fig. 13.17 are significant because the primary cost in this tertiary process is the cost of the heat injected. It can be seen in Fig. 13.17 that the curve with 40 percent vapor shows the largest oil recovery for any amount of heat injection. This is a significant fact in design of the steam-flooding system, since the use of 40 percent vapor results in the largest return in oil for the cost of heating the steam.

To explain why 40 percent is a desirable percentage of vapor, it was necessary to look more carefully at the computer results. It was discovered that the 40 percent vapor leads to a more uniform oil temperature and oil viscosity in the vicinity of the production well. An indication of this is shown in Fig. 13.18a, where the reservoir cross section is drawn with a curve through those points in the reservoir which are at a temperature of 65°C. The portion of the reservoir on the right in Fig. 13.18a is above 65°C; the portion to the left is below 65°C. In this case, the producing well on the left is warmed uniformly from the top to the bottom of the reservoir. The 65°C contours for 20 and 100 percent vapor are shown in Fig. 13.18b. With 20 percent vapor content, the bottom of the producing well is warm but the top is left cool. With 100 percent vapor, the contour at 65°C shows that the top is warm but the bottom cool.

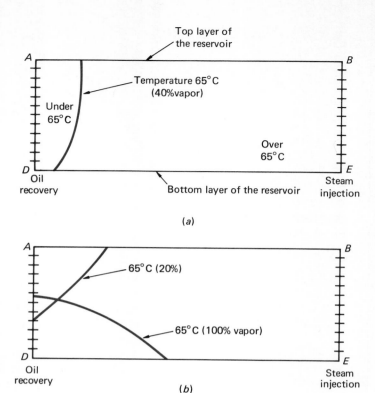

FIGURE 13·18
Temperature contours in the reservoir at 65 °C.

As you can see, the design of a tertiary oil recovery system is much more complicated than are many of the more traditional problems we have previously discussed. It is fortunate that computer technology has come along in time to help engineers in dealing with this problem which so deeply affects our industrial society.

PROBLEMS

13.12 How does steam flooding allow an increase in the oil production from an oil reservoir?

13.13 Is the five-point or seven-point injection approach more effective for steam flooding according to Gomaa's results? Why?

13.14 Are "thin" oil reservoirs most suitable for tertiary recovery by steam flooding? Explain.

13.15 In what way is the graph of Fig. 13.17 more informative than Fig. 13.16?

13.16 Explain why steamflooding with 40% vapor is more effective than steamflooding with substantially lower or higher percentages of vapor content.

REFERENCES

1. Flower, A. R., "World Oil Production," *Scientific American*, vol. 238, no. 3, 1978.
2. Brown, K. E., *Gas Lift Theory and Practice*, Prentice Hall, Englewood Cliffs, N.J., 1967.
3. Gomaa, E. A., "Correlations for Predicting Oil Recovery by Steamflood," *Journal of Petroleum Technology*, February, 1980.

APPENDIX

SUGGESTED GUIDELINES FOR USE WITH THE FUNDAMENTAL CANONS OF ETHICS

1. Engineers shall hold paramount the safety, health and welfare of the public in the performance of their professional duties.
   a. Engineers shall recognize that the lives, safety, health and welfare of the general public are dependent upon engineering judgments, decisions and practices incorporated into structures, machines, products, processes and devices.
   b. Engineers shall not approve nor seal plans and/or specifications that are not of a design safe to the public health and welfare and in conformity with accepted engineering standards.
   c. Should the Engineers' professional judgment be overruled under circumstances where the safety, health, and welfare of the public are endangered, the Engineers shall inform their clients or employers of the possible consequences and notify other proper authority of the situation, as may be appropriate.
      (c.1) Engineers shall do whatever possible to provide published standards, test codes and quality control procedures that will enable the public to understand the degree of safety or life expectancy associated with the use of the design, products and systems for which they are responsible.
      (c.2) Engineers will conduct reviews of the safety and reliability of the design, products or systems for which they are responsible before giving their approval to the plans for the design.
      (c.3) Should Engineers observe conditions which they believe will endanger public safety or health, they shall inform the proper authority of the situation.
   d. Should Engineers have knowledge or reason to believe that another person or firm may be in violation of any of the provisions of these Guidelines, they shall present such information to the proper authority in writing and shall cooperate with the proper authority in furnishing such further information or assistance as may be required.
      (d.1) They shall advise proper authority if an adequate review of the safety and reliability of the products or systems has not been made or when the design imposes hazards to the public through its use.
      (d.2) They shall withhold approval of products or systems when changes or modifications are made which would affect adversely its performance insofar as safety and reliability are concerned.
   e. Engineers should seek opportunities to be of constructive service in civic affairs and work for the advancement of the safety, health and well-being of their communities.
   f. Engineers should be commited to improving the environment to enhance the quality of life.
2. Engineers shall perform services only in areas of their competence.
   a. Engineers shall undertake to perform engineering assignments only when qualified by education or experience in the specific technical field of engineering involved.
   b. Engineers may accept an assignment requiring education or experience outside of their own fields of competence, but only to the extent that their services are restricted to those phases of the project in which they are qualified. All other phases of such project shall be performed by qualified associates, consultants, or employees.
   c. Engineers shall not affix their signatures and/or seals to any engineering plan or document dealing with subject matter in which they lack competence by virtue of education or experience, nor to any such plan or document not prepared under their direct supervisory control.
3. Engineers shall issue public statements only in an objective and truthful manner.
   a. Engineers shall endeavor to extend public knowledge, and to prevent misunderstandings of the achievements of engineering.
   b. Engineers shall be completely objective and truthful in all professional reports, statements, or testimony. They shall include all relevant and pertinent information in such reports, statements, or testimony.
   c. Engineers, when serving as expert or technical witnesses before any court, commission, or other tribunal, shall express an engineering opinion only when it is founded upon adequate knowledge of the facts in issue, upon a background of technical competence in the subject matter, and upon honest conviction of the accuracy and propriety of their testimony.
   d. Engineers shall issue no statements, criticisms, nor arguments on engineering matters which are inspired or paid for by an interested party, or parties, unless they have prefaced their comments by explicitly identifying themselves, by disclosing the identities of the party or parties on whose behalf they are speaking, and by revealing the existence of any pecuniary interest they may have in the instant matters.
   e. Engineers shall be dignified and modest in explaining their work and merit, and will avoid any act tending to promote their own interests at the expense of the integrity, honor and dignity of the profession.
4. Engineers shall act in professional matters for each employer or client as faithful agents or trustees, and

shall avoid conflicts of interest.

a. Engineers shall avoid all known conflicts of interest with their employers or clients and shall promptly inform their employers or clients of any business association, interests, or circumstances which could influence their judgment or the quality of their services.
b. Engineers shall not knowingly undertake any assignments which would knowingly create a potential conflict of interest between themselves and their clients or their employers.
c. Engineers shall not accept compensation, financial or otherwise, from more than one party for services on the same project, nor for services pertaining to the same project, unless the circumstances are fully disclosed to, and agreed to, by all interested parties.
d. Engineers shall not solicit nor accept financial or other valuable considerations, including free engineering designs, from material or equipment suppliers for specifying their products.
e. Engineers shall not solicit nor accept gratuities, directly or indirectly, from contractors, their agents, or other parties dealing with their clients or employers in connection with work for which they are responsible.
f. When in public service as members, advisors, or employees of a governmental body or department, Engineers shall not participate in considerations or actions with respect to services provided by them or their organization in private or product engineering practice.
g. Engineers shall not solicit nor accept an engineering contract from a governmental body on which a principal, officer or employee of their organization serves as a member.
h. When, as a result of their studies, Engineers believe a project will not be successful, they shall so advise their employer or client.
i. Engineers shall treat information coming to them in the course of their assignments as confidential, and shall not use such information as a means of making personal profit if such action is adverse to the interests of their clients, their employers, or the public.
   (i.1) They will not disclose confidential information concerning the business affairs or technical processes of any present or former employer or client or bidder under evaluation, without his consent.
   (i.2) They shall not reveal confidential information nor findings of any commission or board of which they are members.
   (i.3) When they use designs supplied to them by clients, these designs shall not be duplicated by the Engineers for others without express permission.
   (i.4) While in the employ of others, Engineers will not enter promotional efforts or negotiations for work or make arrangements for other employment as principals or to practice in connection with specific projects for which they have gained particular and specialized knowledge without the consent of all interested parties.
j. The Engineer shall act with fairness and justice to all parties when administering a construction (or other) contract.
k. Before undertaking work for others in which Engineers may make improvements, plans, designs, inventions, or other records which may justify copyrights or patents, they shall enter into a positive agreement regarding ownership.
l. Engineers shall admit and accept their own errors when proven wrong and refrain from distorting or altering the facts to justify their decisions.
m. Engineers shall not accept professional employment outside of their regular work or interest without the knowledge of their employers.
n. Engineers shall not attempt to attract an employee from another employer by false or misleading representations.
o. Engineers shall not review the work of other Engineers except with the knowledge of such Engineers, or unless the assignments/or contractual agreements for the work have been terminated.
   (o.1) Engineers in governmental, industrial or educational employment are entitled to review and evaluate the work of other engineers when so required by their duties.
   (o.2) Engineers in sales or industrial employment are entitled to make engineering comparisons of their products with products of other suppliers.
   (o.3) Engineers in sales employment shall not offer nor give engineering consultation or designs or advice other than specifically applying to equipment, materials or systems being sold or offered for sale by them.

5. Engineers shall build their professional reputation on the merit of their services and shall not compete unfairly with others.

a. Engineers shall not pay nor offer to pay, either directly or indirectly, any commission, political contribution, or a gift, or other consideration in order to secure work, exclusive of securing salaried positions through employment agencies.
b. Engineers should negotiate contracts for professional services fairly and only on the basis of demonstrated competence and qualifications for the type of professional service required.
c. Engineers should negotiate a method and rate of compensation commensurate with the agreed upon scope of services. A meeting of the minds of the parties to the contract is essential to mutual confidence. The public interest requires that the cost of engineering services be fair and reasonable, but not the controlling consideration in selection of individuals or firms to provide these services.
   (c.1) These principles shall be applied by Engineers

in obtaining the services of other professionals.
d. Engineers shall not attempt to supplant other Engineers in a particular employment after becoming aware that definite steps have been taken toward the others' employment or after they have been employed.
  (d.1) They shall not solicit employment from clients who already have Engineers under contract for the same work.
  (d.2) They shall not accept employment from clients who already have Engineers for the same work not yet completed or not yet paid for unless the performance or payment requirements in the contract are being litigated or the contracted Engineers' services have been terminated in writing by either party.
  (d.3) In case of termination of litigation, the prospective Engineers before accepting the assignment shall advise the Engineers being terminated or involved in litigation.
e. Engineers shall not request, propose nor accept professional commissions on a contingent basis under circumstances under which their professional judgments may be compromised, or when a contingency provision is used as a device for promoting or securing a professional commission.
f. Engineers shall not falsify nor permit misrepresentation of their, or their associates', academic or professional qualifications. They shall not misrepresent nor exaggerate their degree of responsibility in or for the subject matter of prior assignments. Brochures or other presentations incident to the solicitation of employment shall not misrepresent pertinent facts concerning employers, employees, associates, joint ventures, or their past accomplishments with the intent and purpose of enhancing their qualifications and work.
g. Engineers may advertise professional services only as a means of identification and limited to the following:
  (g.1) Professional cards and listings in recognized and dignified publications, provided they are consistent in size and are in a section of the publication regularly devoted to such professional cards and listings. The information displayed must be restricted to firm name, address, telephone number, appropriate symbol, names of principal participants and the fields of practice in which the firm is qualified.
  (g.2) Signs on equipment, offices and at the site of projects for which they render services, limited to firm name, address, telephone number and type of services, as appropriate.
  (g.3) Brochures, business cards, letterheads and other factual representations of experience, facilities, personnel and capacity to render service, providing the same are not misleading relative to the extent of participation in the projects cited and are not indiscriminately distributed.
  (g.4) Listings in the classified section of telephone directories, limited to name, address, telephone number and specialties in which the firm is qualified without resorting to special or bold type.
h. Engineers may use display advertising in recognized dignified business and professional publications, providing it is factual, and relates only to engineering, is free from ostentation, contains no laudatory expressions or implication, is not misleading with respect to the Engineers' extent of participation in the services or projects described.
i. Engineers may prepare articles for the lay or technical press which are factual, dignified and free from ostentations or laudatory implications. Such articles shall not imply other than their direct participation in the work described unless credit is given to others for their share of the work.
j. Engineers may extend permission for their names to be used in commercial advertisements, such as may be published by manufacturers, contractors, material suppliers, etc., only by means of a modest dignified notation acknowledging their participation and the scope thereof in the project or product described. Such permission shall not include public endorsement of proprietary products.
k. Engineers may advertise for recruitment of personnel in appropriate publications or by special distribution. The information presented must be displayed in a dignified manner, restricted to firm name, address, telephone number, appropriate symbol, names of principal participants, the fields of practice in which the firm is qualified and factual descriptions of positions available, qualifications required and benefits available.
l. Engineers shall not enter competitions for designs for the purpose of obtaining commissions for specific projects, unless provision is made for reasonable compensation for all designs submitted.
m. Engineers shall not maliciously or falsely, directly or indirectly, injure the professional reputation, prospects, practice or employment of another engineer, nor shall they indiscriminately criticize another's work.
n. Engineers shall not undertake nor agree to perform any engineering service on a free basis, except professional services which are advisory in nature for civic, charitable, religious or non-profit organizations. When serving as members of such organizations, engineers are entitled to utilize their personal engineering knowledge in the service of these organizations.
o. Engineers shall not use equipment, supplies, laboratory nor office facilities of their employers to carry on outside private practice without consent.
p. In case of tax-free or tax-aided facilities, engineers should not use student services at less than rates of other employees of comparable competence, including fringe benefits.

6. Engineers shall act in such a manner as to uphold and enhance the honor, integrity and dignity of the profession.
   a. Engineers shall not knowingly associate with nor permit the use of their names nor firm names in business ventures by any person or firm which they know, or have reason to believe, are engaging in business or professional practices of a fraudulent or dishonest nature.
   b. Engineers shall not use association with non-engineers, corporations, nor partnerships as 'cloaks' for unethical acts.
7. Engineers shall continue their professional development throughout their careers, and shall provide opportunities for the professional development of those engineers under their supervision.
   a. Engineers shall encourage their engineering employees to further their education.
   b. Engineers should encourage their engineering employees to become registered at the earliest possible date.
   c. Engineers should encourage engineering employees to attend and present papers at professional and technical society meetings.
   d. Engineers should support the professional and technical societies of their disciplines.
   e. Engineers shall give proper credit for engineering work to those to whom credit is due, and recognize the proprietary interests of others. Whenever possible, they shall name the person or persons who may be responsible for designs, inventions, writings or other accomplishments.
   f. Engineers shall endeavor to extend the public knowledge of engineering, and shall not participate in the dissemination of untrue, unfair or exaggerated statements regarding engineering.
   g. Engineers shall uphold the principle of appropriate and adequate compensation for those engaged in engineering work.
   h. Engineers should assign professional engineers duties of a nature which will utilize their full training and experience insofar as possible, and delegate lesser functions to subprofessionals or to technicians.
   i. Engineers shall provide prospective engineering employees with complete information on working conditions and their proposed status of employment, and after employment shall keep them informed of any changes.

ANSWERS TO SELECTED PROBLEMS

CHAPTER 2

2.1 (a) Undefined: 1,2 or 3; (c) 3; (e) 5; (g) 1; (i) 3.
2.2 (a) 3.00×10^2; (c) 1.15×10^{-8}; (e) 2.0×10^3; (g) 1.0.
2.3 (a) 764; (c) 7.58×10^5; (e) -6.65×10^{-6}.
2.4 (a) 0.66; (c) 4.968×10^{-9}.
2.6 3.1×10^5 J.
2.8 (a) m^3, ft^3; (c) m/s, ft/s; (e) N·m/s (or W), lbf·ft/s or ft·lbf/s; (g) N·m/kg·K (or J/kg·K), (lbf·ft)/(slug·°R); (i) N·s/m^2.
2.9 lbf·s/ft^2.
(a) 1.39 m^2; (c) 33.7 m/s; (e) 999
2.10 kg/m^3; (g) 4.38 kPa; (i) 0.73 J.
(a) 13.4 hp; (c) 4.40 Btu/(h)(ft^2)(°F); (e) 1 kg/(m)(s);
2.12 (g) 2.33×10^4 J/kg.
2.14 25.2 lbf.
2.787 ×
2.16 10^{-4} (m^2)(kPa)/(ft)(yd)(Pa).
2.21 gh/v^2.
(a) $X = 23$, $Y = 0.023$;
2.22 (c) $X = 0.27$, $Y = 0.028$.
2.24 $H = 0.11\ V^{1.9}$.
2.26 $T = 100\ (0.368)^t$.
(a) 11, 11, no modes; (c) 0.115,
2.29 0.113, 0.115; (e) 6.53, 6.53, 6.53. One reasonable set of class intervals

1.5–3.5	1
3.5–5.5	3
5.5–7.5	3
7.5–9.5	3
9.5–11.5	4
11.5–13.5	1
13.5–15.5	3
15.5–17.5	1
17.5–19.5	1

2.30 9.5, 9.75, 4.41.
2.32 (a) 3.59 percent; (b) R = 0.964.
2.34 $y = 0.0708\ x - 29.5$.

CHAPTER 3

3.2 For example, George Boole (1815–1864), Charles Babbage (1792–1871) and Johannes Brahms (1833–1897) all lived at the same time.
3.3 For example, the first hand-held scientific calculators were sold commercially about 3 years after Neil Armstrong's walk on the Moon.
3.4 (b) $(2001)_{16} = 2 \times 16^3 + 0 \times 16^2 + 0 \times 16^1 + 1 \times 16^0$.
3.5 (b) The change in the LSB increases the value by one unit.
3.7 (c) Approximately 10 bits (because $2^{10} - 1 = 1023_{10}$).
3.8 (c) $32g + 8g + 4g + 1g = 45\ g$.
3.10 (b) $(0\ 110)_2 = 6_{10}$
(d) $(1\ 110)_2 = 14_{10}$.
(c) $45_{10} = 0\ 101\ 101_2$.
3.12 (b) $101\ 101_2 = 55_8 = 45_{10}$.
3.13 (c) $1111\ 1111_2 = FF_{16} = 255_{10}$.
3.14 (b) $(0\ 111)_2 + (0\ 111)_2 = (1\ 110)_2$; Check: $7_{10} + 7_{10} = 14_{10}$; (d) $(0\ 010)_2 + (1\ 100)_2 = (1\ 110)_2$; Check: $2_{10} + 12_{10} = 14_{10}$.
3.15 (b) 2's complement of 0 111 is 1001; (d) 2's complement of 0 001 is 1111.
3.16 Remember, you expect a carryout when a sum is zero.
3.17 (c) $(0\ 111)_2 - (0\ 001)_2 = (0111)_2 + (1\ 111)_2 = (0\ 110)_2$; Check: $7_{10} - 1_{10} = 6_{10}$.
3.18 (c) 0.0 to 0.4 V.
3.19 (a) 2.4 to 5.0 V.
3.20 The XOR gate is defective.
3.22 (b) Magnitude = $(0\ 111)_2$; (d) magnitude = $(0\ 001)_2$.
3.23 $Z_0 = 0$, $Z_1 = 1$, $Z_2 = 0$, $Z_3 = 0$; carryout = 1.
3.26 Interchange the FORTRAN IF and PRINT* statements to print results after each cell doubling.
3.28 Instruction stored in location 23 becomes $(71\ 20\ 000\ 040)_8$.

CHAPTER 4

4.1 (a) $F_1 = 10$ N, $F_2 = 50$ N; (b) $F_1 = 400$ N, $F_2 = 200$ N.

4.2 (a) $F_x = 50$ N, $F_y = 86.6$ N;
(b) $F_x = 0.866$ N, $F_y = 0.50$ N.
4.4 $F_{JK} = 866$ lbf, $F_{KL} = 433$ lbf.
4.5 (a) $F_{BC} = 25$ N (compression),
$F_{AB} = 43.3$ N (compression),
$F_{AC} = 21.7$ N (tension).
4.6 (a) $F_{BC} = 76.6$ lbf (compression),
$F_{AB} = 64.3$ lbf (tension),
$F_{AC} = 58.7$ lbf (tension).
4.7 (a) $F_{AC} = F_{CB} = 2.89$ kN (tension), $F_{AD} = F_{FB} = 5.77$ kN (compression), $F_{DC} = F_{FC} = 5.77$ kN (tension), $F_{DE} = F_{EF} = 5.77$ kN (compression), $F_{EC} = 0$.
4.8 (b) $4(2.5) - 1(10) = 0$;
(d) $20(100) - 10(200) = 0$.
4.9 (a) $F_A = 15.6$ N, $F_B = 9.4$ N;
(c) $F_A = 7.5$ kN, $F_B = 2.5$ kN.

CHAPTER 5

5.1 (a) $i = 24$ mA; (c) $v = 0.10$ V.
5.2 (a) $i = 10.8$ mA.
5.3 (a) $R_E = 33.3$ Ω. (c) $R_E = 50$ Ω.
5.4 (a) $v_1 = v_2 = 12$ V, $i_1 = 0.12$ A, $i_2 = 0.06$ A; (c) $v_1 = 1.5$ V, $v_2 = v_3 = 1.91$ V, $i_1 = 150$ mA, $i_2 = 87$ mA, $i_3 = 64$ mA.
5.8 10 V.
5.10 573 °/A.
5.12 (a) 60.0 mA; (b) 59.5 mA.

CHAPTER 6

6.1 (a) 98.1 kPa; (c) 1.33×10^3 lbf/ft^2.
6.4 (a) 3.98×10^4; (c) 3.98×10^4.
6.5 2.91 ft/s, 814 lbf/ft^2.
6.7 (a) 3 kg/s; (c) 15 kg/s.
6.9 0.123 kg CO_2 per minute, 30.6 kW.
6.11 $h = 5.1$ m.
6.13 40 kJ, 10°C.
6.15 365 kW, 5400 m^2.
6.17 (a) 0; (c) 175 W.
6.19 2040 Btu/h.
6.21 (a) 36 Btu/(h)(ft^2); (c) 200 W/m^2.
6.23 $\tau_{\infty i} = 21.7$°C, $\tau_{si} = 15.2$°C, $\tau_{so} = 9.5$°C.

CHAPTER 7

7.2 (a) Possible inputs: water volume, water temperature, duration of wash cycle, duration of rinse cycle, agitator speed, duration and speed of spin dry, amount and type of detergent, etc.
Possible outputs: cost per washload, noise level during operation, machine vibration, quality of the finished washload, etc.
(c) Possible inputs: volume control, channel selection, fine tuning, brightness and contrast settings, quality of the antenna signal, etc.
Possible outputs: sound quality and volume, picture quality including clarity, color, brightness, etc.
7.4 At $t = 25$ s, $v = 35.67$ ft/s; at $t = 50$ s, $v = 45.90$ ft/s; at $t = 75$, $v = 13.15$ ft/s; at $t = 100$, $v = 3.77$ ft/s; at $t = 125$, $v = 1.08$ ft/s; at $t = 150$, $v = 0.025$ ft/s.
7.6 $\tau = 50$ s.
7.7 (a) $\tau = 1$, $v_o = 100$; (c) $\tau = 2.5$, $v_o = 50$.
7.11 1.0 Hz, 0.45 Hz.
7.12 (a) 0.16, 0.50, 100; (c) 0.16, 1.0, 100.
7.15 At $t = 0$ s, $v_c = 50$ V; at $t = 0.1$ s, $v_c = 81.6$ V; at $t = 0.2$ s, $v_c = 93.2$ V; at $t = 0.3$ s, $v_c = 97.5$ V; equilibrium, $v_c = 100$ V.
7.16 $f_n = 15.9$ Hz, $\zeta = 0.05$.
7.18 (a) 16 V; (c) 8.5 V.
7.20 (a) 50 V; (c) 10 V.

CHAPTER 9

9.1 102 kN, 186 kN.
9.3 (a) 6.8 mm; (c) 0.40 in.
9.6 (a) 18.9 kg; (b) 6.8 kg.
9.8 (a) 0.7 mm, 8.83 mm^2; (c) 0.05 in, 0.0318 in^2; (e) 0.0357 in, 0.0364 in^2.
9.9 (a) 1 mm; (c) 0.50 mm.
9.10 (a) 0.0357 in; (c) 0.0179 in.
9.11 273 in·lbf.
9.12 $\tan \psi = 0.058$.

CHAPTER 10

10.2 For each hundred connections, five failures are expected in a 5-year period.
10.4 $c = 5.0$.
10.5 $c = 7.9$.
10.6 The resistance of 4 in (10 cm) of #12 AWG aluminum wire is 0.82×10^{-3} Ω.
10.7 The stress of 113 Mpa is smaller than the yield stress of wrought aluminum alloy, yellow brass, or carbon steel.
10.8 In doing this problem, it is reasonable to assume that the compressive strength of aluminum (a ductile material) is approximately equal to the tensile strength.
10.9 Using Fig. 10.5, you can estimate that the force falls below 1000 N in approximately one hour.
10.10 Torque range 96 to 14 in·lbf. This 7-to-1 range greatly simplifies the craftsperson's task.

CHAPTER 11

11.1 (a) 22.2 kN, buckling; (c) 727 N, buckling.
11.2 (a) 1800 lbf/in^2; (c) 900 lbf/in^2.
11.3 Yes, largest stress only 109 MPa.
11.5 Cable diameter = 0.29 in; square rod size = 1.32 in.
11.7 25-mm square.
11.8 Buckling of AB or BD at 123 kN
11.10 0.25-in square.

CHAPTER 12

12.3 (a) 12.7 pF; (c) 225 pF.
12.5 (a) amplification = $A/A_o = 5$.
12.8 (a) 0.001 s; (c) 0.003 s.
12.9 (a) 0.8 s; (c) 0.8 ms.

CHAPTER 13

13.1 (a) 1986
13.2 (b) 1998
13.5 (a) 8.34 MPa gage; (b) 737 lbf/in^2 gage.
13.5 (a) 5600 lbf/in^2 gage; (b) 14 MPa, gage.

INDEX

AAES (American Association of Engineering Societies, Inc.), 38, 44
Abacus, 118
ABET (Accreditation Board for Engineering and Technology, Inc.), 38, 41–42
Abscissa, 87
Absolute pressure, 228–229
Acceptance tests, 339–340
Accreditation of engineering schools, 41–42
Accuracy:
 of calculations, 51–54
 of measuring devices, 53–54
Aerospace engineering, 20
Agricultural engineering, 21–22
AIAA (American Institute of Aeronautics and Astronautics, Inc.), 38
AIChE (American Institute of Chemical Engineers), 38
AIIE (American Institute of Industrial Engineers, Inc.), 38
Aiken, Howard, 127
AIME (American Institute of Mining, Metallurgical, and Petroleum Engineers), 38

Alumina, 393–394
Aluminum:
 electric contacts using, 393–394
 oxide of, 393–394
 stress relaxation in, 395–396
Aluminum conductors, problems with, 353–363
Aluminum connectors, life tests on, 399
Aluminum contacts:
 design of, 396–400
 life tests on, 399–400
Aluminum house wiring, problems with, 359–363
Aluminum transmission cables, 354–359
AM (amplitude modulation) radio, 422–426
American units, 65–68
American wire gage (AWG), 360
Ammeters, 219–220
Ampere (unit), 63, 198
Amplitude modulation (AM) radio, 422–426
Analog computers, 116, 431
Analytical engines, 120
ANS (American Nuclear Society), 38

ANSI (American National Standards Institute), 40–41
Arch bridges, 417–419
Architectural engineering, 23–24
Arithmetic operations, digital hardware for, 157–162
 (See also Computer arithmetic)
Armatures, 221
ASAE (American Society of Agricultural Engineers), 38
ASCE (American Society of Civil Engineers), 38
ASEE (American Society for Engineering Education), 38
ASHRAE (American Society of Heating, Refrigerating and Air-Conditioning Engineers), 38
ASME (American Society of Mechanical Engineers), 38
ASTM (American Society for Testing and Materials, Inc.), 38
Audio signals (see Tuned circuits)
Audio tapers, 213
Automatic riveting machines, 345–348
 computerization of, 348–351

Automatic riveting machines:
design of, 352–353
Automobiles:
controls of, simulation of, 430–432
drag and propulsive forces on, 173–175
Avogadro's number, 63

Babbage, Charles, 120–123
Balsa wood, properties of, table, 413
Bar graphs, 83, 84
BASIC (computer language), 141, 162, 165
Basic units, 63
Bending, failure of structures due to, 406–407
Bimodal data, 99
Binary digits, physical representation of, 153–154
Binary systems, 118, 141–146
Binary-to-decimal conversions, 143–144
Bioengineering, 23
Bolts, 376
Boole, George, 126
Bowden, F. P., 388
Brainstorming sessions, 329
Bridges, examples of, 415–419
British thermal unit (Btu), 68, 248
Brittleness, 372–373
Buckingham pi (π) theorem, 76–81
Buckling, 373–374
failure of structures due to, 405–406
Byron, Ada Augusta, 122

Cables:
static forces on, 176–178
transmission, 354–359
Calculations, 51–111
expressing results of, 56–58
Candela (unit), 64
Capacitance, electric (see Electric capacitance)
Capacitors (see Electric capacitors)

Carbon microphones, 213–214
Careers for engineers:
engineering positions, 26–33
other positions, 33–35
role of engineering societies, 45–46
Celsius scale, 63
Census data, Hollerith machine for, 123–126
Chemical engineering, 18–19
Circle graphs, 83, 84
Circuits (see Electric circuits; Tuned circuits)
Civil engineering, 17–18
COBOL (computer language), 141
Code of Ethics for Engineers, 44, 460
Combustion:
example of, gasoline combustion, 243–244
theory of, 240–242
Combustion products, 243–244
Communication:
human, 422
radio, 422–426
Commutators, 222
Compression, 373–375
failures of structures due to, 404–405
Computer arithmetic:
addition and subtraction, 148–152
complements in binary, 150–152
complements in decimal, 149–150
hardware fundamentals, 152–156
Computer engineering, 20
Computer languages, 166, 169–171
(See also names of specific languages)
Computer programs:
early type, 121–122
preparation of, 162–171
(See also names of specific programs)
Computers, digital, 115–171
engineering uses of, 116–117
future of, 133–138

Computers:
history of, 117–133
mass-produced, 131–133
memories, 131
for CYBER, 166–167
number systems for, 141–146
(See also Computer arithmetic)
operation and design of, 139
programming of, 162–171
use of, in oil recovery calculations, 455–458
Conduct of engineers, 11
Conduction of heat, 259, 261–264
Conductivity, thermal, table, 262
Conductors, electric (see Electric conductors)
Connectors (see Aluminum connectors; Electric connections)
Conservation of energy (see Energy, conservation of)
Conservation of matter, 237–244
Construction engineers, 28
Consulting as a career, 34–35
Consumer Products Safety Commission, 354
Contacts (see Aluminum contacts; Electric contacts)
Control Data Corporation, 132
Convection, 260, 264–265
Conversion of units, 68–73
table, 69
Coplanar forces, 194
Copper transmission cables, 356–358
Costs, relation to engineering, 6–7
Coulomb (unit), 198
Countess of Lovelace, 122
Creep of materials, 369
Current, electric (see Electric current)
CYBER, 137, 166–169

Damping, viscous, 287–293
Damping ratio, 293–298, 305–306, 314–318
Decimal-to-binary conversions, 143–144

Demodulation process, 425
Density of fluids, 232–233
Derived dimensons, 61
Derived units, 64
Design (*see* Engineering design)
Design engineers, 31–32
Development engineers, 32
Difference engines, 120
Digital computers (*see* Computers, digital)
Digital displays, 53–54
Digital Equipment Corporation, 133
Dimensional analysis, 73–81
Dimensionless groups, 76–81
Dimensions, 60–62
   derived, 61
   fundamental, 61
Drafters, 26
Ductility, 372–373
Dynamic systems, 273–320

Eckert, J. Presper, 127, 128
EDSAC, 130
Education in engineering technology, 25–26
Education of engineers, 10–11, 24
   enrollments in schools, 25
   nature of, 35
   role of engineering societies in, 41–42
   (*See also* Engineering schools)
EDVAC, 129
Efficiency, 255
Einstein's equation for mass and energy, 237
Elastic region of materials, 369
Electric capacitance, 299–300
   in filter circuits, 434–437
   in tuned circuits, 426–429
Electric capacitors, 299–300
Electric circuits, 197–224
   dynamics in, 301–306
   (*See also specific types of circuits*)
Electric conductors, 216–217
   aluminum, problems with, 353–363

Electric conductors:
   forces on, due to magnetic fields, 216–217
   voltage differences created by movement of, 216
Electric connections:
   reliability of, 387–388
   screws for, 387–400
   wire-binding screws for, 387–400
   (*See also* Electric contacts)
Electric contacts:
   with aluminum, 393–394
   measurements of resistance of, 391–393
   with metal surfaces, 388–396
   (*See also* Aluminum contacts; Electric connections)
Electric current, 198
   laws for, 200–202
   magnetic fields created by, 216
Electric filters, 421–439
   design of, 433–437
   resistance and capacitance in, 434–437
Electric inductance, 300–301
   in tuned circuits, 427–429
Electric motors, 221–223
Electric resistance, 198–199
   of contacts, measurement of, 391–393
   in filter circuits, 434–437
   variable, applications of, 212–214
Electric resistivity, 198
Electric resistors:
   in parallel, 204–209
   in series, 202–203
   variable, 212–214
Electric systems, 299–306
Electrical engineering, 14–15
Electromagnetic waves, communication using, 422
Electromagnets, 216
Electromechanical devices, 215–223
Electronic engineering, 22
Employment of engineers:
   engineering positions, 26–33
   other positions, 33–35

Employment of engineers:
   role of engineering societies, 45–46
Endurance tests, 338
Energy:
   conservation of, 246–256
      applying equation for, 248–251
      steady-flow processes, 251–255
   dimensions of, 62, 248
   mass and, relationship between, 237
   relation to work, 247
   units of, 248
Engineering:
   as an art, 7
   benefits from, for society, 4–5
   as a career, paths to, 3–46
   computers used in, 116–117
   definition of, 4
   nature of, 1–46
   as a profession, 8–11
   relation to values and costs, 6–7
   reliance on mathematics and science, 5–6
Engineering design, 321–365
   applications of, 367–458
   evolution of, 344–353
   one-of-a-kind designs, 335–336
   process of, 325–332
      application of, 332–341
   testing of, 337–341
Engineering disciplines, nature of, 13–24
Engineering education (*see* Education of engineers; Engineering schools)
Engineering schools:
   accreditation of, 41–42
   growth of, 36
   (*See also* Education of engineers)
Engineering science, 20–21
Engineering societies:
   growth of, 36–37
   legal functions of, 42
   list of, 37–38
   role of, 38–46
      in education, 41–42

Engineering technicians, 26–27
Engineering technology, education for, 25–26
Engineers:
  computers used by, 116–117
  functions of, 26–35
  salaries of, 11–13
ENIAC, 127–128
Environmental tests, 338
Equations, development of, from graphs, 92–95
Equilibrium temperature, calculation of, 268–270
Equilibrium values of first-order systems, calculation of, 284–285
Ethical standards for engineers, 43–45
Euler, Leonhard, 120
Eyes (threaded), 380–382

Fahrenheit scale, 68
Failures of structures, modes of, 403–407
Fasteners, threaded, 376–382
Field evaluation of aluminum house wiring, 362–363
Filters (see Electric filters)
First-order systems, 282–285
  FORTRAN programs for, 279–281, 310–314
  frequency response of, 310–314
Flow, steady, 251–255
Flowcharts, 162
  use of, to analyze problems, 163–164
Fluid columns, pressure in, 229–230
Fluid mechanics, 228–236
Fluids, 228
Foot-pounds-force (unit), 248
Force (dimension):
  Newton's laws of, 64, 66–67, 72
  relation to mass, 66–67
Force, creation of, by magnetic fields, 216–217
Forrester, J. W., 130
FORTRAN (computer language), 132, 141, 162, 164–165, 169–171

FORTRAN (computer language):
  for first-order systems, 279–281, 310–314
  for second-order systems, 291–293, 314–318
Fourier series, 308
Free-body diagrams, 174
Freedom versus responsibilities in engineering, 10
Frequency response, 306
  background of, 307–310
  of first-order systems, 310–314
  of second-order systems, 314–318
Full adders, 158–160
Fundamental dimensions, 61

Gage pressure, 228–229
Galvanometers, 217–219
Gasoline, combustion of, 243–244
Gates, logic, 156
Gauss, Karl Friedrich, 120
Gomaa, E. A., 454, 455
Graphs, 83–93
  bar, 83, 84
  development of equations from, 92–95
  logarithmic scales for, 89–92
  pie, 83, 84
  rules for plotting of, 87
Groups, dimensionless, 76–81

Half-adders, 158
Heat transfer, 259–270
Heat-transfer coefficient, 264–265
  experimental determination of, 265–268
Hertz, Heinrich, 295
Hertz (unit), 295
Hewlett Packard Corporation, 133
Hexadecimal system, 141
  conversions to and from, 144–145
High-frequency signals, removal of, 430–437
Histograms, 101–102

History:
  of computers, 117–133
  of engineering, 36–37
Hollerith, Herman, 123–126
Holm, Ragnar, 390
Hooke's law, 369
House wiring, aluminum, 359–362
  field evaluation of, 362–363
Human communication, 422
Human factors tests, 338
Hydraulic systems, 239–240

IBM (International Business Machines Corporation), 127
  computer production by, 131–132
IEEE (Institute of Electrical and Electronic Engineers), 38
Inductance, electric (see Electric inductance)
Industrial engineering, 19–20
Instruction format for CYBER, 168–169
Integration of digital hardware, levels of, 160–161
Intellectual nature of engineering, 8–9
International System of Units (Système International d'Unités, SI), 60, 62–65

Jacquard looms, 121
JOHNIAC, 130
Joints, pinned, 180–182, 415
Joules (unit), 248
Journals for engineers, 39
Judgment involved in engineering, 9–10

Kelvin (unit), 63
Kilogram (unit), 63
Kinetic energy, 247
Kirchhoff's laws, 200–201

Lagrange, Joseph Louis, 120
Laminar flow, 234, 236

Laplace, Pierre Simon de, 120
LC (tuning) circuits, design of, 428–429
Least-squares fit, 106, 108, 110–111
Legal functions of engineering societies, 42
Leibniz, Gottfried Wilhelm, 118
Life tests, 338
Liquids, flow of, in pipes, 231–236
Logarithmic scale of graphs, 89–92
Logic, 154–155
Logic circuits, 156
Logic gates, 156

Machine-language programs, 166, 169–171
(See also names of specific languages)
Magnetic fields, creation of, by electric currents, 216
Management positions for engineers, 34
MANIAC, 130
Manometers, 230–231
Marine engineering, 22
Mass (dimension), relation to force, 66–67
Mass and energy, relation between, 237
Materials, 367–375
mechanical properties of, 368–375
Mathematics, relation to engineering, 5–6
Matter, conservation of, 237–244
Mauchly, John W., 127
Maxwell, James Clerk, 390
Mean, 99–100
Mean deviation, 120
Measurements, significant figures in, 52–53
Mechanical computation, origins of, 118–119
Mechanical engineering, 15–17
Mechanical properties of materials, 368–375
Mechanics of fluids, 228–236

Median, 98–99
Meetings for engineers, 39
Memories for computers, 131
Metal surfaces, electric contacts using, 388–396
Metallurgical engineering, 22
Meter (unit), 63
Metric units, 60, 62–65
Mining engineering, 22
Mode, 99
Modulation process, 423
Modulus of elasticity, 369
Mole (unit), 63
Moments (statics), 189–192
Motor vehicles (see Automobiles)
Motors, electric, 221–223
Multimodal data, 99

Nader, Ralph, 46
NASA (National Aeronautics and Space Agency), 5, 80, 137
Natural frequency, 293–298, 305–306, 314–318
Natural logarithms, 103
NCEE (National Council of Engineering Examiners, Inc.), 38
Newton, Isaac, 119
Newton (unit), 64
Newton's laws:
of cooling, 265
of force, 64, 66–67, 72
NICE (National Institute of Ceramic Engineers), 38
Normal distribution, 102–106
Notation, scientific, 54–56
NSPE (National Society of Professional Engineers), 38
Nuclear engineering, 22–23
Number systems for computers, 141–146
(See also Computer arithmetic)
Nusselt number, 266

Octal system, 141
conversions to and from, 144–145
Ohm's law, 200

Oil:
depletion of supplies of, 443
production of, 441–458
recovery of, 442–443
computer calculations for, 455–458
tertiary methods for, 453–458
Oil wells:
mixtures of oil, gas, and water in, 450–452
pressure required for, 446–450
tubing for, design calculations, 446–452
Operations engineers, 30–31
Ordinate, 87
Oscilloscope, 432–436

Parallel resistors, 204–209
Pascal, Blaise, 118
Pascal (computer language), 162
PE (professional engineer), requirements for, 42–43
Pension plans, 45
Performance evaluation, role of safety in, 363–365
Performance tests, 337
Periodic functions, 308
Perpetual motion machines, 255
Petroleum (see Oil)
Petroleum engineering, 21
Photoresistive cells, 214
Pie graphs, 83, 84
Pinned joints, static forces on, 180–182, 415
Pipes, liquid flow in, 231–236
Plant engineers, 30–31
Plastic deformation, 371
Pollution due to combustion products, 243–244
Positional number systems, 141
Potential energy, 247
Potentiometers, 212–213
Pound (unit), 66–67
Power (dimension), 62
Prandtl number, 266
Precision of figures, 53–54
Pressure:
absolute, 228–229
in fluid columns, 229–230
gage, 228–229

Primary dimensions, 61
Probability, 103
Production engineers, 27–28
Professional conduct, nature of, 8–11
Professional engineer (PE), requirements for, 42–43
Programming of computers, 162–171
    sample calculations preceding programming, 164
    (See also Computer programs)
Prototypes, 330

Quality control tests, 339–340

Radiation of heat, 260
Radio, amplitude modulation (AM), 422–426
Random variations, 97–98
Rankine scale, 68
Rate of change of first-order systems, calculations of, 284–285
Raw data, 98
Registers for computers, 167
Registration of engineers, 42–43
Reliability, 106
    for electric connections, 387–388
Research engineers, 32–33
Resistance (see Electric resistance)
Resistivity (see Electric resistivity)
Resistors (see Electric resistors)
Reynolds number, 81, 234–236, 266
Right-hand rule, 216
Rivets (see Automatic riveting machines)
Rotation due to static forces, 189–192

SAE (Society of Automotive Engineers), 38
Safety, role of, in performance evaluation, 363–365

Salaries of engineers, 11–13
    comparison with other occupations, 12–13
Sales engineers, 30
Science, relation to engineering, 5–6
Scientific notation, 54–56
Screw threads, 376–378
    standards for, 377
Screws, 376–382
    for electric connections, 387–400
    threads of, 376–378
    torque for turning, 378–379
SEAC, 130
Second (unit), 63
Second-order differential equations, 289
Second-order systems, 287–289
    FORTRAN program for, 291–293, 314–318
    frequency response of, 314–318
    general nature of, 293–298
    transient response of, 289–293
Secondary dimensions, 61
Self-locking screws, 379
Semilogarithmic graphs, 92
Series resistors, 202–203
SES (Society of Engineering Science), 39
Shannon, Claude, 126–127
SI units (Système International d'Unités), 60, 62–65
Significant figures, 52–53, 56–58
Simulators of automobile controls, 430–432
Skewed distributions, 102
Slide rule, 116
Slug (unit), 67
SME (Society of Manufacturing Engineers), 39
Societies for engineers (see Engineering societies)
Society, benefits received by, from engineering, 4–5
Specific heat, table, 247
Standard deviation, 102
Standards for engineering, 39–40
Statically indeterminate problems, 194–196
Statics, 173–196

Statistics, concepts of, 97–111
Steady flow, 251–255
Steam flooding:
    computer calculations for, 455–458
    recovery of oil using, 453–458
Stibitz, George, 127
Strain, 369, 370
Stress, 369, 370
Stress relaxation in aluminum, 395–396
Structural steel, 404
Structures, 403–411
    failure modes of, 403–407
Suspension bridges, 416–417
Système International d'Unités (International System of Units, SI), 60, 62–65
Systems, dynamic, 273–320

Tabor, D., 388
Tapped holes for eyes (threaded), 381
Technical journals, 39
Technical meetings, 39
Technical sales engineers, 30
Technical standards, 39–40
Tensile strength, table, 370
Tensile tests, 368–370
Tension, failures of structures due to, 404–405
Tertiary recovery of oil, 453–458
Test engineers, 28–30
Testing of designs, 337–341
Texas Instruments, 133
Thermal conductivities, table, 262
Thermal-fluid sciences, 227–272
Thermal radiation, 260
Thermistors, 214
Thermodynamic temperature, 63
Thermodynamics:
    equation for, 247
    first law of, 246
    second law of, 246
Thread pitch, 377
Threaded connections, 380–382
Threaded fasteners, 376–382
Threads of screws, 376–378

Time constant, 283, 316, 434–437
Torque for screws, 378–379
Transient response problems, 274–275
　mathematical description, 275–277
　numerical solution, 277–281
Transistor-transistor logic (TTL), 154
Transmission cables, 354–359
Truss bridges, 415–416
Truss structures, 409–411
　design of, 409–411
Trusses, static forces on, 182–185

Tubing for oil wells, design calculations for, 446–452
Tuned circuits, 422–429
　analysis of, 427–428
　capacitance and inductance in, 426–429
　design of, 426–427
Turbulent flow, 234, 236

Underwriters Laboratories, Inc., 361
Unimodal data, 99
U.S. Army, support of ENIAC by, 127–128
USCS (U.S. Customary System) units, 60, 65–68
Units:
　American, 60, 65–68
　basic, 63
　conversion of, 68–73
　table, 69
　derived, 64
　SI (metric), 60, 62–65

Variable resistance (electric), applications of, 212–214
Vectors, 194
Vested rights in pensions, 45
Viscosity, 232
　tables, 233, 454

Viscous damping, 287–293
Voltage, 198
　laws for, 200–202
Voltage differences, creation of, by moving electric conductors, 216
Voltmeters, 220–221
Von Neumann, John, 128–130
　architecture of, for computers, 140

Warnings, 364–365
Whirlwind I, 130
"Whistle blowing," 45, 46
Work (dimension), 61–62
Work, relation to energy, 247

Yield strength, 371

Zuse, Konrad, 127